Trademarks

All terms noted in this publication that are believed to be registered trademarks or trademarks are listed below:

- PC-DOS, IBM, IBM PC/XT, IBM PC/AT, IBM PS/2 are registered trademarks of International Business Machines Corporation.
- Microsoft, MS-DOS and Windows are registered trademarks of Microsoft Corporation.
- Intel is a registered trademark of the Intel Corporation
- Allen- Bradley, Siemens , Omron, Arduino, Raspberry pi,ARM are trademark of respective Corporations.

Disclaimer

Whilst all reasonable care has been taken to ensure that the description, opinions, listings and diagrams are accurate and workable, Authors does not accept any legal responsibility or liability to any person, organization or other entity for any direct loss, consequential loss or damage, however caused, that may be suffered because of the use of this publication.

About Authors

Madhukar Varshney

(B.E. Electronics, MBA)

A qualified engineer, a seasoned Electronics and Telecom professional, entrepreneur, motivational speaker and thinker, with a successful track record of over 21 years, involved in Electronics, Telecom and IT in multiple telecom technologies and business environments.

An expert in network technology with a strong understanding of Instrumentation, Automation, Cloud Computing, IoT and emerging technologies such as 5G.

You can reach Madhukar thru: madhuvarsha@gmail.com

Sanjay Galhan

(Diploma and B.E. Electronics & Communications)

A qualified Instrumentation and Industrial Automation Engineer. Expert in Automation Design, planning and troubleshooting.

You can reach Sanjay thru: skgalhan@gmail.com

Contents

1 Chapter: Definition of Terminology

Below is a list of terms and their definitions that are used throughout this manual for Instrumentation and Automation.

Accuracy

How precise or correct the measured value is to the actual value. Accuracy is an indication of the error in the measurement.

Ambient

The surrounds or environment about a point or object.

Attenuation

A decrease in signal magnitude over a period.

Calibrate

To configure a device so that the required output represents (to a defined degree of accuracy) the respective input.

Closed loop

Relates to a control loop where the process variable is used to calculate the controller output.

Coefficient, temperature

A coefficient is typically a multiplying factor. The temperature coefficient defines how much change in temperature there is for a given change in resistance (for a temperature dependent resistor).

Cold junction

The thermocouple junction which is at a known reference temperature.

Compensation

A supplementary device used to correct errors due to variations in operating conditions.

Controller

A device which operates automatically to regulate the control of a process with a control variable.

Elastic

The ability of an object to regain its original shape when an applied force is removed. When a force is applied that exceeds the elastic limit, then permanent deformation will occur.

Excitation

The energy supply required to power a device for its intended operation.

Gain

This is the ratio of the change of the output to the change in the applied input. Gain is a special case of sensitivity, where the units for the input and output are identical and the gain is unitless.

Hunting

Generally, an undesirable oscillation at or near the required setpoint. Hunting typically occurs when the demands on the system performance are high and possibly exceed the system capabilities. The output of the controller can be overcontrolled due to the resolution of accuracy limitations.

Hysteresis

The accuracy of the device is dependent on the previous value and the direction of variation. Hysteresis causes a device to show an inaccuracy from the correct value, as it is affected by the previous measurement.

Ramp

Defines the delayed and accumulated response of the output for a sudden change in the input.

Range

The region between the specified upper and lower limits where a value or device is defined and operated.

Reliability

The probability that a device will perform within its specifications for the number of operations or time period specified.

Repeatability

The closeness of repeated samples under exact operating conditions.

Reproducibility

The similarity of one measurement to another over time, where the operating conditions have varied within the time span, but the input is restored.

Resolution

The smallest interval that can be identified as a measurement varies.

Resonance

The frequency of oscillation that is maintained due to the natural dynamics of the system.

Response

Defines the behavior over time of the output as a function of the input. The output is the response or effect, with the input usually noted as the cause.

Self-Heating

The internal heating caused within a device due to the electrical excitation. Self heating is primarily due to the current draw and not the voltage applied and is typically shown by the voltage drop as a result of power (I2R) losses.

Sensitivity

This defines how much the output changes, for a specified change in the input to the device.

Setpoint

Used in closed loop control, the setpoint is the ideal process variable. It is represented in the units of the process variable and is used by the controller to determine the output to the process.

Span Adjustment

The difference between the maximum and minimum range values. When provided in an instrument, this changes the slope of the input-output curve.

Steady state

Used in closed loop control where the process no longer oscillates or changes and settles at some defined value.

Stiction

Shortened form of static friction and defined as resistance to motion. More important is the force required (electrical or mechanical) to overcome such a resistance.

Thermal shock
An abrupt temperature change applied to an object or device.
Time constant
Typically, a unit of measure which defines the response of a device or system. The time constant of a first order system is defined as the time taken for the output to reach 63.2% of the total change, when subjected to a step input change.
Transducer
An element or device that converts information from one form (usually physical, such as temperature or pressure) and converts it to another (usually electrical, such as volts or millivolts or resistance change). A transducer can be considered to comprise a sensor at the front end (at the process) and a transmitter.
Transient
A sudden change in a variable which is neither a controlled response nor long lasting.
Transmitter
A device that converts from one form of energy to another. Usually from electrical to electrical for the purpose of signal integrity for transmission over longer distances and for suitability with control equipment.
Variable
Generally, this is some quantity of the system or process. The two main types of variables that exist in the system are the measured variable and the controlled variable. The measured variable is the measured quantity and is also referred to as the process variable as it measures process information. The controlled variable is the controller output which controls the process.
Vibration
This is the periodic motion (mechanical) or oscillation of an object.
Zero adjustment
The zero in an instrument is the output provided when no, or zero input is applied. The zero adjustment produces a parallel shift in the input-output curve.

IoT Related Terms and Definitions

Below is a list of IoT terminology, that anyone interested in the Internet of Things should know and understand.

API – Application Programming Interface
Software programs that incorporate protocols, tools, and additional resources used by developers to build interoperability between and across programs which are running in the same environment. APIs are frequently needed when 2 or more IoT companies or devices integrate. For example, when a smart light bulb works with voice-controlled platforms such as Amazon's Alexa.
Algorithm
An extremely structured set of instructions developed to perform a specific task that contains a finite set of steps. Algorithms are used for computational problem-solving.

Artificial Intelligence
Software which uses an extremely complex set of rule-based instructions. The goal is to boost computational decision making at or above a human's level and ability. Some fear that as artificial intelligence improves there is a chance technology will overpower the human race because we have programmed them to learn and become extremely intelligent.

Augmented Reality
Using technology to enhance the physical world. Any device that works in conjunction with your surroundings creates an augmented reality. For example, google glass creates a sense of AR because of the screen which allows you to view a screen while simultaneously interacting with your surrounding area.

Autonomous Vehicle
A computer-controlled vehicle (car, truck, train, etc.) that is equipped with sensors and cameras that allow the vehicle to navigate without the aid a human being. Telsa (example) and the google self-driving car (example) are splendid examples of this.

Big Data
When large data sets are analyzed to determine trends, events, or activities on a very grand scale. For example, data collected from a city's traffic camera determine when and where the most traffic accidents occur. Internet of Things devices have the ability to record massive amounts of useful data.

Bluetooth
An open standard for wireless digital communications over short distances. Radio frequency technology which allows devices to transfer and receive digital audio, video, text, signals, and much more. Bluetooth is constantly improving which is great for IoT. The latest advancement will be the release of Bluetooth 5 which is scheduled for late 2016 or early 2017.

Byte
A standard unit of measurement for computing. Bytes are comprised of 8 binary digits incorporating alphanumeric characters. Types of storage include bytes, kilobytes, megabytes, gigabytes, terabytes, petabytes, exabytes, and zettabytes.

Cloud Computing
When remote servers are used to store data and host applications. Storing data in the cloud is becoming the industry standard because of the flexibility and convenience. Cloud applications can be accessed from anywhere in the world.

Connected Devices
A term used to explain when any device, whether industrial or personal, is connected through a network. The network connection can come in various forms such as the internet or Bluetooth. All Internet of Things devices are also considered to be connected devices.

Contextual Awareness

The capability of a machine to recognize and adapt to environmental factors. These factors include user behavior, surroundings, and many other variables. All of the information collected is processed to determine the next action. For example, many laptop computers will adjust the screen's brightness depending on the environment that the computer is in. Another example is when the Nest smart thermostat will adjust depending on the temperature within the home as well as patterns of occupancy.

Cybersecurity

Security that relates to online systems, applications, and devices. Cyber security is one of the main concerns relating to the Internet of Things. Every new IoT product to market must consider cyber security very seriously during the development process. The ability for hackers to overtake your smart home, smart car, or smart device is very concerning for overall public safety.

DIY – Do it Yourself

The process of building, tinkering or experimenting without the aid of professionals. Do it Yourself innovation comes in many different forms. For example, 3D printing can be used when developing DiY IoT devices.

Encryption

When data becomes scrambled for the sole purpose of security. When data is encrypted, only the person sending or receiving the information is able to unscramble (read) it. Encryption is an important part of cyber security as it aids in preventing someone from hacking your device or information.

Ethernet

A local area network (LAN) that is connected with a cable that allows data to travel through it.

Geolocation

Specific coordinates used to communicate a location. The use of satellites, cellular data, WiFi, and other systems are used to provide a general location and information.

GPS – Global Positioning System

A system that uses satellites in space to determine the specific location of objects on earth. Almost all smartphones today contain GPS technology.

H2M – Human to Machine Communication

The interaction between people and computers.

Industrial Internet

A term coined by General Electric (GE) used to explain the connectivity between humans and machines.

Internet

Infrastructure used to connect computers, smartphones, machines, and other devices to each other, as well as over a common network.

IoE – Internet of Everything

A term coined by Cisco Systems used to describe all things connected. This includes the Internet of Things (IoT).

IoT – Internet of Things

The IoT acronym stands for Internet of Things, which is any device that connects and shares data over the internet. The Internet of Things has become a buzzword for the ever-increasing technology industry which includes connected devices and more and more 'things'.

IP – Internet Protocol

A communications protocol which is used as a networking standard for the Internet. Allowing for computers to handle packet switching, routing, addressing, as well as other functions.

LAN – Local Area Network

A group of locally connected devices including computers, scanners, phones, and more that communicate in real time over a cable or wireless infrastructure.

M2M – Machine to Machine Communication

The ability for computing devices and other machines to exchange data and perform functions without human involvement. M2M derived from telemetry communication but has since improved.

NFC – Near Field Communication

Wireless communication technology that allows objects with NFC to exchange data with little to no human intervention.

PAN – Personal Area Network

An interconnected set of devices used by a single person within a restricted area. Typically, around 10 meters.

RFID – Radio Frequency Identification

A wireless technology that uses either passive or active tags and readers with antennas to identify objects. Passive means non-powered and active means powered. The data from these objects can determine location, condition, and any changes to computers. RFID chips are key to developing some IoT devices and applications.

RTLS – Real Time Location System

A system which uses radio frequency tags to automate tracking on a continuous basis.

Robotics

A branch of computer science and engineering that involves developing and building complex machines which can perform complex tasks. Robots are continuing to get smarter with the advancement of Artificial Intelligence.

Sensor

A device which detects changes to the surrounding area and environment. Sensors are increasingly able to communicate with smartphones and other computing devices.

Sensors are a key component of the Internet of Things. Embedding sensors inside of objects are what creates connected devices.

Smart Home

A connected house, or living area, which uses technology to improve the quality of life. Smart homes will typically learn and adapt to your daily habits. For example, a smart home outfitted with a smart thermostat will learn your heating/cooling preferences and update accordingly without the need for human input.

Smartphone

A mobile phone which incorporates sophisticated sensors and a variety of digital computing capabilities. These capabilities include cameras, GPS, microphone, and many other features. There are currently over 2 Billion smartphones worldwide.

Telemetry

The ability for machines to communicate with one another and exchange data with computers and other systems through advanced telecommunication features.

UAV – Unmanned Air Vehicles

An aerial vehicle that operates without the aid of an onboard crew. These are often referred to as drones and can be used for military, business, or personal reasons. Drones are controlled remotely, usually from the ground and within sight of the UAV.

Wearable Computing

Also known as wearables, wearable computing is when small computing devices are worn and used on the body. For example, smart watches, fitness trackers, smart glasses, or any other device similar device is considered a wearable. Wearable computing devices typically have built in sensors which exchange data with other devices such as smartphones.

2 Chapter: Introduction

The rapid evolution of technology in industrial automation systems requires tighter integration between devices on the plant floor and the rest of the enterprise. This integration requires a secure network infrastructure, smart devices for efficient data collection, and the ability to turn data into actionable information.

The integration of control and information across the enterprise enables our customers to optimize their operations by connecting the plant, site, facility, and people.

In a time of constant and rapid technological development, it would be quite ambitious to develop and present a course that claimed to cover each industrial measuring type of equipment. This course is not intended to be an encyclopedia of Instrumentation, Control Valves, Industrial Automation, Ethernet or any upcoming modern technologies i.e. IoT, LoRa, 5G, IoT protocols and Project Management but rather a training guide for gaining experience in this fast-changing environment.

This handbook is aimed at providing engineers, technicians and any other personnel involved with Process Management, Industrial Automation and upcoming technologies with more experience in that field. It is also designed to give readers the fundamentals on analyzing the process requirements and selecting suitable solutions for their applications.

3 Chapter: Instrumentation

Instrumentation is defined as the art and science of measurement and control of process variables within a production or manufacturing area.

3.1 Instrument

An instrument is a device that measures a physical quantity such as flow, temperature, level, distance, angle, or pressure. Instruments may be as simple as direct reading thermometers or may be complex multi-variable process analyzers. Instruments are often part of a control system in refineries, factories, and vehicles.

Classification of instruments

Analog instrument

The measured parameter value is display by the moveable pointer. The pointer will have moved continuously with the variable parameter/analog signal which is measured. The reading is inaccurate because of parallax error (parallel) during the skill reading. E.g: ampere meter, voltage meter, ohm meter etc.

Digital instrument

The measured parameter value is display in decimal (digital) for m which the reading can be read thru in numbers form. Therefore, the parallax error is not existed and terminated.

The concept used for digital signal in a digital instrument is logic binary '0 'and '1'.

Characteristic of Instruments

Figure below presents a generalized model of a simple instrument. The physical process to be measured is in the left of the figure and the measurand is represented by an observable physical variable X.

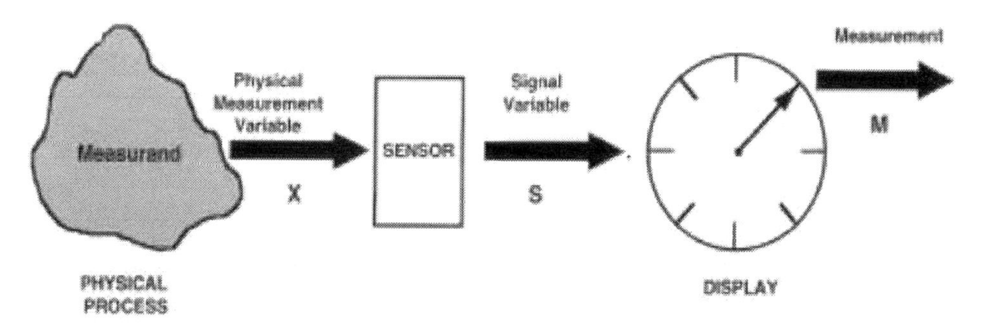

Figure: Characteristic of Instruments

For example, the mass of an object is often measured by the process of weighing, where the measurand is the mass but the physical measurement variable is the downward force the mass exerts in the Earth's gravitational field. There are many possible physical measurement variables.

The key functional element of the instrument model shown in Figure above is the sensor, which has the function of converting the physical variable input into a signal variable output.

Signal variables have the property that they can be manipulated in a transmission system, such as an electrical or mechanical circuit. Because of this property, the signal variable can be transmitted to an output or recording device that can be remote from the sensor. In electrical circuits, voltage is a common signal variable. In mechanical systems, displacements or force are commonly used as signal variables. Other examples of signal variable are shown in Table below.

Common physical variables	Typical signal variables
• Force • Length • Temperature • Acceleration • Velocity • Pressure • Frequency • Capacity • Resistance • Time • …	• Voltage • Displacement • Current • Force • Pressure • Light • Frequency

Table: Example physical variables

Instruments attached to a control system may provide signals used to operate solenoids, valves, regulators, circuit breakers, or relays. These devices control a desired output variable and provide either remote or automated control capabilities. These are often referred to as final control elements when controlled remotely or by a control system.

The signal output from the sensor can be displayed, recorded, or used as an input signal to some secondary device or system. In a basic instrument, the signal is transmitted to a Display or recording device where the measurement can be read by a human observer. The observed output is the measurement **M**.

Measurement instruments have three traditional classes of use:

- Monitoring of processes and operations
- Control of processes and operations
- Experimental engineering analysis

3.2 Measurement Parameters

Instrumentation is used to measure many parameters (physical values). These parameters include:

• Pressure, either differential or static • Flow • Temperature • Levels of liquids, etc. • Density • Viscosity	• Other mechanical properties of materials • Properties of ionizing radiation • Frequency • Current • Voltage • Inductance	• Capacitance • Resistivity • Chemical composition • Chemical properties • Properties of light • Vibration • Weight

Process of measurement

Measurement is essentially the act, or the result, of a quantitative comparison between a given quantity and a quantity of the same kind chosen as a unit. The result of measurement is expressed by a number representing the ratio of the unknown quantity to the adopted unit of measurement.

3.3 Pressure Transducer

A pressure transducer is used to convert a certain value of pressure into its corresponding mechanical or electrical output. Measurement if pressure is of considerable importance in process industries.

3.3.1 Pressure

Pressure is the force per unit area applied in a direction perpendicular to the surface of an object. Gauge pressure is the pressure relative to the local atmospheric or ambient pressure. Pressure is an effect which occurs when a force is applied on a surface. Pressure is the amount of force acting on a unit area. The symbol of pressure is P.

Types of Pressure Measurement:

- Gage pressure (psig) quantifies fluid pressure relative to ambient air pressure.
- Absolute pressure (psia) measures pressure relative to a vacuum.
- Sealed reference pressure (psis) (bars) is measured relative to a situation pressure whose magnitude is at, or close to, standard atmospheric pressure.
- Differential pressure (psid) (bard) quantifies the pressure difference between two points within a system.

The measurement also must consider the magnitude of the system's line pressure. Measurements usually are taken from two dissimilar fluid inputs within the system using a transducer designed specifically for differential-pressure calculations.

The types of pressure sensors are differentiated according to the amount of differential pressure they are able to measure. For low differential pressure measurement Liquid Column Manometers are used. Elastic type pressure gauges are also used for pressure measurement up to 700 MPa. Some of the common elastic/mechanical types are:

3.3.1.1 Bourdon Tubes

Bourdon Tubes are known for its very high range of differential pressure measurement in the range of almost 100,000 psi (700 MPa). It is an elastic type pressure transducer. The bourdon pressure gauges used today have a slight elliptical cross-section and the tube is generally bent into a C-shape or arc length of about 27 degrees. The detailed diagram of the bourdon tube is shown below.

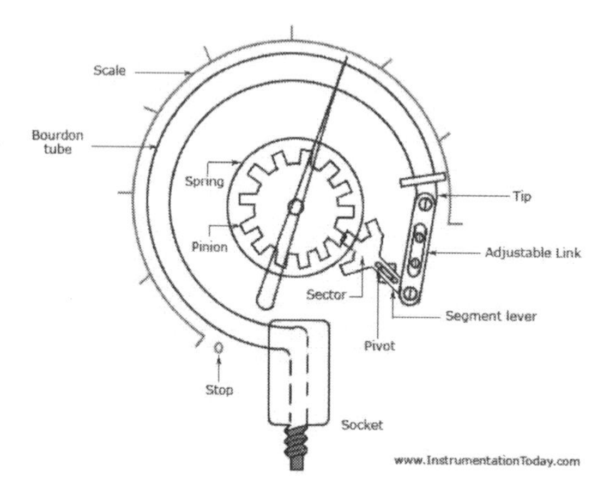

Bourdon Tube Pressure Gauge

Figure: Bourdon Tube Pressure Gauge

Working Principle:

As the fluid pressure enters the bourdon tube, it tries to be reformed and because of a free tip available, this action causes the tip to travel in free space and the tube unwinds. The simultaneous actions of bending and tension due to the internal pressure make a non-linear movement of the free tip. This travel is suitable guided and amplified for the measurement of the internal pressure. But the main requirement of the device is that whenever the same pressure is applied, the movement of the tip should be the same and on withdrawal of the pressure the tip should return to the initial point.

3.3.1.2 Diaphragm

A diaphragm pressure transducer is used for low pressure measurement. They are commercially available in two types – metallic and non-metallic.

Metallic diaphragms are known to have good spring characteristics and non-metallic types have no elastic characteristics. Thus, non-metallic types are used rarely, and are usually opposed by a calibrated coil spring or any other elastic type gauge. The non-metallic types are also called slack diaphragm.

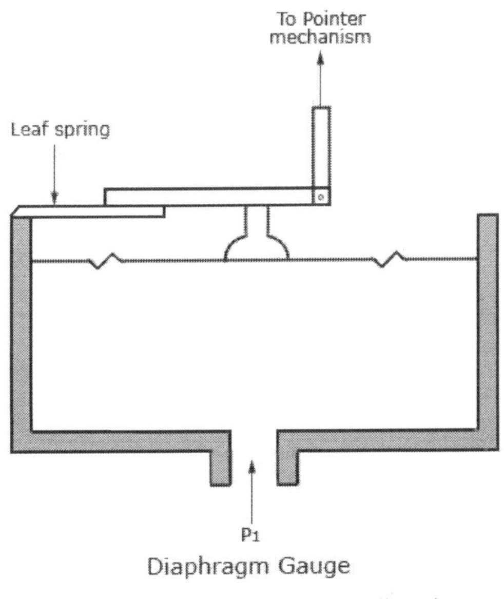

Figure: Diaphragm Gauge

Working Principle:
The diagram of a diaphragm pressure gauge is shown below. When a force acts against a thin stretched diaphragm, it causes a deflection of the diaphragm with its centre deflecting the most.

Since the elastic limit has to be maintained, the deflection of the diaphragm must be kept in a restricted manner. This can be done by cascading many diaphragm capsules as shown in the figure below. A main capsule is designed by joining two diaphragms at the periphery. A pressure inlet line is provided at the central position. When the pressure enters the capsule, the deflection will be the sum of deflections of all the individual capsules. As shown in figure (3), corrugated diaphragms are also used instead of the conventional ones.

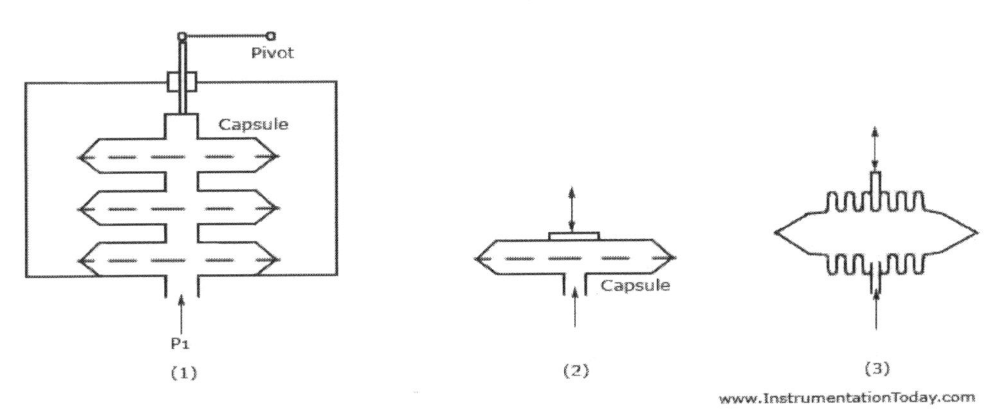

Figure: Diaphragm Pressure Transducer

Diaphragm Pressure Transducer

Corrugated designs help in providing a linear deflection and also increase the member strength. The total amount of deflection for a given pressure differential is known by the following factors:

- Number and depth of corrugation
- Number of capsules
- Capsule diameter
- Shell thickness
- Material characteristics

3.3.1.3 Bellows

Like a diaphragm, bellows are also used for pressure measurement, and can be made of cascaded capsules. The basic way of manufacturing bellows is by fastening together many individual diaphragms. The bellows element, basically, is a one piece expansible, collapsible and axially flexible member. It has many convolutions or fold. It can be manufactured form a single piece of thin metal. For industrial purposes, the commonly used bellow elements are:

- By turning from a solid stock of metal
- By soldering or welding stamped annular rings
- Rolling a tube
- By hydraulically forming a drawn tubing

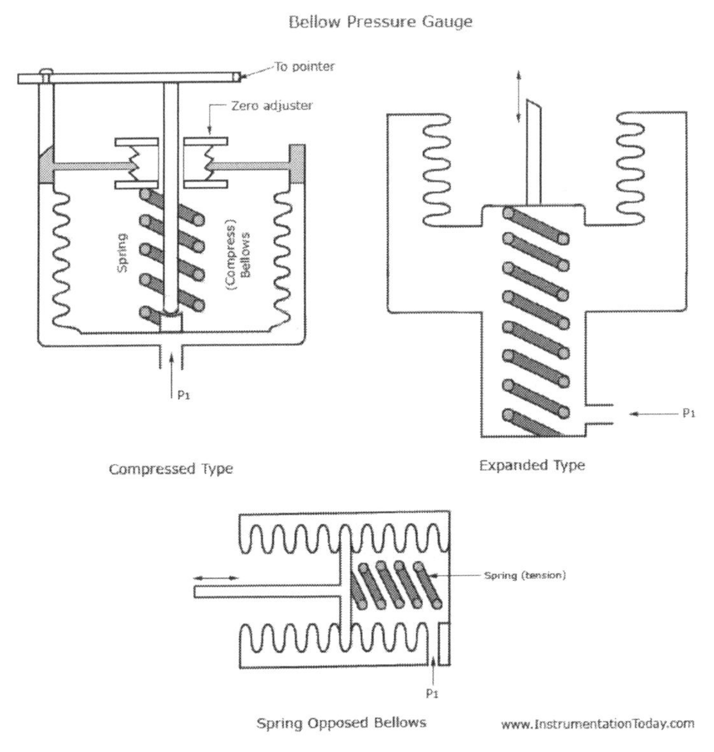

Figure: Bellows

Working Principle:

The action of bending and tension operates the elastic members. For proper working, the tension should be least. The design ideas given for a diaphragm is applied to bowels as well. The manufacturer describes the bellows with two characters – maximum stroke and maximum allowable pressure. The force obtained can be increased by increasing the diameter. The stroke length can be increased by increasing the folds or convolutions.

3.3.1.4 Piston Type Pressure Transducer

As shown in the figure below, the input pressure is given to the piston. This moves the piston accordingly and causes the spring to be compressed. The piston position will be directly proportional to the amount of input pressure exerted. A meter is placed outside the piston and spring arrangement, which indicates the amount of pressure exerted. As the device has the ability to withstand shock, sudden pressure changes, and vibrations, it is commonly used in hydraulic applications. Mostly, the output of the piston and spring arrangement is given to a secondary device to convert movement into an electrical signal.

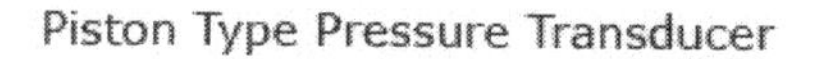

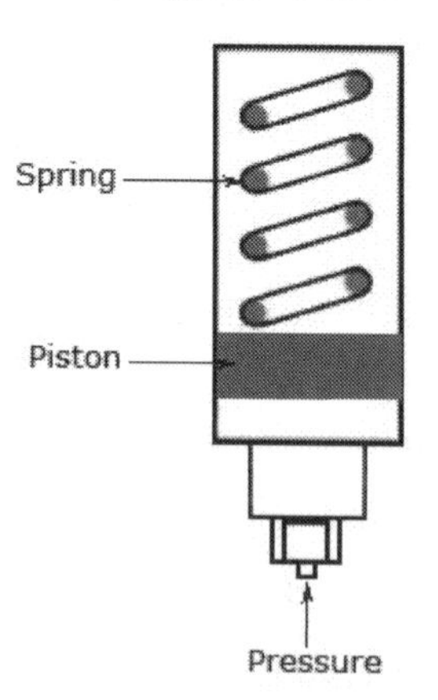

Figure: Piston Type Pressure Transducer

3.3.1.5 Electric Pressure Transducers

In general, “a transducer is a device which converts one form of energy into another form of energy.” However, in the field of electrical instrumentation, a transducer is a tool that converts a physical quantity, a physical condition or mechanical output to an electrical signal. Most of method converting mechanical output into an electrical signal works equally well for the bellows, the diaphragm, the bourdon tube. In this conversion a mechanical motion is first converted into a change in electrical resistance and then change in resistance is transformed into change in electrical current or voltage. Generally electrical pressure transducers consist of three elements:

- Pressure sensing element such as bellow, a diaphragm or a bourdon tube
- Primary conversion element e.g. resistance or a voltage
- Secondary conversion element

Types of Electric Pressure Transducers:

Strain Gauge Pressure Transducers:

Strain gauge is passive type of resistance pressure transducer, whose electrical resistance changes when it is stretched or compressed. It can be attached to a pressure sensing diaphragm.

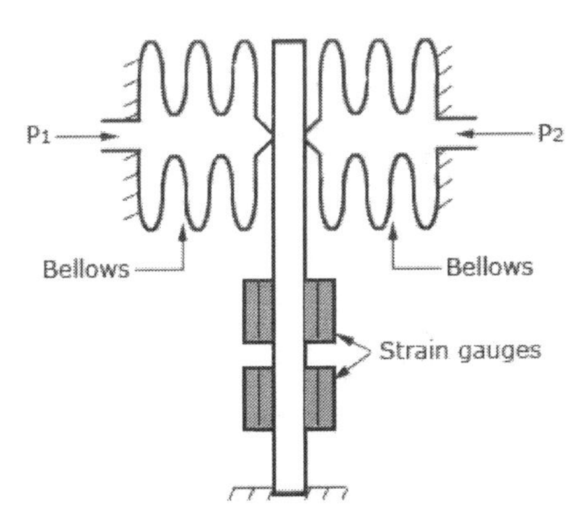

Figure: Pressure Measurement with Strain Gauge on Bellows

Working Principle:

The strain gauge is a fine wire which changes its resistance when mechanically strained, due to physical effects. A strain gauge may be attached to the diaphragm so that when the diaphragm flexes due to the process pressure applied on it, the strain gauge stretches or compresses. This deformation of the strain gauge causes the dissimilarity in its length and cross-sectional area due to which its resistance also changes. The resistance change of a strain gauge is usually converted into voltage by linking one, two, or four similar gauges as of Wheatstone bridge (known as strain gauge bridge) and applying excitation to the bridge. The bridge output voltage is then a measure of the pressure sensed by strain gauges.

Advantages:

- They are small and easy to install
- They have good accuracy
- They are available for wide range of measurements (from vacuum to 200000 psig)
- They possess good stability
- They have high output signal strength
- They have high over range capacity
- They are simple to maintain
- They contain no moving parts
- They possess good shock and vibration characteristics
- They are readily adoptable to electronic systems
- They possess fast speed of response

Disadvantages:

- Their cost is moderate to high (could be offset by reduced installation cost)
- Electrical readout is necessary in these transducers
- They are requiring constant voltage supply

- They require temperature compensation due to problem presented by temperature variations

3.3.1.6 Potentiometric Pressure Transducers

In this type of pressure transducer, there is a potentiometer (basically a variable resistance) which is made by winding resistance wire around an insulated cylinder. A moveable electrical contact, called a wiper, slides along the cylinder, touching the wire at one point on each turn. The position of the wiper determines how much wire, and therefore, how much resistance, is between the end of the wire and the wiper. A mechanical linkage from the pressure sensing element (such as bellows, a diaphragm, and the bourdon tube) controls the position of the wiper on the potentiometer. Some potentiometer is made curved so that the wiper can pivot in a circular motion rather than moving along a straight line. The location of wiper determines the confrontation of the potentiometer which in turn determines the pressure.

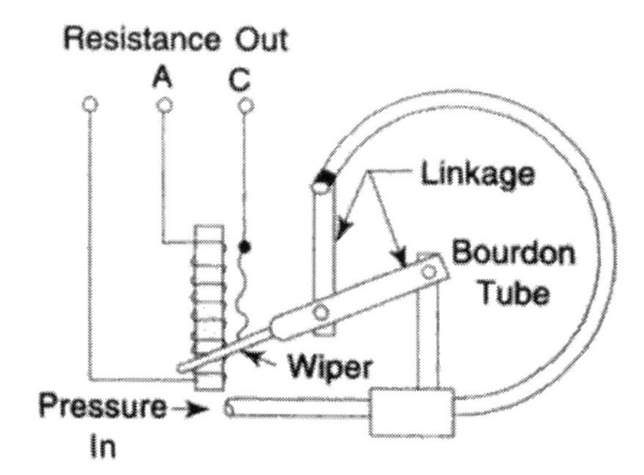

Figure: Potentiometric Pressure Transducers

Working Principle:

A Potentiometer pressure transducer in which the sensing element is bourdon tube. An increase in pressure makes the bourdon tube relax out partially. This motion causes the linkage to move the wiper across the winding on potentiometer as the wiper moves; it increases the resistance between terminals A and B which is equal to the pressure sensed by the bourdon tube.

Advantages:

- Provide strong output so no need of additional speaker
- Potentiometer pressure transducers have high range
- Potentiometer pressure transducers have high starkness
- Potentiometer pressure transducers have simple instrumentation
- Potentiometer pressure transducers have high electrical efficiency
- Potentiometer pressure transducers are in expensive

- Potentiometer pressure transducers are best suitable for measurements in the systems with least requirements

Disadvantages:

- Potentiometer pressure transducers have finite resolution
- Potentiometer pressure transducers have limited life
- Potentiometer pressure transducers are of large size
- Potentiometer pressure transducers have poor frequency response
- Potentiometer pressure transducers have tendency to develop noise
- Potentiometer pressure transducers have susceptibility to vibration

3.3.1.7 Reluctance Pressure Transducers

Reluctance in magnetic circuit is corresponding to resistance in an electrical circuit. Whenever coupling between two magnetic coils changes, then reluctance between them also changes. So, pressure sensor can be used to change the spacing between the coils by moving one part of the magnetic circuit. This motion changes the reluctance between the coils which causes the change voltage induced by one coil in other. This induced voltage can be interpreted as a change in pressure. Reluctance pressure transducer of several types
One of them is discussed below.

Linear variable differential transformer (LVDT)
It is widely used inductive transducer that translates linear motion into an electrical signal.

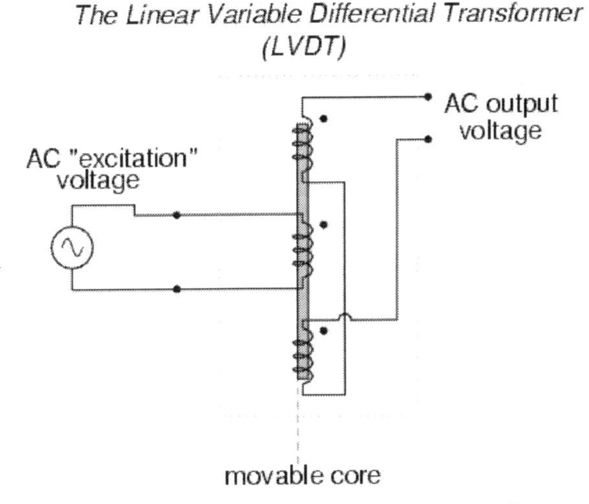

Figure: Linear variable differential transformer

Construction and Working:

It consists of one primary and two secondary coils. These windings are arranged concentrically next to each other. They are wound over a hollow bobbin which is usually a non-magnetic and insulating material. A ferromagnetic armature is attached to the transducers sensing shaft. A core is generally made of a high permeability alloy and has a shape of cylinder or rod. A.C excitation is supplied to a primary coil and moveable core is varies the coupling between it and two secondary coils. When the core is in center position then coupling with secondary coil is equal. As the core moves away from the center position then the coupling the one secondary, and hence its output voltage, increases while the coupling and the output voltage of other secondary decreases.

Any change in pressure makes the bellows expand or contract. This motion moves the magnetic core inside hollow portion of bobbin. It causes the voltage of one secondary winding to increases, while simultaneously reducing voltage in other secondary winding. The difference of this voltage that appear across the output terminals of the transducers and gives the measure of physical position of the core and hence the pressure.

Advantages:

- It possesses a high sensitivity
- It is very rugged in construction
- It tolerates high degree of shock and vibration without any adverse effect
- It is stable and easy to align and maintain due to simplicity of construction, small size and light body
- The output voltage of this transducer is practically linear for the displacement of about 5 mm
- It has infinite resolution
- It shows a low hysteresis, hence repeatability is excellent under all condition

Disadvantages:

- Relatively large core displacements are required for appreciable amount of differential output
- They are sensitive to stray magnetic fields, but shielding is possible
- Temperature effects the performance of transducers

3.3.1.8 Piezoelectric Pressure Transducers

These devices utilize the piezoelectric characteristics of certain crystalline and ceramic materials (such as quartz) to generate an electrical signal.

Basic Principle:

These transducers depend upon the principle that “when pressure is applied on piezoelectric crystals (such as a quartz), an electrical charge is generated”. These are about 40 crystal-line material that, when subjected to “squeeze”, generate an electric charge. Some of piezoelectric materials are barium titanate sintered powder, crystal of quartz, tourmaline, Rochelle salts.

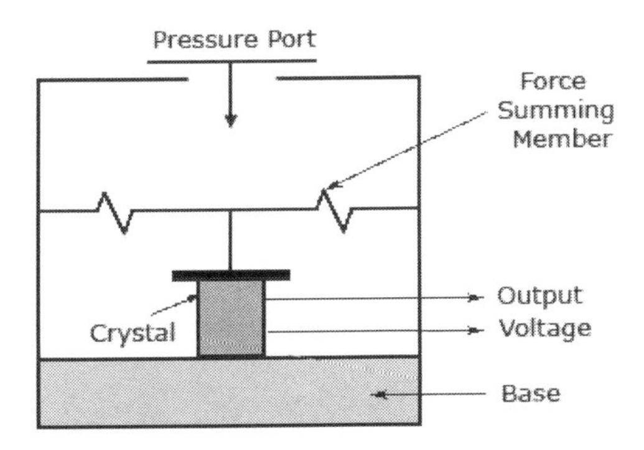

Figure: Piezoelectric Pressure Transducers

Construction and Working:
The X-Y axis of a piezoelectric crystal and its cutting technique is shown in the figure below.

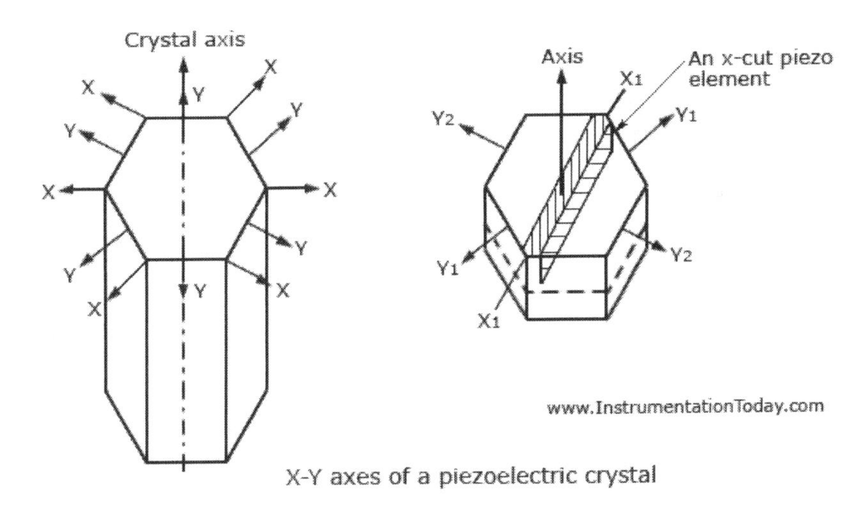

X-Y axes of a piezoelectric crystal

The direction, perpendicular to the largest face, is the cut axis referred to.

If an electric stress is applied in the directions of an electric axis (X-axis), a mechanical strain is produced in the direction of the Y-axis, which is perpendicular to the relevant X-axis. Similarly, if a mechanical strain is given along the Y-axis, electrical charges will be produced on the faces of the crystal, perpendicular to the X-axis which is at right angles to the Y-axis.

It consists of diaphragm by which pressure is transmitted to the piezoelectric crystal Y1. This crystal generates electrical signal which is amplified by a charge amplifier. A second piezoelectric crystal Y2 is included to compensate for any acceleration of the device during use. This compensation is needed because rapid acceleration of transducers creates additional pressure on piezoelectric crystal. Vibration is major a source high, rapidly changing acceleration.

Signal from the compensating crystal are amplified by second charge amplifier. A different amplifier which are used to substrates pressure alone; all effects of acceleration are removed. Piezoelectric pressure transducers are used to measure very high pressure that change very rapidly. For example, the pressure inside the cylinder of gasoline engine changes very rapidly from less than atmospheric pressure to many thousands of PSI.

Similar pressure changes in compressors, rocket motors, etc. It is impossible for ordinary pressure transducers to measure such a high-pressure change over such short time periods. They do not respond fast enough. But piezoelectric materials produced electrical voltage when they or squeezed suddenly. The voltage disappears when the pressure stops changing. Piezoelectric pressure transducer may be used to measure a pressure over a range up to 0-50,000 psi. However, piezoelectric transducers cannot measure steady pressure. They respond only to changing pressures.

Advantages:

- The transducers no needs external power and is therefore self-generating (active type)
- It has a good frequency response
- Simple to use as they have small dimensions and large measuring range.
- Barium titanate and quartz can be made in any desired shape and form. It also has a large dielectric constant. The crystal axis is selectable by orienting the direction of orientation.

Disadvantages:

- This type of transducers cannot measure static pressure
- Output of the transducers is affected by changes in temperature. Therefore, temperature-compensating devices have to be used.

3.3.1.9 Capacitive Pressure Transducers

A capacitance pressure transducer is based on the fact that dielectric constants of liquids, solids and gases change under pressure. The figure below shows an arrangement of a cylindrical capacitor that can withstand large pressure. As the change in dielectric constant is quite small (only about ½ percent change for a pressure change of about 10 MPa), it is usable only at large change in pressure. Besides, the capacitance-pressure relation is non-linear and is affected by temperature variation. The measurement of this capacitance is done by a resonance circuit. The schematic is shown below. The oscillogram giving the variation of the output voltage with capacitance is also shown below.

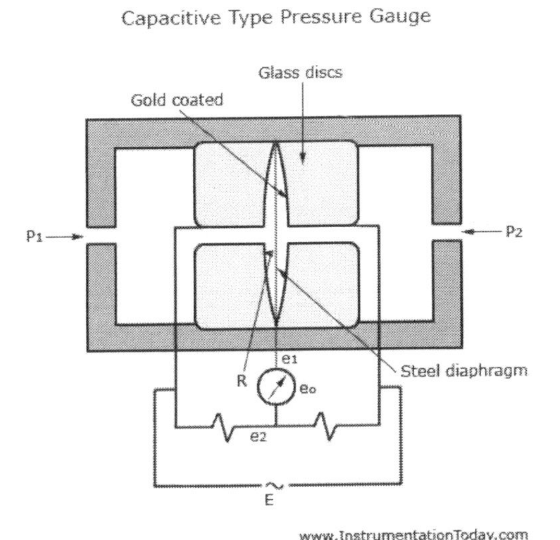

Figure: Capacitive Type Pressure Gauge

The term capacitor is defined as two metal plates are separated by a distance d. A dielectric medium is placed between the plates. When voltage or potential difference is applied to them, equal and opposite charges is getting developed on the plates. A capacitive transducer works on the principle of capacitance of parallel plate capacitor.

$C=\varepsilon 0\ \varepsilon r\ A/d$

Where,

C = the capacitance of a capacitor in farad

A = area of each plate in m2

d = distance between two plates in m

εr= dielectric constant

ε0 = 8.854*10-12 farad/m2

Thus, capacitance can be varied by changing distance between the plates, area of the plate or value of the dielectric medium between the plates. Any change in these factors cause change in capacitance. In capacitive transducers, pressure is utilized to vary any of the above-mentioned factors which will cause change in capacitance and that is a measurable by any suitable electric bridge circuit and is proportional to the pressure.

Working Principle:

It uses the principle of change in capacitance due to change in distance between two plates when pressure is applied to the diaphragm. A capacitive pressure transducer consist of an airtight housing in which metallic diaphragm is placed. At the inner side, fixed plate is placed. In between diaphragm and fixed plate, the dielectric medium is placed. When pressure is applied to the diaphragm, it moves towards a fixed plate resulting change in the capacitance. This capacitance is calibrated into voltage proportional to the applied pressure.

Advantages:

- As these transducers have high input impedance, the loading effect is minimum
- They require small force for operation, hence is useful for small displacement, pressure measurements and require small power
- These transducers have good frequency response
- They are less affected by stray magnetic fields

Disadvantages:

- These transducers require good quality isolation so as protect the transducer (metal plates) from stray capacitance
- Guard rings are necessaries so as to minimize stray electric fields
- Sometime frequency response affected by loading effects from connecting links and cables
- The performance may be affected by parameters like dust, temperature, moisture content variations and so fourth
- They require complex circuit arrangement like bridge, amplifier, etc. for measurement purpose

What are Pressure Transmitters?

Pressure transducers are devices that convert the mechanical force of applied pressure into electrical energy. This electrical energy becomes a signal output that is linear and proportional to the applied pressure. Pressure transducers are very similar to pressure sensors and transmitters. In fact, transducers and transmitters are nearly synonymous. The difference between them is the kind of electrical signal each sends. A transducer sends a signal in volts (V) or millivolt per volt (mV/V), and a transmitter sends signals in milliamps(mA).

Both transmitters and transducers convert energy from one form to another and give an output signal. This signal goes to any device that interprets and uses it to display, record or alter the pressure in the system. These receiving devices include computers, digital panel meters, chart recorders and programmable logic controllers (PLC). There are a wide variety of industries that use pressure transducers and transmitters for various applications. These include, but are not limited to, medical, air flow management, factory automation, HVAC and refrigeration, compressors and hydraulics, aerospace and automotive

Signal connection and programming standards vary somewhat between different models of PLC, but they are similar enough to allow a "generic" introduction to PLC programming here. The following illustration shows a simple PLC, as it might appear from a front view. Two screw terminals provide connection to 120 volts AC for powering the PLC's internal circuitry, labeled L1 and L2. Six screw terminals on the left-hand side provide connection to input devices, each terminal representing a different input "channel" with its own "X" label. The lower-left screw terminal is a "Common" connection, which is generally connected to L2 (neutral) of the 120 VAC power source.

3.4 Sensor

All types of sensors can be classed as two kinds, passive and active.
A sensor (also called detectors) is a converter that measures a physical quantity and converts it into a signal which can be read by an observer or by an (today mostly electronic) instrument. For example, a mercury-in-glass thermometer converts the measured temperature into expansion and contraction of a liquid which can be read on a calibrated glass tube. A thermocouple converts temperature to an output voltage which can be read by a voltmeter. For accuracy, most sensors are calibrated against known standards.

Active sensors

Active sensors require some form of external power to operate. For example, a strain gauge is a pressure-sensitive resistor. It does not generate any electrical signal, but by passing a current through it. its resistance can be measured by detecting variations in the current and/or voltage across it relating these changes to the amount of strain or force.

Passive sensor

A passive sensor does not need any additional energy source. and directly generates an electric signal in response to an external stimulus. For example, a thermocouple or photodiode. Passive sensors are direct sensors which change their physical properties, such as resistance, capacitance or inductance etc. As well as analogue sensors, Digital Sensors produce a discrete output representing a binary number or digit such as a logic level "0" or a logic level "1

Proximity Sensor

Proximity sensors detect the presence of objects without physical contact. Since 1983 Fargo Controls' proximity sensors have been of the highest quality, durability & repeatability to meet today's tough industrial requirements. Typical applications include the detection, position, inspection and counting on automated machines and manufacturing systems. They are also used in the following machinery: packaging, production, printing, plastic molding, metal working, food processing, etc

Inductive & Capacitive Proximity Sensor

Their operating principle is based on a high frequency oscillator that creates a field in the close surroundings of the sensing surface. The presence of a metallic object (inductive) or any material (capacitive) in the operating area causes a change of the oscillation amplitude. The rise or fall of such oscillation is identified by a threshold circuit that changes the output state of the sensor. The operating distance of the sensor depends on the actuator's shape and size and is strictly linked to the nature of the material (Table 1 & Table 2.). A screw placed on the back of the capacitive sensor allows regulation of the operating distance. This sensitivity regulation is useful in applications, such as detection of full containers and non-detection of empty containers.

***Table 1:* INDUCTIVE SENSORS**
Sensitivity when different metals are present.
Sn = operating distance.

Fe37 (iron)	1 x Sn
Stainless steel	0.9 x Sn
Brass- bronze	0.5 x Sn
Aluminum	0.4 x Sn
Copper	0.4 x Sn

***Table 2:* CAPACITIVE SENSORS**
Sensitivity when different materials are present.
Sn = operating distance.

Material	Distance
Metal	1 x Sn
Water	1 x Sn
Plastic	0.5 x Sn
Glass	0.5 x Sn
Wood	0.4 x Sn

Photoelectric

These sensors use light sensitive elements to detect objects and are made up of an emitter (light source) and a receiver. Three types of photoelectric sensors are available. Direct Reflection - emitter and receiver are housed together and uses the light reflected directly off the object for detection. Reflection with Reflector - emitter and receiver are housed together and requires a reflector. An object is detected when it interrupts the light beam between the sensor and reflector. Thru Beam - emitter and receiver are housed separately and detects an object when it interrupts the light beam between the emitter and receiver.

Magnetic

Magnetic sensors are actuated by the presence of a permanent magnet. Their operating principle is based on the use of reed contacts, which consist of two low reluctance ferro-magnetic reeds enclosed in glass bulbs containing inert gas. The reciprocal attraction of both reeds in the presence of a magnetic field, due to magnetic induction, establishes an electrical contact.

3.5 Temperature Measurement

Temperature measurement relies on the transfer of heat energy from the process material to the measuring device. The measuring device therefore needs to be temperature dependent.

There are two main industrial type of temperature sensors:

- Contact
- Non- Contact

Contact

Contact is the more common and widely used form of temperature measurement. The three main types are:

- Thermocouples
- Resistance Temperature Detectors (RTD's)
- Thermistors

These types of temperature devices all vary in electrical resistance for temperature change. The rate and proportion of change is different between the three types, and also different within the type classes.

Another less common device relies on the expansion of fluid up a capillary tube. This is where bulk of the fluid is exposed to the process materials temperature.

Non-Contact

Temperature measuremnt by non-contact means is more specialised and can be performed with the following technologies:

- Infrared
- Acoustics

Thermocouple Basics

Let's start with T. J. Seebeck, who in 1821 discovered what is now termed the thermoelectric effect. He noted that when two lengths of dissimilar metal wires (such as iron and Constantan) are connected at both ends to form a complete electric circuit, an emf is developed when one junction of the two wires is at a different temperature than the other junction. Basically, the developed emf (actually a small millivoltage) is dependent upon two conditions: (1) the difference in temperature between the hot junction and the cold junction. Note that any change in either junction temperature can affect the emf value and (2) the metallurgical composition of the two wires.

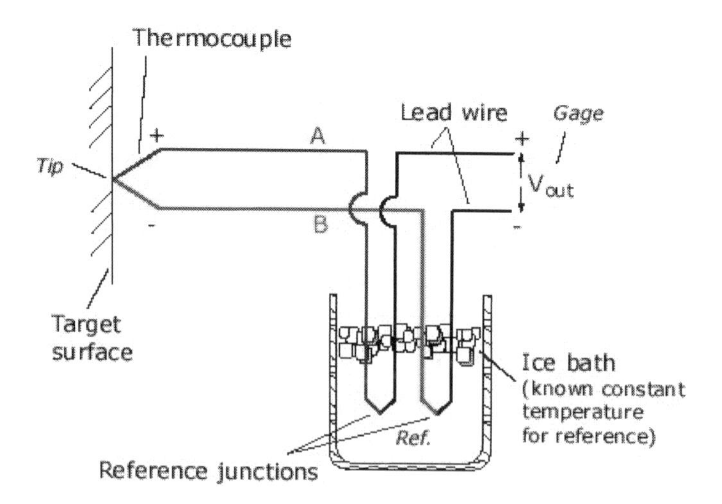

The Law of Intermediate Temperatures provides a means of relating the emf generated under ordinary conditions to what it should be for the standardized constant temperature (e.g., 32¡F). Referring to Figure, which shows thermocouples A and B made of the same two dissimilar metals; this diagram will provide an example of how the law works.

Resistance temperature detectors

Resistance temperature detectors (RTDs) operate on the principle of changes in the electrical resistance of pure metals and are characterized by a linear positive change in resistance with temperature. Typical elements used for RTDs include nickel (Ni) and copper (Cu), but platinum (Pt) is by far the most common because of its wide temperature range, accuracy, and stability.

RTDs are popular because of their excellent stability and exhibit the most linear signal with respect to temperature of any electronic temperature sensor. They are generally more expensive than alternatives, however, because of the careful construction and use of platinum. RTDs are also characterized by a slow response time and low sensitivity, and because they require current excitation, they can be prone to self-heating.

RTDs are commonly categorized by their nominal resistance at 0 °C. Typical nominal resistance values for platinum thin-film RTDs include 100 W and 1000 W. The relationship between resistance and temperature is very linear and follows the equation

For < 0°C $R_T = R_0 [1 + aT + bT^2 + cT^3 (T - 100)]$

For > 0°C $R_T = R_0 [1 + aT + bT^2]$

Where R_T = resistance at temperature T

R_0 = nominal resistance

a, b, and c are constants used to scale the RTD

The most common RTD is the platinum thin-film with an a of 0.385%/°C and is specified per DIN EN 60751. The a value depends on the grade of platinum used, and commonly include 0.3911%/°C and 0.3926%/°C. The a value defines the sensitivity of the metallic element but is normally used to distinguish between resistance/temperature curves of various RTDs.

2 Wire PT 100 provides one connection to each end of the sensor. This construction is suitable particularly where the changes in lead resistance due to ambient temperature changes can be ignored.

3 Wire PT 100 provides one connection to one end and two to the other end of the sensor. Connected to an instrument designed to accept three-wire input, compensation is achieved for lead resistance and temperature change in lead resistance. This is the most commonly used configuration

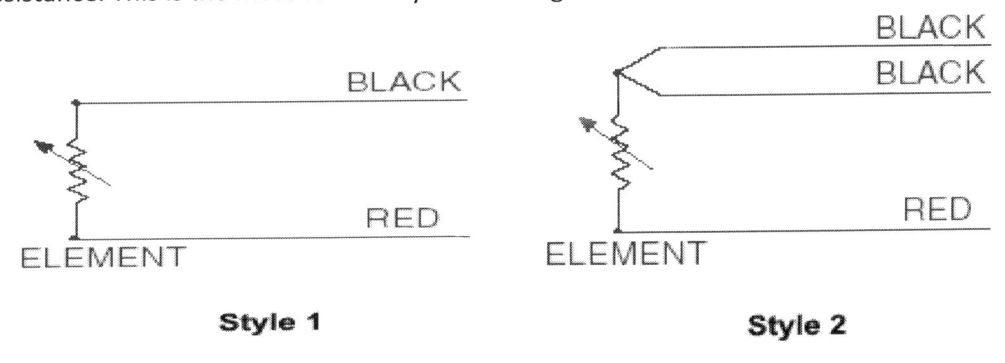

Figure: Temperature Sensor

Attributes of the Temperature Sensor		
Parameter/Criteria	**Thermocouple**	**RTD**
Typical Measurement Range	-450 °F (-267 °C) to +4200 °F (2316 °C)	-400 °F (-240 °C) to +1200 °F (649 °C)
Interchangeability	Good	Excellent
Long-term Stability	Poor to Fair	Excellent
Accuracy	Medium	High
Repeatability	Poor to Fair	Excellent
Sensitivity (output)	Low	Good
Response	Medium to Fast	Good
Linearity	Fair	Good
Self Heating	No	Low
Tip (end) Sensitivity	Excellent	Fair
Lead Effect	High	Medium
Size/Packaging	Small to Large	Medium to Small

Advantages and Disadvantages		
Sensor	**Advantages**	**Disadvantages**
Thermocouple	• Inexpensive • No resistance leadwire problems • Fastest response • Simple and rugged • High temperature operation • Tip (end) temperature sensing	• Least sensitive • Non-linear • Low voltage • Least stable, repeatable
RTD	• Good stability • Excellent accuracy • Contamination resistant • Good linearity • Area temperature sensing • Very repeatable temperature measurement	• Marginally higher cost • Current source required • Self-heating • Slower response time • Medium sensitivity to small temperature changes

Choosing the Right

Criteria	Thermocouples vs RTDs	TC	RTD
Range	Although new and improved manufacturing techniques have increased the range of RTDs, this category belongs to Thermocouples. Better than 95% of RTDs are used in temperatures below 1000° F. Thermocouples can be used up to 2700° F.	*	
Sensitivity	Grounded Thermocouples are inherently tip sensitive; while RTD elements are isolated from	*	

	their sheaths. A grounded Thermocouple will respond to a 63% step change in temperature nearly 3 times faster than a RTD counterpart.		
Cost	Comparing a 12 inch, SS sheath .25", Type J grounded Thermocouple, with a 100 Ohm platinum RTD.00385 Alpha, prices the thermocouple at 2.5 to 3 times less than an RTD. Installed cost make up some of this difference since RTDs use inexpensive copper lead wire to transmit the signal back to the DCS.	*	
Accuracy	There are many factors to determine accuracy; linearity, stability, and repeatability to name a few that can affect accuracy. While a Thermocouple's standalone accuracy can approach that of an RTD, the superior advantages in these other areas make the RTD the choice.		*
Linearity	Temperature vs. resistance nearly plot a straight line for an RTD, while a Thermocouple shows an almost "S" like curve.		*
Ruggedness	Thermocouples can essentially be one piece. RTD elements both thin film and wire wound must be connected to copper wire.	*	
Stability	Due to their linearity and virtually drift free output, RTDs are more stable than Thermocouples.		*

Thermistor

A thermistor is an input transducer (sensor) which converts temperature (heat) to resistance. Almost all thermistors have a negative temperature coefficient (NTC) which means their resistance decreases as their temperature increases. It is possible to make thermistors with a positive temperature coefficient (resistance increases as temperature increases) but these are rarely used. Always assume NTC if no information is given.

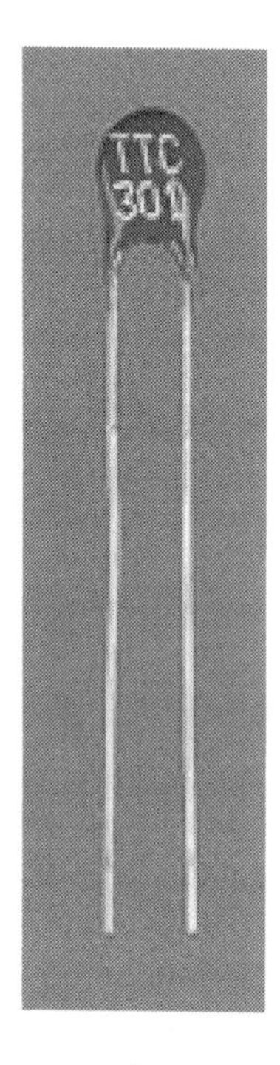

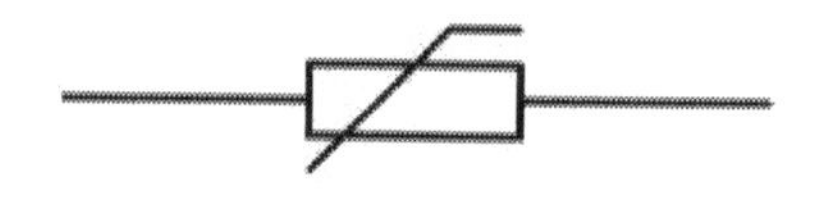

Thermistor circuit symbol

A multimeter can be used to find the resistance at various temperatures, these are some typical readings for example:
Icy water 0°C: high resistance, about 12kΩ.
Room temperature 25°C: medium resistance, about 5kΩ.
Boiling water 100°C: low resistance, about 400Ω.
Suppliers usually specify thermistors by their resistance at 25°C (room temperature). Thermistors take several seconds to respond to a sudden temperature change; small thermistors respond more rapidly.
A thermistor may be connected either way round and no special precautions are required when soldering. If it is going to be immersed in water the thermistor and its connections should be insulated because water is a weak conductor; for example, they could be coated with polyurethane varnish.

Non-Contact Pyrometers

Non-contact temperature sensors work in the infrared portion of the spectrum. The infrared range falls between 0.78 microns and 1000 microns in wavelength and is invisible to the naked eye. The infrared is region can be divided into three regions: near-infrared (0.78-3.0 microns); middle infrared (3-30 microns); and far infrared (30-300 microns). The range between 0.7 microns and 14 microns is normally used in infrared temperature measurement. The divisions have been related to the transmission of the atmosphere for several types of applications.

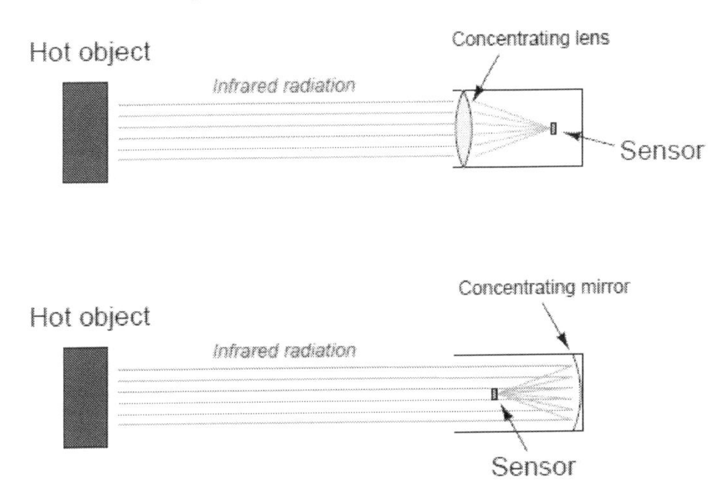

Figure: Two designs of non-contact pyrometer

Infrared Pyrometers

Infrared pyrometer is derived from the Greek root pyro, meaning fire. The term pyrometer was originally used to denote a device capable of measuring temperatures of objects above incandescence, objects bright to the human eye. The original pyrometers were non-contacting optical devices which intercepted and evaluated the visible radiation emitted by glowing objects. A modern and more correct definition would be any non-contacting device intercepting and measuring thermal radiation emitted from an object to determine surface temperature.

Infrared Thermometers: Function and Operation

Infrared thermometers are adaptable to most temperature measurement applications. Capable of measuring a broad spectrum of temperature ranges from approximately -4 to 7,200°F (-20 to 4,000°C), they are available in designs as simple as hand-held point-and-shoot guns with laser sighting or as sophisticated as integrated process control systems that can output temperature measurements in real time.

All infrared thermometers function utilizing the same basic design principles (figure 4). Using various infrared filters, an optical lens system focuses energy to an infrared detector, which converts the energy to an electrical signal. This electrical signal is compensated for emissivity, typically manually. Through linearization and amplification in the instrument's processor, an analog signal (typically, 1 to 5 VDC or 4 to 20 mA) is output. Electronics can be incorporated to convert the analog output to digital signals that can be transmitted at high speeds, allowing for extremely fast data acquisition rates. Ambient temperature compensation electronics ensure that temperature variations inside the infrared thermometer do not impact its output. Some applications with extremely hot operating environments can require water-jacket cooling at the infrared sensor.

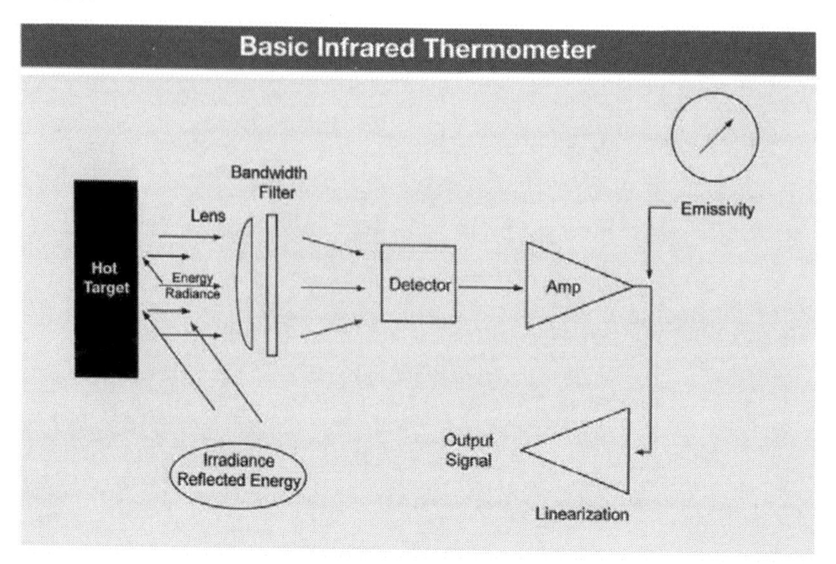

Figure: Basic Infrared Thermometer

Infrared thermometers usually operate in either very broad or very narrow bandwidths. Typically, broad-bandwidth infrared thermometers have a much wider temperature range than narrow-bandwidth units. The drawback to broad-bandwidth infrared thermometers is that emissivity influences them to a greater degree. Keep in mind that emissivity is dynamic and can have different values at different wavelengths. If an infrared thermometer operating at 3 to 5 µm has an emissivity value of 0.6 at 3 µm and 0.9 at 5 µm, what is the true emissivity value? Without knowing the true emissivity value, the true target temperature cannot be established. Narrow-bandwidth infrared thermometers are not constrained by this limitation because the emissivity value typically will not vary over the narrow bandwidth utilized. Still, the target's emissivity value must be determined as it significantly impacts temperature measurement. Narrow-band infrared thermometers are somewhat more expensive than broadband units as more sophisticated electronics are required to accommodate the lower energy levels found at these bandwidths. Additionally, lower temperature ranges (under 482°F [250°C]) are not readily achievable with narrow-band infrared thermometers due to the reduced energy being measured.

Acoustic Pyrometer

Acoustic Pyrometer is a non-contact measurement device that obtains highly accurate instantaneous gas temperature data in any area of the boiler. The system utilizes the principle that the velocity of sound through a medium is related to the temperature of the medium. Instantaneous temperature data is spatially averaged and can provide zonal temperature maps.

Thermowells

Thermowells are tubular fittings used to protect temperature sensors installed in industrial processes. A thermowell consists of a tube closed at one end and mounted in the process stream. A temperature sensor such as a thermometer, thermocouple or resistance temperature detector is inserted in the open end of the tube, which is usually in the open air outside the process piping or vessel and any thermal insulation. The process fluid transfers heat to the thermowell wall, which in turn transfer heat to the sensor. Since more mass is present, the sensor's response to process temperature changes is delayed. If the sensor fails, it can be easily replaced without draining the vessel or piping.

The most common types of thermowells are (1) threaded, (2) socket weld, and (3) flanged welded. Thermowells are classified according to their connection to a process. For example, a threaded Thermowell is screwed into the process. A socket weld Thermowell is welded into a weldalet and a weld in Thermowell is welded directly into the process. A flanged Thermowell has a flange collar which is attached to a mating flange.

Typically, a Thermowell consists of (1) a process connection, (2) shank construction, (3) a "Q dimension", (4) bore size, (5) immersion ("U) length, and (6) lagging extension ("T") length.

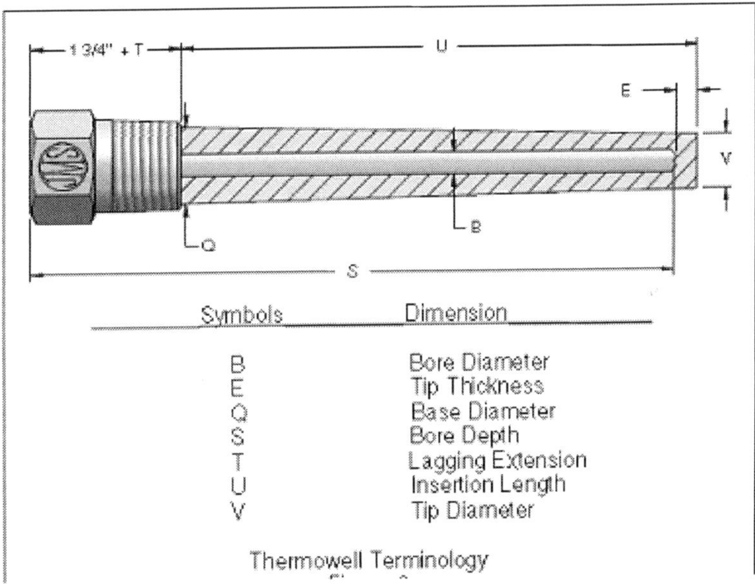

Thermowell Terminology

Figure : Thermowell

1. **Thermowell process connections:**

Thermowells are inserted into and connected into a process in a pressure tight manner. The most common process connections for thermowells include threaded, socket weld, and flanged connections.

2. **Thermowell Shank construction:**

The most common shank constructions for thermowells are (1) straight, (2) step, and (3) tapered. A straight shank Thermowell is the same size all along the immersion length of the Thermowell. A step shank Thermowell has an outer diameter of ½" at the end of the thermowell immersion length to provide a quicker response time. In a tapered Thermowell the outside diameter of the Thermowell decreases gradually along the immersion length of the Thermowell.

A heavy duty tapered Thermowell is typically used for high velocity applications.

3. **Q Dimension:**

The "Q" dimension of a Thermowell is the thickest part of the shank of the Thermowell that is on the hot side of the process connection or flange. The size of a Thermowell Q dimension is, of course, related to the bore size of the Thermowell and the process connection size.

4. **Bore size:**

The inside diameter of a Thermowell. Standard Thermowell bore sizes are .260" and .385". These sizes are intended to accept a quarter or three eights inch diameter sensor.

5. **Thermowell Immersion ("U") Length:**

Thermowell immersion lengths are often called the "U" length. The U length is the measurement of the Thermowell from the bottom of the process connection to the tip of the Thermowell. The U length establishes the length of the Thermowell that is actually in the process being measured.

6. **Lagging Extension ("T") Length:**

The lagging extension of a thermowell is often referred to as the Thermowell's "T" length. The lagging extension or T length is located on the cold side of the process connection and is usually an extension of the hex length of the Thermowell. Typically, the T length enables the probe and thermowell to extend through insulation or walls.

3.6 Flow Measurements

Flow measurement is the quantification of bulk fluid movement. Flow can be measured in a variety of ways. Positive-displacement flow meters accumulate a fixed volume of fluid and then count the number of times the volume is filled to measure flow. Other flow measurement methods rely on forces produced by the flowing stream as it overcomes a known constriction, to indirectly calculate flow. Flow may be measured by measuring the velocity of fluid over a known area.

There are several types of flow meter that rely on Bernoulli's principle, either by measuring the differential pressure within a constriction, or by measuring static and stagnation pressures to derive the dynamic pressure.

Orifice Plate & Venturi Tube

In a differential pressure (or differential head) flowmeter, a restriction is used to create a pressure drop. Measuring the pressure drop across the restriction can be used to determine the flowrate. Types of differential pressure flowmeter include orifice plates, flow nozzles and venturi tubes.

In an orifice flowmeter, the restriction is simply a plate with a hole (or orifice) drilled through it. The smaller the hole size, the larger the pressure drop across it for a given flowrate. Pressure tappings upstream and downstream of the orifice plate measure the pressure drop. Venturi flowmeters work on a similar principle but the orifice plate is replaced with a venturi.

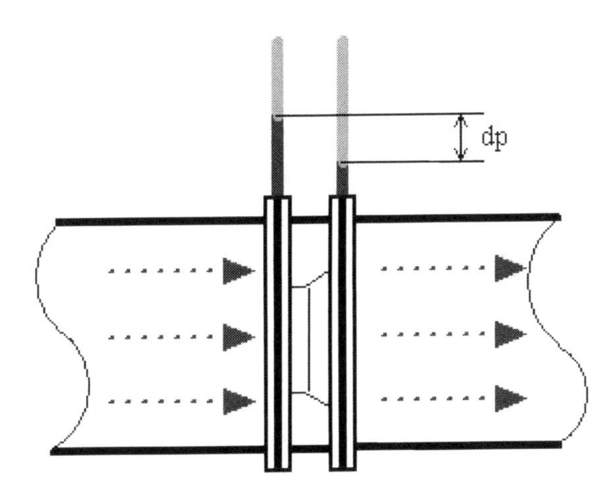

Figure: Orifice Plate

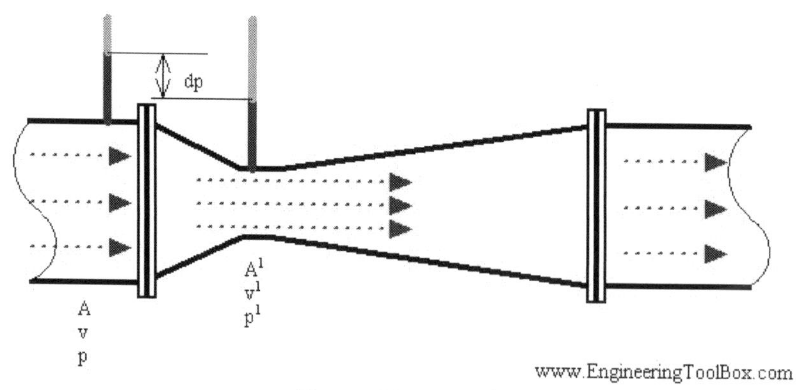

Figure: Venturi Plate

Pitot tubes

The pitot tube are one the most used (and cheapest) ways to measure fluid flow, especially in air applications as ventilation and HVAC systems, even used in airplanes for the speed measurement.

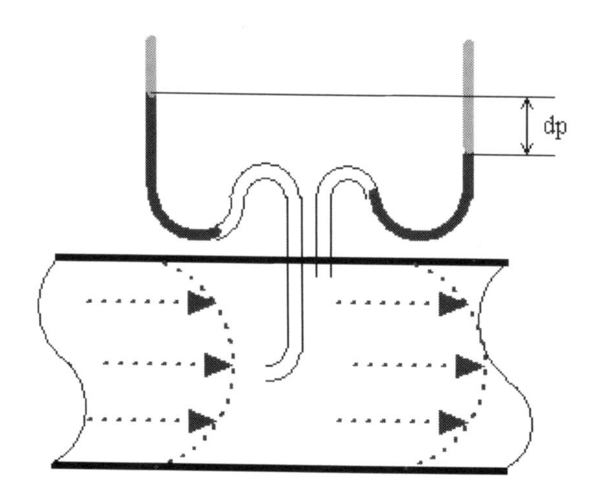

Figure: Pitot tubes

The pitot tube measures the fluid flow velocity by converting the kinetic energy of the flow into potential energy.

Note that a square root relationship exists between velocity and pressure drop. This limits the accuracy to a small turndown range.

$$u_1 = \sqrt{\frac{2\,\Delta P}{\rho}}$$

u1 = The fluid velocity in the pipe Δp = Dynamic pressure - Static pressure ρ = Density.

Inferential Type Meters

The flow of the fluid is inferred from some effect produced by the flow. Usually this is a rotor which is made to spin and the speed of the rotor is sensed mechanically or electronically. The main types are :

Turbine Type

The turbine type shown has an axial rotor which is made to spin by the fluid and the speed represents the flow rate. This may be sensed electrically by coupling the shaft to a small electric tachometer. Often this consists of a magnetic slug on the rotor which generates a pulse of electricity each time it passes the sensor.

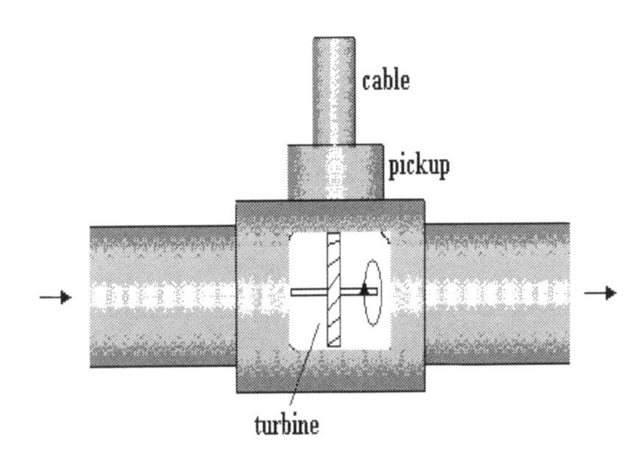

Figure: Turbine Type

Rotating Vane Type

The jet of fluid spins around the rotating vane and the speed of the rotor is measured mechanically or electronically.

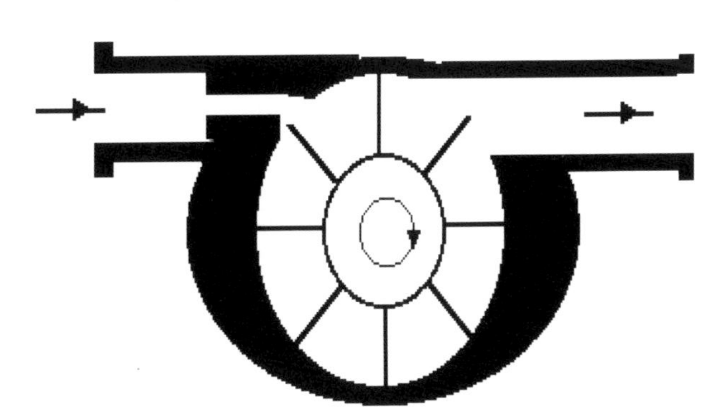

Figure: Rotating Vane Type

Speed Transducers

Speed transducers are widely used for measuring the output speed of a rotating object. There are many types using different principles and most of them produce an electrical output.

Optical Types

These use a light beam and a light sensitive cell. The beam is either reflected or interrupted so that pulses are produced for each revolution. The pulses are then counted over a fixed time and the speed obtained. Electronic processing is required to time the pulses and turn the result into an analogue or digital signal.

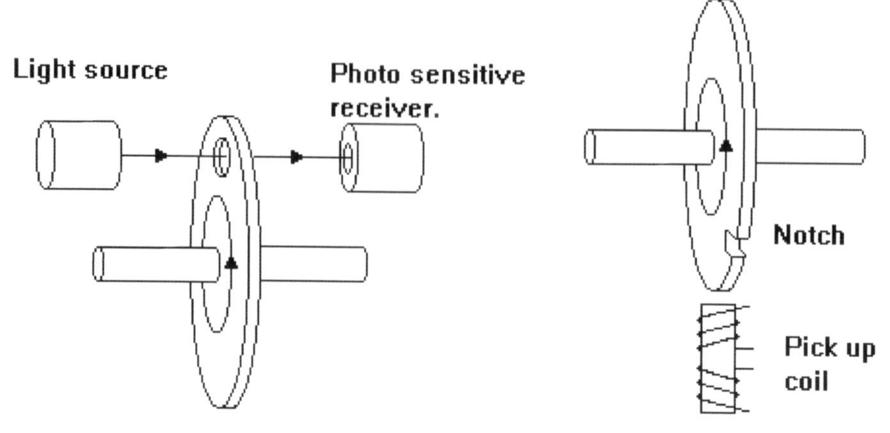

Figure: Optical Type

Magnetic Pick Ups

These use an inductive coil placed near to the rotating body. A small magnet on the body generates a pulse every time it passes the coil. If the body is made of ferrous material, it will work without a magnet. A discontinuity in the surface such as a notch will cause a change in the magnetic field and generate a pulse. The pulses must be processed to produce an analogue or digital output.

Tachometers

There are two types, A.C. and D.C. The A.C. type generates a sinusoidal output. The frequency of the voltage represents the speed of rotation. The frequency must be counted and processed. The D.C. type generates a voltage directly proportional to the speed. Both types must be coupled to the rotating body. very often the tachometer is built into electric motors to measure their speed.

Figure: Tachometer Generator

3.7 *Level Measurement*

Ultrasonic Level Sensor

The working principle of a typical Ultrasonic level sensor is illustrated in the figure below.

In this design, the level sensor is located at the top of the tank in such a way that it sends out the sound waves in the form of bursts in downward direction to the fluid in the tank under level measurement. As soon as the directed sound waves hits the surface of the fluid, sound echoes gets reflected and returned back to the sensor.

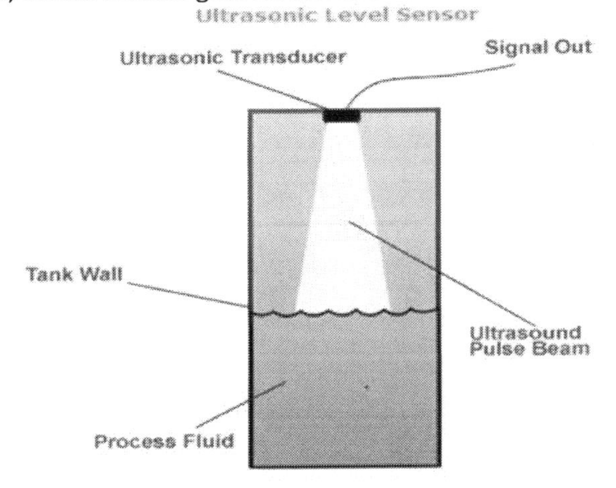

Figure: Ultrasonic Level Sensor

The time taken by the sound wave to return is directly proportional to the distance between the piezo electric sensor and the material in the tank. This time duration is measured by the sensor which is then further used to calculate the level of liquid in the tank. The speed of the sound waves can sometimes be affected due to variations in temperature for which appropriate compensations need to be provided in the sensor design.

Depth Gauges

Depth gauges measure the depth of liquids and powder in tanks. They use a variety of principles and produce outputs in electrical and pneumatic forms. The type to use depends on the substance in the tank. Here are a few.

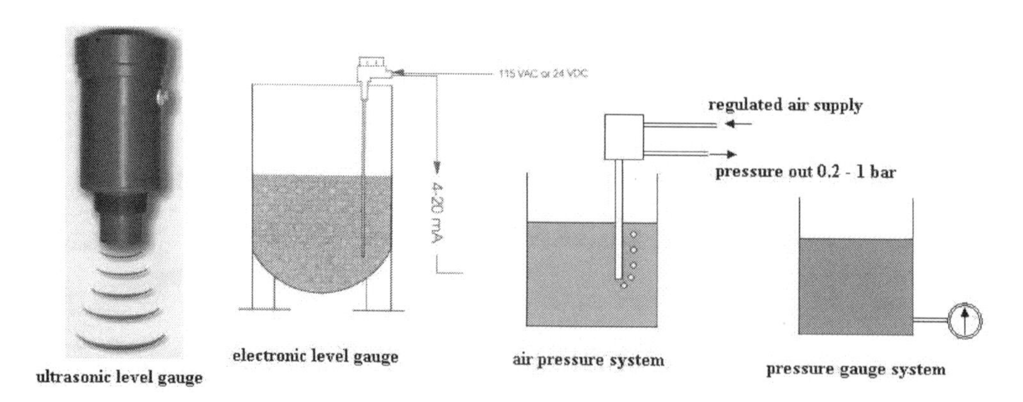

Figure: Depth Gauges

The ultrasonic system reflects sound waves from the surface and determines the depth from the time taken to receive the reflected sound. The electronic version uses a variety of electrical affects including conduction of the fluid and capacitance. The pneumatic version bubbles air through the liquid and the pressure of the air is related to the depth. A simple pressure gauge attached to a tank is also indicates the depth since depth is proportional to pressure.

Optical Types

Optical types are mainly used for producing digital outputs. A common example is found on machine tools where they measure the position of the work table and display it in digits on the gauge head. Digital micrometers and verniers also use this idea. The basic principle is as follows:
Light is emitted through a transparent strip or disc onto a photo electric cell. Often reflected light is used as shown. The strip or disc has very fine lines engraved on it which interrupt the beam. The number of interruptions are counted electronically and this represents the position or angle. This is very much over simplified and you should refer to more advanced text to find out how very accurate measurements are obtained and also the direction of movement.

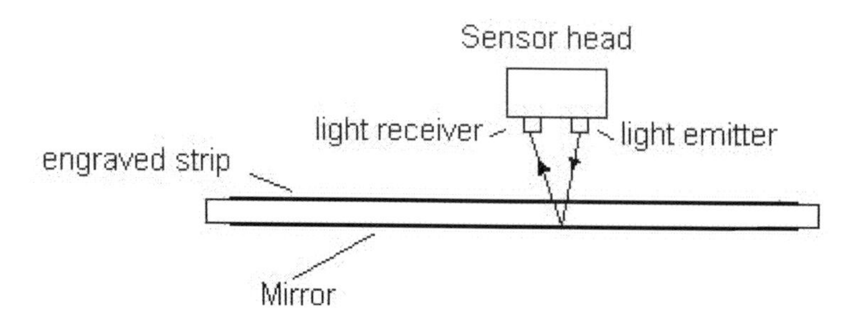

Figure: Optical Type

3.8 Control Valve

Control valves are imperative elements in any system where fluid flow must be monitored and manipulated. Selection of the proper valve involves a thorough knowledge of the process for which it will be used. Involved in selecting the proper valve is not only which type of valve to use, but the material of which it is made and the size it must be to perform its designated task.

The basic valve is used to permit or restrain the flow of fluid and/or adjust the pressure in a system. A complete control valve is made of the valve itself, an actuator, and, if necessary, a valve control device. The actuator is what provides the required force to cause the closing part of the valve to move. Valve control devices keep the valves in the proper operating conditions; they can ensure appropriate position, interpret signals, and manipulate responses.

Valve Types

Although many different types of valves are used to control the flow of fluids, the basic valve types can be divided into two general groups: stop valves and check valves.

Besides the basic types of valves, many special valves, which cannot really be classified as either stop valves or check valves, are found in the engineering spaces. Many of these valves serve to control the pressure of fluids and are known as pressure-control valves. Other valves are identified by names that indicate their general function, such as thermostatic recirculating valves. The following sections deal first with the basic types of stop valves and check valves, then with some of the more complicated special valves.

Stop Valves

Stop valves are used to shut off or, in some cases, partially shut off the flow of fluid. Stop valves are controlled by the movement of the valve stem. Stop valves can be divided into four general categories: globe, gate, butterfly, and ball valves. Plug valves and needle valves may also be considered stop valves.

Globe Valves

Globe valves are probably the most common valves in existence. The globe valve derives its name from the globular shape of the valve body. However, positive identification of a globe valve must be made internally because other valve types may have globular appearing bodies. Globe valve inlet and outlet openings are arranged in several ways to suit varying requirements of flow. Figure shows the common types of globe valve bodies: straight-flow, angle-flow, and cross flow. Globe valves are used extensively throughout the engineering plant and other parts of the ship in a variety of systems.

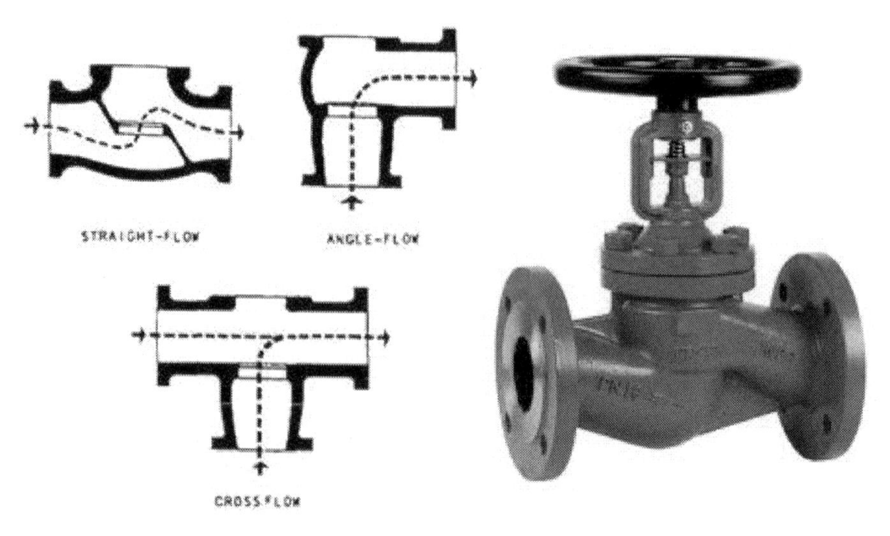

Figure: Types of globe valve bodies

Gate Valves

Gate valves are used when a straight-line flow of fluid and minimum restriction is desired. Gate valves are so named because the part that either stops or allows flow through the valve acts somewhat like the opening or closing of a gate and is called, appropriately, the gate. The gate is usually wedge shaped. When the valve is wide open, the gate is fully drawn up into the valve, leaving an opening for flow through the valve the same size as the pipe in which the valve is installed. Therefore, there is little pressure drop or flow restriction through the valve. Gate valves are not suitable for throttling purposes since the control of flow would be difficult due to valve design and since the flow of fluid slapping against a partially open gate can cause extensive damage to the valve. Except as specifically authorized, gate valves should not be used for throttling.

Gate valves are classified as either RISINGSTEM or NONRISING-STEM valves. On the nonrising-stem gate valve shown in figure the stem is threaded on the lower end into the gate. As the handwheel on the stem is rotated, the gate travels up or down the stem on the threads, while the stem remains vertically stationary. This type of valve almost always has a pointer-type indicator threaded onto the upper end of the stem to indicate valve position.

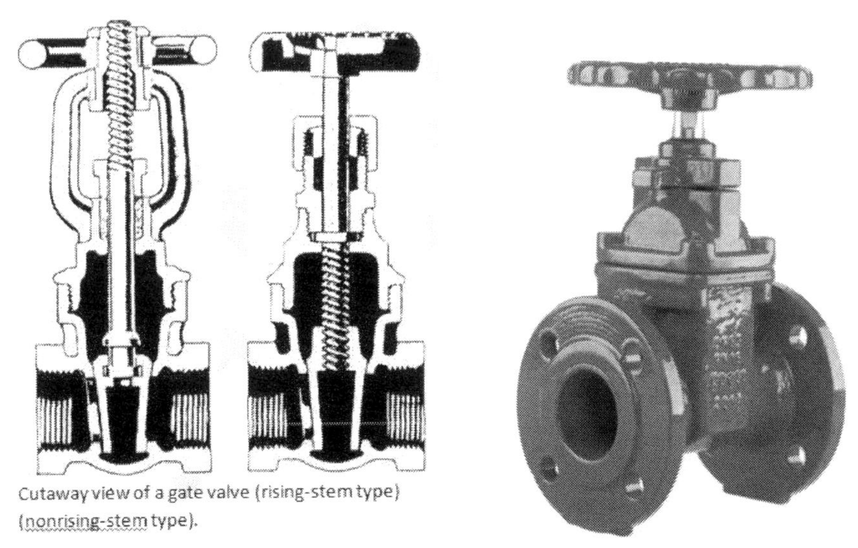

Cutaway view of a gate valve (rising-stem type) (nonrising-stem type).

Figure: Gate Valve

The rising-stem gate valve, shown in figure has the stem attached to the gate; the gate and stem rise and lower together as the valve is operated.

Gate valves used in steam systems have flexible gates. The reason for using a flexible gate is to prevent binding of the gate within the valve when the valve is in the closed position. When steam lines are heated, they will expand, causing some distortion of valve bodies. If a solid gate fits snugly between the seat of a valve in a cold steam system, when the system is heated, and pipes elongate, the seats will compress against the gate, wedging the gate between them and clamping the valve shut. This problem is overcome by use of a flexible gate (two circular plates attached to each other with a flexible hub in the middle). This design allows the gate to flex as the valve seat compresses it, thereby preventing clamping.

Butterfly Valves

The butterfly valve, one type of which is shown in figure may be used in a variety of systems aboard ship. These valves can be used effectively in freshwater, saltwater, JP-5, F-76 (naval distillate), lube oil, and chill water systems aboard ship. The butterfly valve is light in weight, relatively small, relatively quick-acting, provides positive shut-off, and can be used for throttling.

The butterfly valve has a body, a resilient seat, a butterfly disk, a stem, packing, a notched positioning plate, and a handle. The resilient seat is under compression when it is mounted in the valve body, thus making a seal around the periphery of the disk and both upper and lower points where the stem passes through the seat. Packing is provided to form a positive seal around the stem for added protection in case the seal formed by the seat should become damaged.

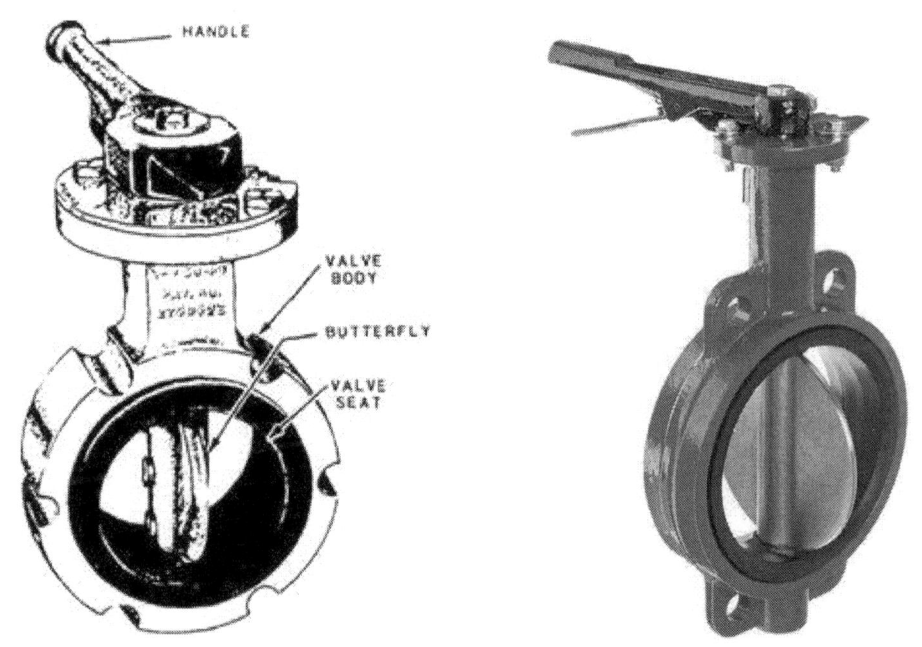

Figure: Butterfly Valve

To close or open a butterfly valve, turn the handle only one quarter turn to rotate the disk 90°. Some larger butterfly valves may have a hand wheel that operates through a gearing arrangement to operate the valve. This method is used especially where space limitation prevents use of a long handle.

Butterfly valves are relatively easy to maintain. The resilient seat is held in place by mechanical means, and neither bonding nor cementing is necessary, because the seat is replaceable, the valve seat does not require lapping, grinding, or machine work.

Ball Valves

Ball valves, as the name implies, are stop valves that use a ball to stop or start the flow of fluid. The ball (fig) performs the same function as the disk in the globe valve. When the valve handle is operated to open the valve, the ball rotates to a point where the hole through the ball is in line with the valve body inlet and outlet. When the valve is shut, which requires only a 90-degree rotation of the handwheel for most valves, the ball is rotated so the hole is perpendicular to the flow openings of the valve body, and flow is stopped.

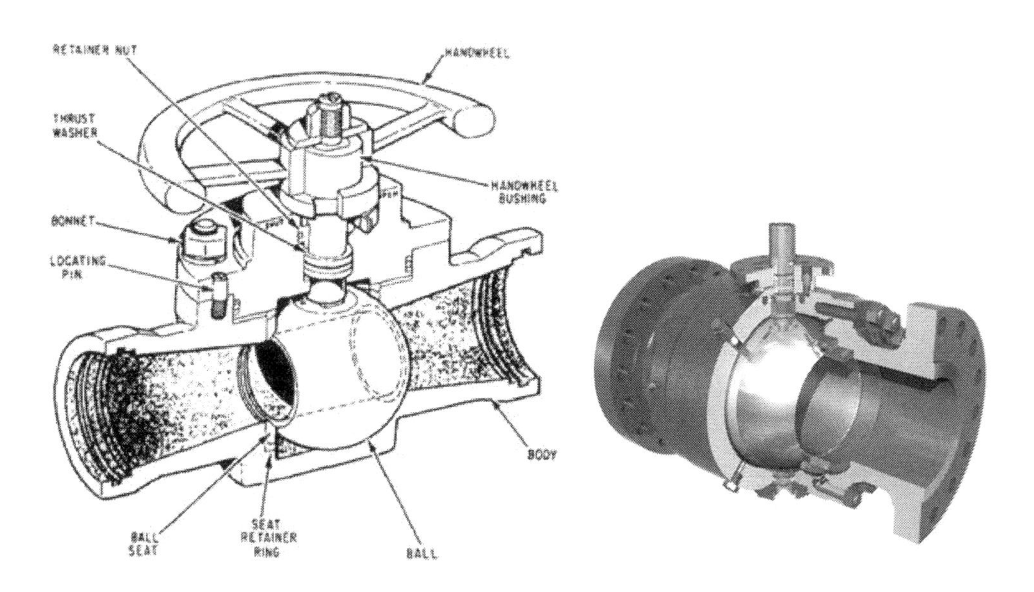

Figure-Typical seawater ball valve

Most ball valves are of the quick-acting type (requiring only a 90-degree turn to operate the valve either completely open or closed), but many are planetary gear operated. This type of gearing allows the use of a relatively small handwheel and operating force to operate a fairly large valve. The gearing does, however, increase the operating time for the valve. Some ball valves contain a swing check located within the ball to give the valve a check valve feature. Ball valves are normally found in the following systems aboard ship: seawater, sanitary, trim and drain, air, hydraulic, and oil transfer.

3.9 Load Measurement

3.9.1 Strain Gauges

A strain gauge is a device used to measure strain on an object. Invented by Edward E. Simmons and Arthur C. Ruge in 1938, the most common type of strain gauge consists of an insulating flexible backing which supports a metallic foil pattern. The gauge is attached to the object by a suitable adhesive, such as cyanoacrylate. As the object is deformed, the foil is deformed, causing its electrical resistance to change. This resistance change, usually measured using a Wheatstone bridge, is related to the strain by the quantity known as the gauge factor.

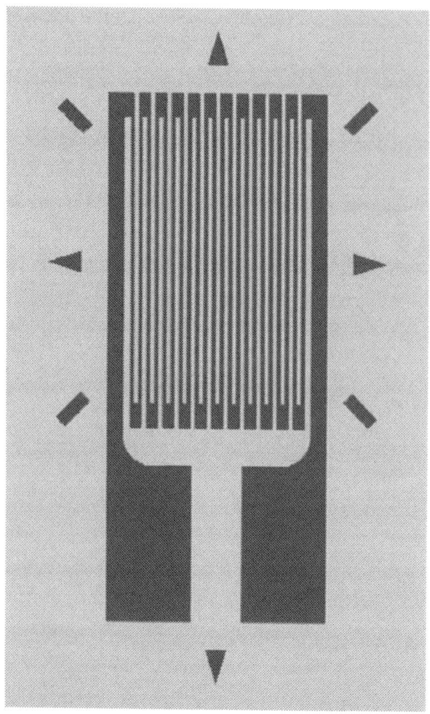

Figure: Typical foil strain gauge

Typical foil strain gauge; the blue region is conductive, and resistance is measured from one large blue pad to the other. The gauge is far more sensitive to strain in the vertical direction than in the horizontal direction. The markings outside the active area help to align the gauge during installation.

The gauge factor GF is defined as: where R G is the resistance of the un deformed gauge , ΔR is the change in resistance caused by strain, and ε is strain.

Wheatstone Bridge

A Wheatstone bridge is a measuring instrument invented by Samuel Hunter Christie in 1833 and improved and popularized by Sir Charles Wheatstonein1843. It is used to measure an unknown electrical resistance by balancing two legs of a bridge circuit, one leg of which includes the unknown component. Its operation is similar to the original potentiometer except that in potentiometer circuits the meter used is a sensitive galvanometer. Wheat stone's bridge circuit diagram. In the circuit at right, Rx is the unknown resistance to be measured; R1, R2 and R3 are resistors of known resistance and the resistance of R2 is adjustable. If the ratio of the two resistances in the known leg (R2/ R1) is equal to the ratio of the two in the unknown leg (R x/ R3), then the voltage between the two midpoints (B and D) will be zero and no current will flow through the galvanometer Vg . R2 is varied until this condition is reached. The current direction indicates whether R2 is too high or too low.

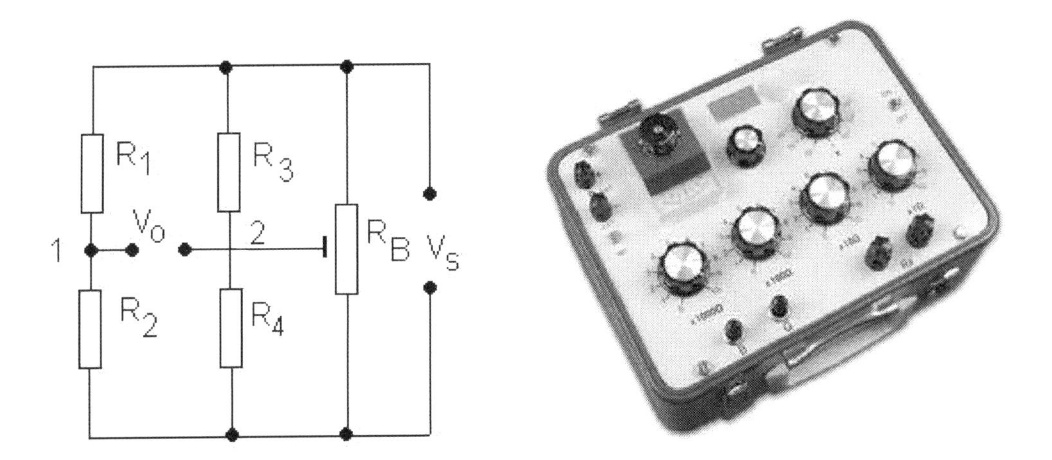

Figure: Wheatstone Bridge

Detecting zero current can be done to extremely high accuracy (see galvanometer). Therefore, if R1, R2 and R3 are known to high precision, then R x can be measured to high precision. Very small changes in R x disrupt the balance and are readily detected. At the point of balance, the ratio of R2/ R1= R x/ R3 Therefore, Alternatively, if R1, R2, and R3 are known, but R2 is not adjustable, the voltage or current flow through the meter can be used to calculate the value of R x, using Kirchhoff's circuit laws(also known as Kirchhoff's rules). This setup is frequently used in strain gauge and Resistance.

Measuring Circuits

For measurement of strain via a bonded resistance strain gage, it must be connected to an electrical measuring circuit which can measure even the minute changes in resistance corresponding to strain. Modern strain-gage transducers usually employ a grid of four strain elements electrically connected to form a Wheatstone bridge measuring circuit. A Wheatstone bridge is a divided bridge circuit employed for the measurement of static or dynamic electrical resistance. The output voltage of the Wheatstone bridge is expressed in millivolts output per volt input. Besides, this bridge circuit is appropriate for temperature compensation. A quarter bridge strain gauge circuit is shown in the figure below:

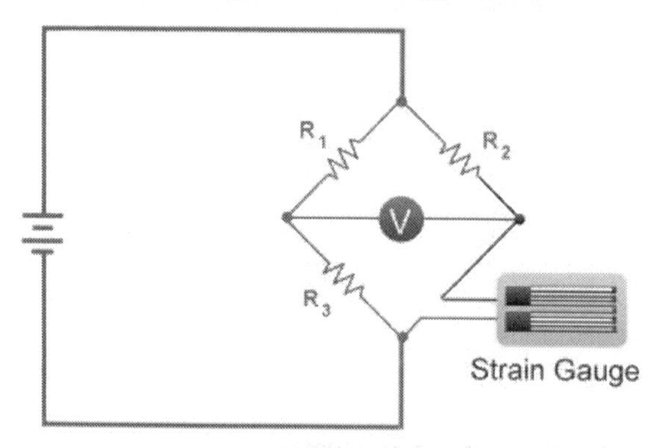

Figure: Quarter- bridge Strain Gauge Circuit

3.9.2 Instrument Transformers

Instrument Transformers are used in AC system for measurement of electrical quantities i.e. voltage, current, power, energy, power factor, frequency. Instrument transformers are also used with protective relays for protection of power system. Basic function of Instrument transformers is to step down the AC System voltage and current. The voltage and current level of power system is very high. It is very difficult and costly to design the measuring instruments for measurement of such high level voltage and current. Generally measuring instruments are designed for 5 A and 110 V.

Figure: Instrument Transformer

The measurement of such very large electrical quantities, can be made possible by using the Instrument transformers with these small rating measuring instruments. Therefore, these instrument transformers are very popular in modern power system.

Advantages of Instrument Transformers

- The large voltage and current of AC Power system can be measured by using small rating measuring instrument i.e. 5 A, 110 – 120 V.
- By using the instrument transformers, measuring instruments can be standardized. Which results in reduction of cost of measuring instruments. More ever the damaged measuring instruments can be replaced easy with healthy standardized measuring instruments.
- Instrument transformers provide electrical isolation between high voltage power circuit and measuring instruments. Which reduces the electrical insulation requirement for measuring instruments and protective circuits and also assures the safety of operators.
- Several measuring instruments can be connected through a single transformer to power system.
- Due to low voltage and current level in measuring and protective circuit, there is low power consumption in measuring and protective circuits.

Instrument transformers are of two types: Current Transformer (C.T.) & Voltage or Potential Transformer (P.T.)

Current Transformers

Current transformer is used to step down the current of power system to a lower level to make it feasible to be measured by small rating Ammeter (i.e. 5A ammeter). A typical connection diagram of a current transformer is shown in figure below.

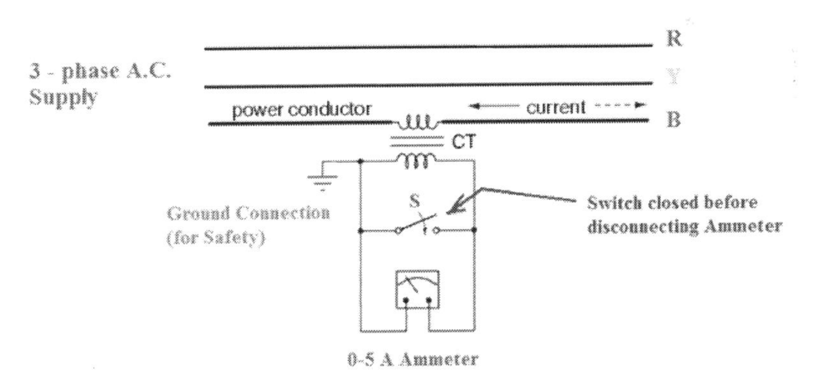

Figure: Current Transformer (C.T.)

A current transformer is a type of "instrument transformer" that is designed to provide a current in its secondary which is accurately proportional to the current flowing in its primary.

Current transformers are commonly used in metering and protective relaying to facilitate the measurement of large currents and isolation of high voltage systems which would be difficult to measure more directly.

Figure: Current Transformer

Current transformers are often constructed by passing a single primary turn (either an insulated cable or an uninsulated conductor (copper or aluminum are typical in electric utility applications) through a well-insulated toroidal core wrapped with many turns of wire. Current transformers (CTs) are used extensively in the electrical power industry for monitoring of the power grid. The CT is described by its current ratio from primary to secondary. Common secondaries are 1 or 5 amperes. The secondary winding can be single ratio or multi ratio, with five taps being common for multi ratio CTs. Typically, the secondary connection points are labeled as X1, X2 and so on. The multi ratio CTs are typically used for current matching in current differential protective relaying applications. Often, multiple CTs will be installed as a "stack" for various uses (for example, protection devices and revenue metering may use separate CTs). For a three-stacked CT application, the secondary winding connection points are typically labeled Xn, Yn, Zn.

Specially constructed "wideband current transformers" are also used (usually with an oscilloscope) to measure waveforms of high frequency or pulsed currents. One type of specially constructed wideband transformer provides a voltage output that is proportional to the measured current. Another type (called a Rogowski coil) requires an external integrator in order to provide a voltage output that is proportional to the measured current.

Care must be taken that the secondary of a current transformer is not disconnected from its load while current is flowing in the primary, as this will produce a dangerously high voltage across the open secondary.

Voltage Transformers

Voltage transformers (also called potential transformers) are another type of instrument transformer. They are used by the electricity supply industry to accurately measure high voltages for metering and protective relay purposes. They are designed to present negligible load to the voltage being measured and to have a precise turns ratio to accurately step down dangerously high voltages so that metering and protective relay equipment can be operated at a lower, and safer, potential. Voltage transformer is used to step down the voltage of power system to a lower level to make is feasible to be measured by small rating voltmeter i.e. 110 – 120 V voltmeter. A typical connection diagram of a potential transformer is showing figure below.

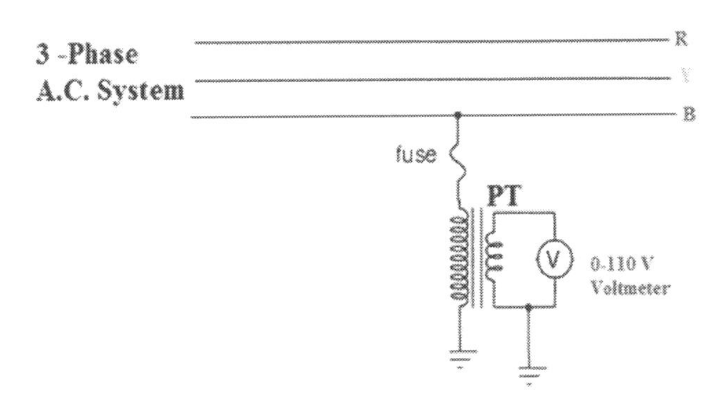

Potential Transformer (P.T.)

Figure : Voltage Transformer

This safer voltage is typically 66 to 120 volts phase to ground. 69 volts phase to ground is equivalent to 120 volts phase to phase (69√3) which is required by some protective relays.

Primary of Voltage Transformer is having large no. of turns. Primary is connected across the line (generally between on line and earth). Hence, sometimes it is also called the parallel transformer. Secondary of Voltage Transformer is having few turns and connected directly to a voltmeter. As the voltmeter is having large resistance. Hence the secondary of a Voltage Transformer operates almost in open circuited condition. One terminal of secondary of Voltage Transformer is earthed to maintain the secondary voltage with respect to earth. Which assures the safety of operators.

Circuit Breaker

A circuit breaker is an automatically operated electrical switch designed to protect an electrical circuit from damage caused by overload or short circuit. Its basic function is to detect a fault condition and, by interrupting continuity, to immediately discontinue electrical flow. a circuit breaker can be reset (either manually or automatically) to resume normal operation.

Earth Leakage Circuit Breaker

An Earth Leakage Circuit Breaker (ELCB) is a safety device used in electrical installations with high earth impedance to prevent shock. An ELCB is a specialized type of latching relay that has a building's incoming mains power connected through its switching contacts so that the ELCB disconnects the power in an earth leakage (unsafe) condition. There are two types of ELCB:

- voltage operated and,
- current operated.

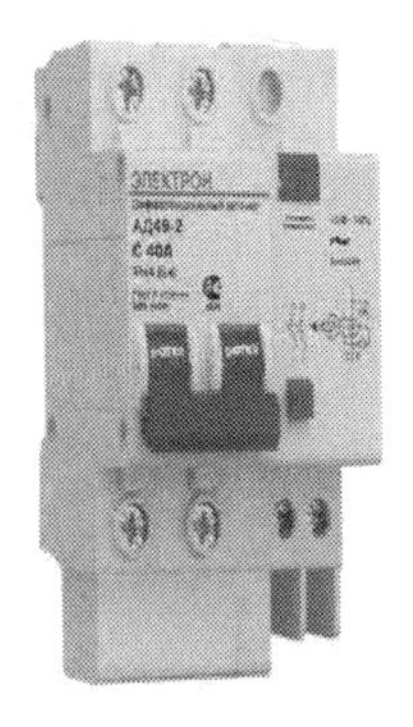
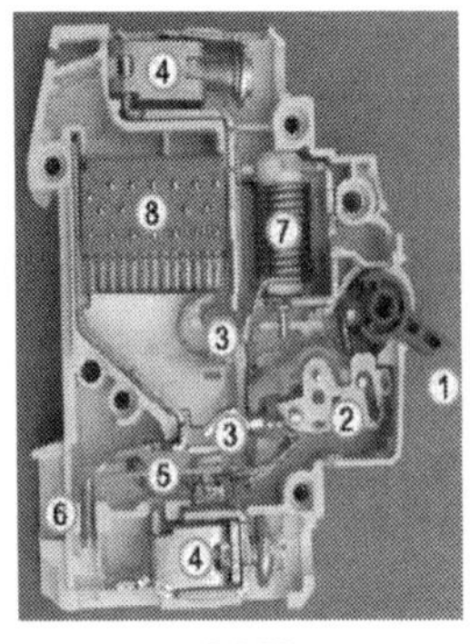

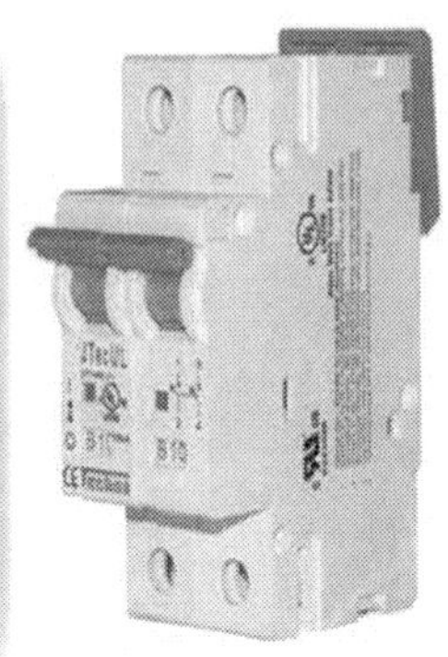

ELCB MCB RCCB MCCB

MCB Circuit Breaker

A MCB Circuit Breaker is an automatically operated electrical switch designed to protect an electrical circuit from damage caused by overload or short circuit. Its basic function is to detect a fault condition and, by interrupting continuity, to immediately discontinue electrical flow. Unlike a fuse, which operates once and then has to be replaced, a circuit breaker can be reset (either manually or automatically) to resume normal operation. Circuit breakers are made in varying sizes, from small devices that protect an individual household applianca up to large switchgear designed to protect high voltage circuits feeding an entire city.

RCCB

The Residual Current Circuit Breaker (RCCB), is an electrical circuit breaker that disconnects a circuit whenever it detects imbalance of the electric current in the phase ("hot" or "live"). These electrical circuit breakers are installed in the circuit to prevent shocks and provide protection against fire hazards.

MCCB (Molded Case Circuit Breaker)

A MCCB has a housing that is made from nonconductive material. Available for low voltage applications and comes in amperages from the single poles in the panel in your house to 3 phase and about 4000A. MCCB's are not meant to come apart. This circuit breaker applies to the circuit with AC 50/60Hz, rated insulating voltage 660V, max rated current 630A as on-off operation, and takes effect, it also can be used as infrequent conversion of line and infrequent start of motor.

Air Circuit breaker

Air Circuit Breaker (ACB) is an electrical device used to provide Overcurrent and short-circuit protection for electric circuits over 800 Amps to 10K Amps. These are usually used in low voltage applications below 450V. We can find these systems in Distribution Panels (below 450V). Air circuit breaker is circuit operation breaker that operates in the air as an arc extinguishing medium, at a given atmospheric pressure.

Air circuit breakers operate with their contacts in free air. Their method of arc quenching control is entirely different from that of oil circuit-breakers. They are always used for a low-voltage interruption and now tends to replace high-voltage oil breakers. The below-shown figure illustrates the principle of air breaker circuit operation.

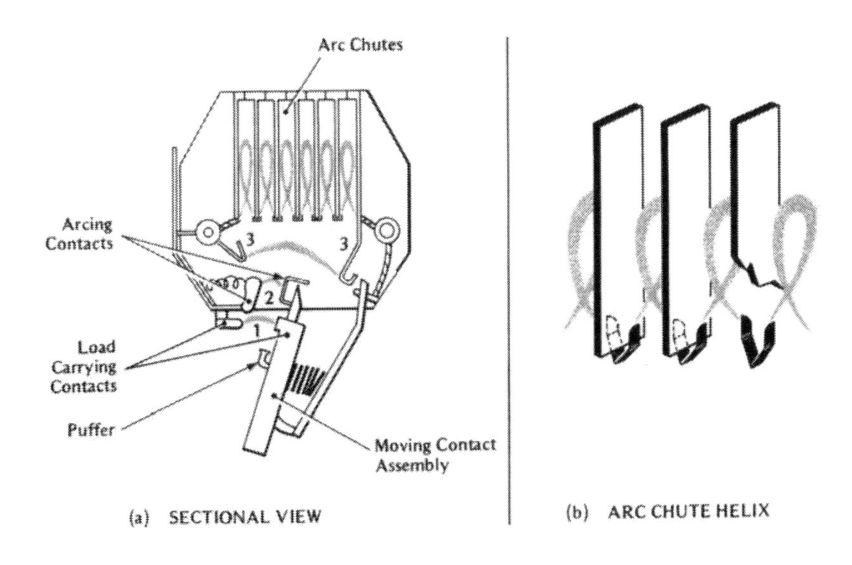

Figure: Air Circuit Breaker (Sectional view)

Air Circuit breakers generally have two pairs of contacts. The main pair of contacts (1) carries the current at normal load and these contacts are made of copper metal. The second pair is the arcing contact (2) and is made of carbon. When the circuit breaker is being opened, the main contacts open first. When the main contacts opened the arcing contacts are still in touch with each other. As the current gets a parallel low resistive path through the arcing contact. During the opening of main contacts, there will not be any arcing in the main contact. The arcing is only initiated when finally the arcing contacts are separated. The each of the arc contacts is fitted with an arc runner which helps. The arc discharge to move upward due to both thermal and electromagnetic effects as shown in the figure. As the arc is driven upward it enters in the arc chute, consisting of splatters.

The arc in the chute will become colder, lengthen and split hence arc voltage becomes much larger than the system voltage at the time of operation of air circuit breaker, and therefore the arc is extinguished finally during the current zero.

The air brake circuit box is made of insulating and fireproof material and it is divided into different sections by the barriers of the same material, as shown above, figure (a). At the bottom of each barrier is a small metal conducting element between one side of the barrier and the other. When the arc, driven upwards by the electromagnetic forces, enters the bottom of the chute, it is split into many sections by the barriers, but the each metal piece ensures electrical continuity between the arcs in each section, the several arcs are consequently in the series.

The electromagnetic forces within each and every section of the chute cause the arc in that section to start the form of a helix, as shown above, figure (b). All these helices are in series so that the total length of the arc has been greatly extended, and its resistance is abundantly increased. This will affect the current reduction in the circuit.

Figure (a) shows the development of the arc from the time it leaves the main contacts until it is within the arc chute. When the current next ceases at a current zero, the ionised air in the path of where the arc had been being in parallel with the open contacts and acts as a shunt resistance across both the contacts and the self-capacitance C, shown in below figure with red as a high resistance R.

Oil Circuit Breaker

Oil circuit breakers have their contacts immersed in insulating oil. They are used to open and close high-voltage circuits carrying relatively large currents in situations where air circuit breakers would be impractical because of the danger of the exposed arcs that might be formed (as per below figure).

When the contacts are drawn apart, the oil covering them tends to quench the arc by its cooling effect and by the gases thereby generated, which tend to "blow out" the arc. At the instant the contacts part, the arc formed at each contact not only displaces the oil but decomposes it, creating gas and a carbon residue. If these carbon particles were to remain in place, as a conductor they would tend to sustain the arc formed. However, the violence of the gas and the resulting turbulence of the oil disperse these particles and they eventually settle to the bottom of the tank.

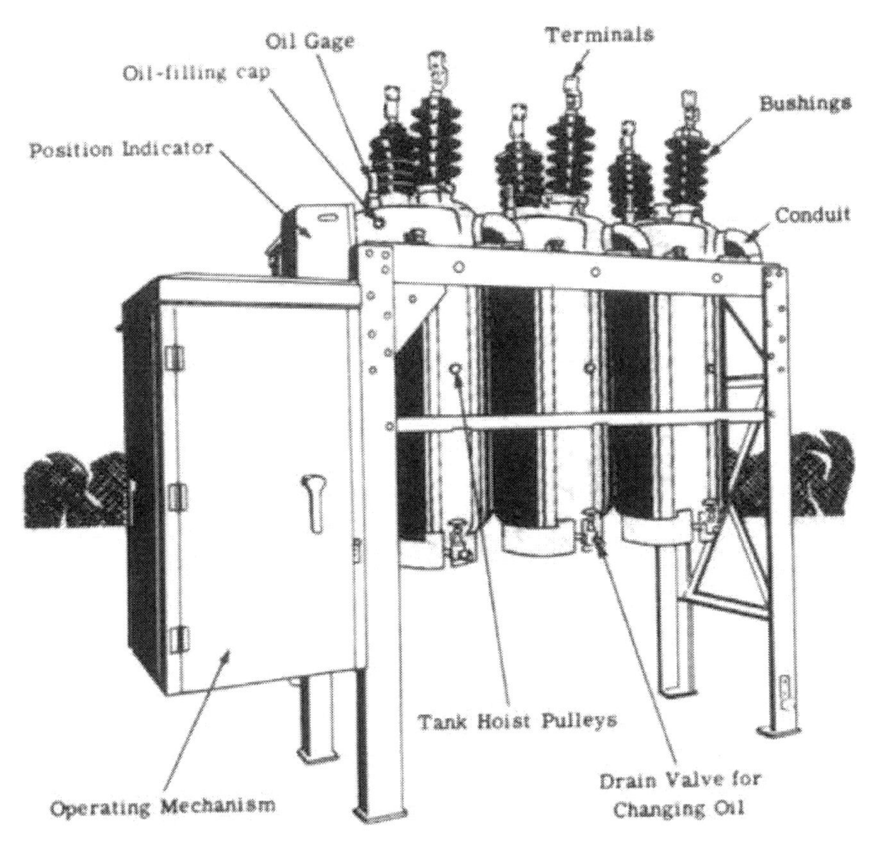

Figure: Oil Circuit Breaker

The insulating oil normally used as a dielectric strength of around 30 kV per one tenth of an inch (compared to a similar value of 1 kV for air). Oil is also an effective cooling medium.

Vaccum Circuit Breaker

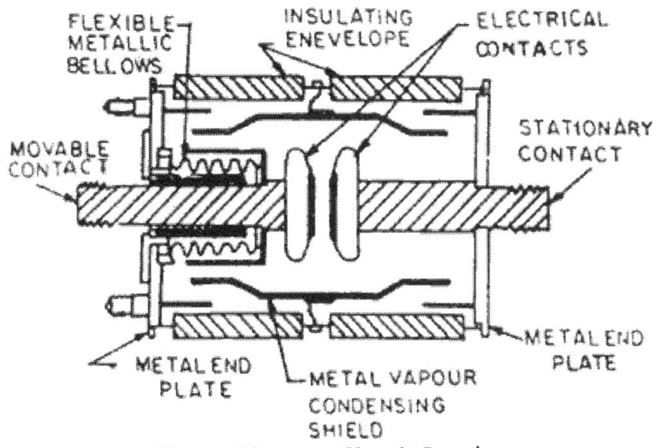

Figure: Vaccum Circuit Breaker

Here the contacts are drawn apart in a chamber from which air has been evacuated. The electric arc is essentially an electric conductor made up of ionized air. Thus, if there is no air, theoretically the arc cannot form. In practice, however, a perfect vacuum is not likely to be obtained. The small residual amount of air that may exist permits only a small arc to be formed and one of only a very short direction. The same vacuum however, will not dissipate the heat generated as readily as other insulating media. This type of breaker has certain advantages in terms of its size and simplicity.

SF6 (Sulphur hexafluoride) Circuit Breaker

In SF6 circuit breakers, the same principle is employed, with SF6 as the medium instead of air. In the "puffer" SF6 breaker, the motion of the contacts compresses the gas and forces it to flow through an orifice into the neighborhood of the arc. Both types of SF6 breakers have been developed for EHV (extra high voltage) transmission systems.

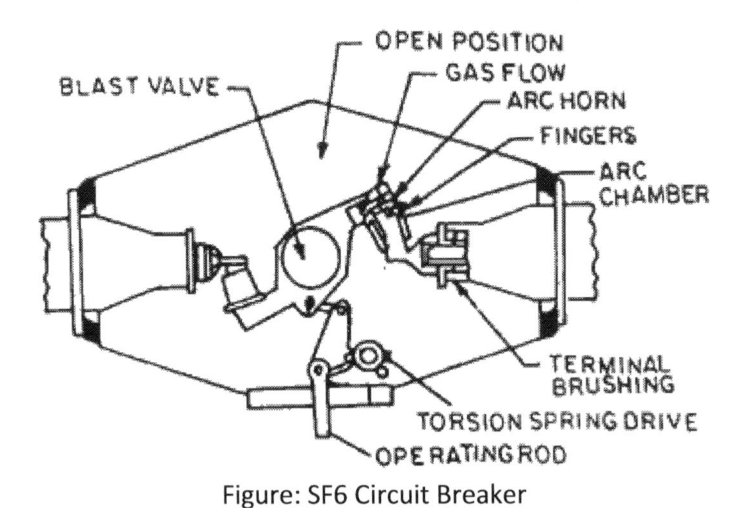

Figure: SF6 Circuit Breaker

The breakers are constructed to modules capable of operation at voltages from 34.4 kV at gas pressure of 45 psi to 362 kV at 240 psi. By connecting two or three such modules in series, breaker capable of operating at 800 kV at 240 psi can be constructed with two-three-cycle interrupting time. Features of sulphur hexafluoride interrupting module.

4 Chapter: Tube and tube fittings

Tube, like pipe, is a hollow structure designed to provide an enclosed pathway for fluids to flow. In the case of tubing, it is usually manufactured from rolled or extruded metal (although plastic is a common tube material for many industrial applications). This section discusses some of the more common methods for joining tubes together (and joining tube ends to equipment such as pressure instruments).

One of the fundamental differences between tube and pipe is that tube is never threaded at the end to form a connection. Instead, a device called a tube fitting must be used to couple a section of tube to another tube, or to a section of pipe, or to a piece of equipment (such as an instrument).

Unlike pipes which are thick-walled by nature, tubes are thin-walled structures. The wall thickness of a typical tube is simply too thin to support threads. Tubes are generally favored over pipe for small-diameter applications. The ability for skilled workers to readily cut and bend tube with simple hand tools makes it the preferred choice for connecting instruments to process piping. When used as the connecting units between an instrument and a process pipe or vessel, the tube is commonly referred to as an impulse tube or impulse line (Impulse lines are alternatively called gauge lines or sensing lines).

4.1 Compression tube fittings

By far the most common type of tube fitting for instrument impulse lines is the compression-style fitting, which uses a compressible ferrule to perform the task of sealing fluid pressure. The essential components of a compression tube fitting are the body, the ferrule, and the nut. The ferrule and body parts have matching conical profiles designed to tightly fit together, forming a pressure-tight metal-to-metal seal. Some compression fitting designs use a two-piece ferrule assembly, such as this tube fitting shown here (prior to full assembly, this happens to be a Swagelok brass instrument tube fitting being installed on a 3/8 inch copper tube):

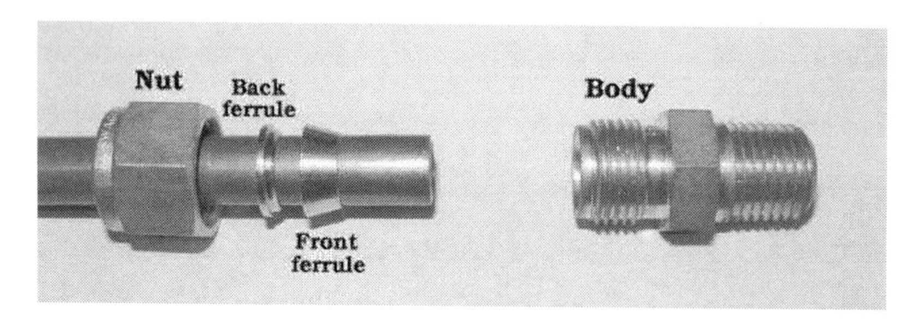

Just prior to assembly, we see how the nut will cover the ferrule components and push them into the conical entrance of the fitting body:

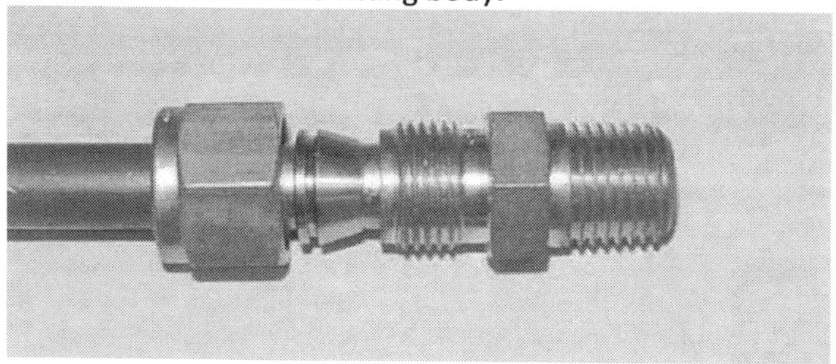

After properly tightening the nut, the ferrule(s) will compress onto the outside circumference of the tube, slightly crimping the tube in the process and thereby locking the ferrules in place:

When initially assembling compression-style tube fittings, you should always precisely follow the manufacturer's instructions to ensure correct compression. For Swagelok-brand instrument tube fittings 1 inch in size and smaller, the general procedure to "swage" a new connector to a tube is to tighten the nut 1-1/4 turns past finger-tight. Insufficient turning of the nut will fail to properly compress the ferrule around the tube, and excessive turning will over-compress the ferrule, resulting in leakage. After this initial "swaging," the connector may be separated by loosening the nut until it no longer engages with the body, then the connection may be re-made by threading the nut back on the body until finger-tight and then gently tightening with a wrench until snug (no additional 1-1/4 turns!!!).

Swagelok provides special gauges which may be used to measure proper ferrule compression during the assembly process. The design of the gauge is such that its thickness will fit between the nut and fitting shoulder if the nut is insufficiently tightened, but will not fit if it is sufficiently tightened. Thus the gauge has the ability to reveal an under-tightened fitting, but not an overtightened fitting. These gauges fit easily in the palm of one's hand:

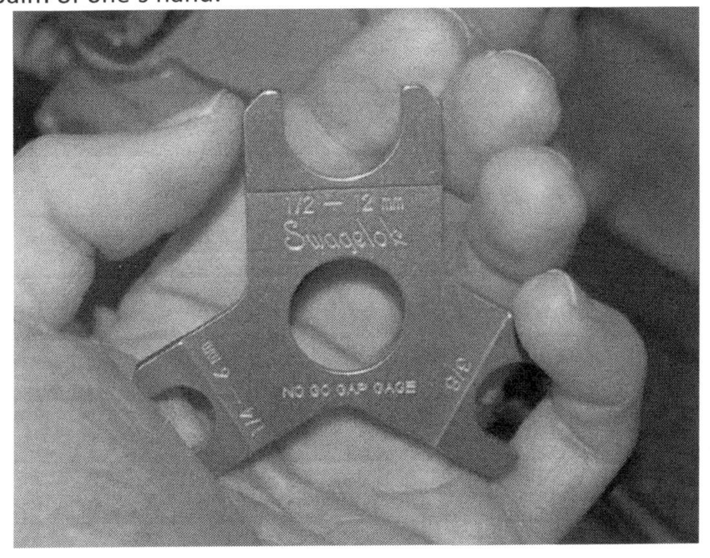

Such gauges are referred to in the industry as no-go gap gauges, because their inability to fit between the nut and body shoulder of a tube fitting indicates a properly-tightened fitting. In other words, the gauge fit will be "no-go" if the tube fitting has been properly assembled.

Photographs showing one of these gauges testing a properly-tightened fitting (left) versus an under-tightened fitting (right) appear here:

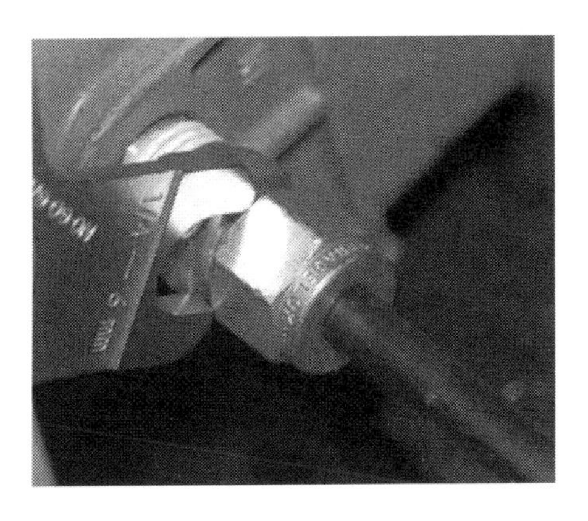

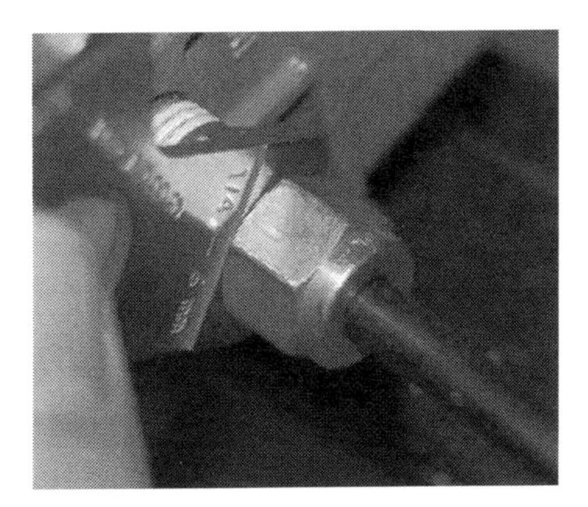

Parker is another major manufacture *(So is Gyrolok, Hoke, and a host of others. No intention to advertise for different manufacturers in this textbook, but merely to point out some of the more common brands an industrial instrument technician might encounter on the job)* of instrument tube fittings, and their product line uses a single-piece ferrule instead of the two-piece design preferred by Swagelok. Like Swagelok fittings, Parker instrument fitting sized 1/4 inch to 1 inch require 1-1/4 turns past hand tight to properly compress the ferrule around the circumference of the tube. Parker also sells gauges which may be used to precisely determine when the proper amount of ferrule compression is achieved.

Regardless of the brand, compression-style instrument tube fittings are incredibly strong and versatile. Unlike pipe fittings, tube fittings may be disconnected and reconnected with ease. No special procedures are required to "re-make" a disassembled instrument fitting connection: merely tighten the nut "snug" to maintain adequate force holding the ferrule to the fitting body, but not so tight that the ferrule compresses further around the tube than it did during initial assembly.

A very graphic illustration of the strength of a typical instrument tube fitting is shown in the following photograph, where a short section of 3/8 inch stainless steel instrument tube was exposed to high liquid pressure until it ruptured. Neither compression fitting on either side of the tube leaked during the test, despite the liquid pressure reaching a peak of 23000 PSI before rupturing the tube*(It should be noted that the fitting nuts became seized onto the tube due to the tube's swelling. The tube fittings may not have leaked during the test, but their constituent components are now damaged and should never be placed into service again.):*

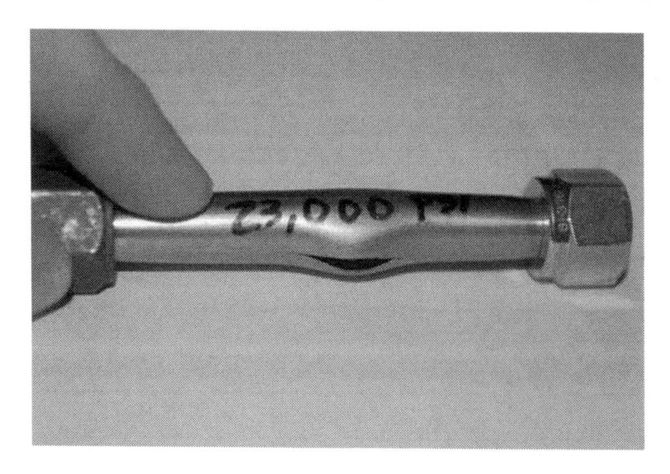

6

4.1.1 Common tube fitting types and names

Tube fittings designed to connect a tube to pipe threads are called connectors. Tube fittings designed to connect one tube to another are called unions:

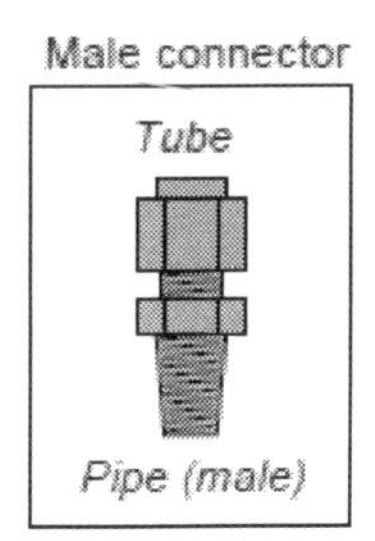

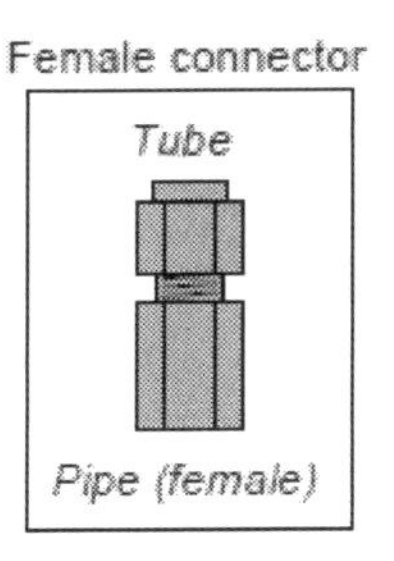

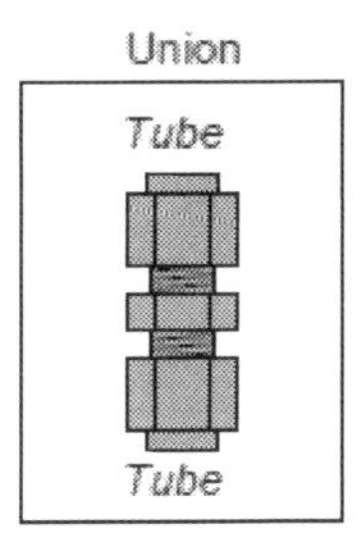

If a tube union joins together different tube sizes rather than tubes of the same size, it is called a reducing union.

A variation on the theme of tube connectors and unions is the bulkhead fitting. Bulkhead fittings are designed to fit through holes drilled in panels or enclosures to provide a way for a fluid line to pass through the wall of the panel or enclosure. In essence, the only difference between a bulkhead fitting and a normal fitting is the additional length of the fitting "barrel" and a special nut used to lock the fitting into place in the hole. The following illustration shows three types of bulkhead fittings:

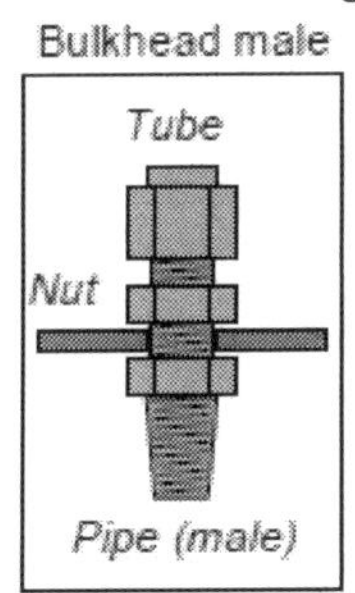

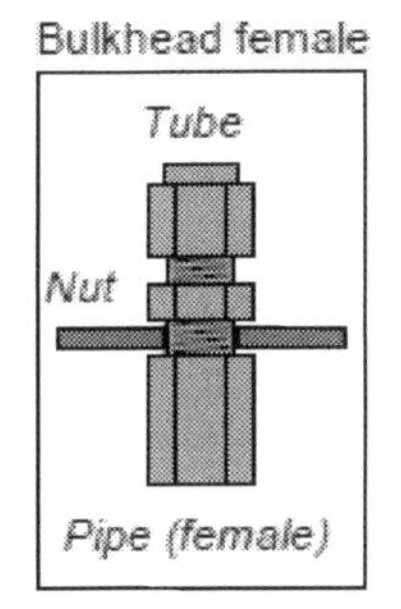

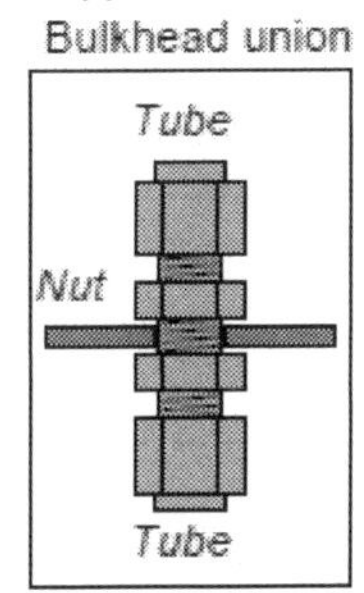

4.1.2 Tube and tube fittings

Tubing elbows are tube connectors with a bend. These are useful for making turns in tube runs without having to bend the tubing itself. Like standard connectors, they may terminate in male pipe thread, female pipe threads, or in another tube end:

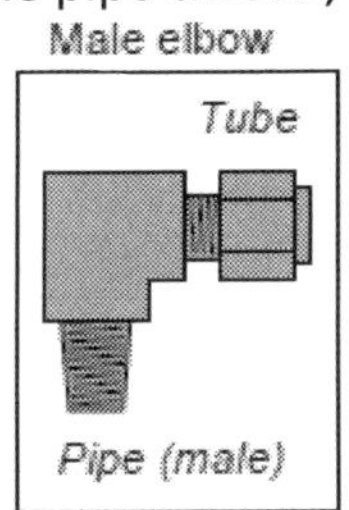

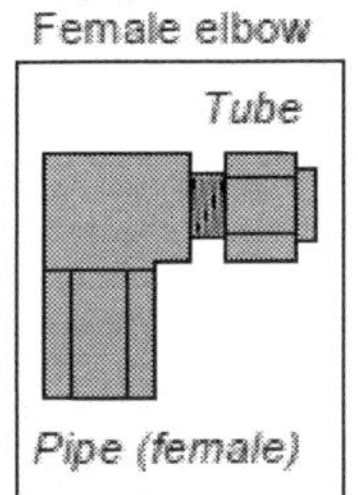

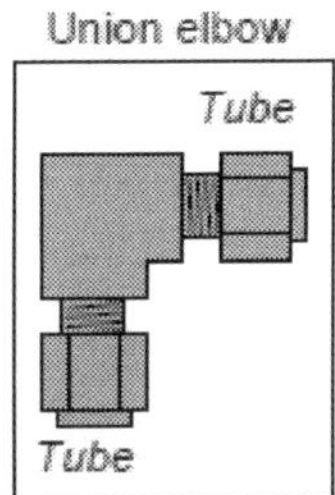

These elbows shown in the above illustration are all 90o, but this is not the only angle available. 45o elbows are also common.

Tee fittings join three fluid lines together. Tees may have one pipe end and two tube ends (branch tees and run tees), or three tube ends (union tees). The only difference between a branch tee and a run tee is the orientation of the pipe end with regard to the two tube ends:

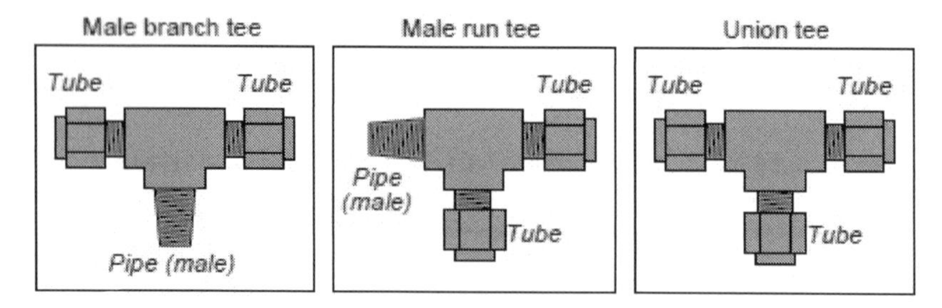

Of course, branch and run tee fittings also come in female pipe thread versions as well. A variation of the theme of union tees is the cross, joining four tubes together:

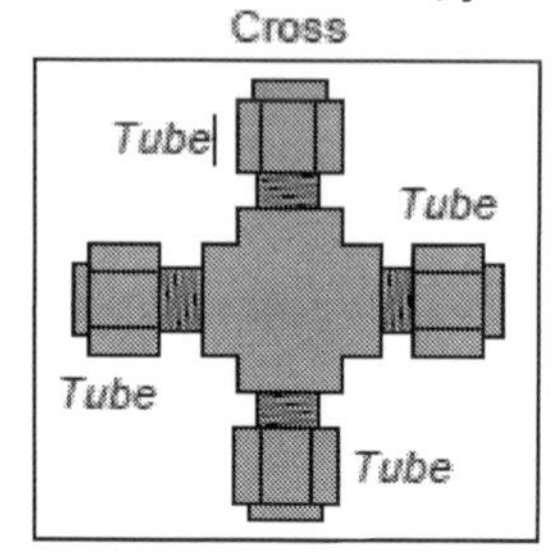

Special tube fittings are made to terminate tube connections, so they are sealed up instead of open. A piece designed to seal off the open end of a tube fitting is called a plug, while a piece designed to seal off the end of an open tube is called a cap:

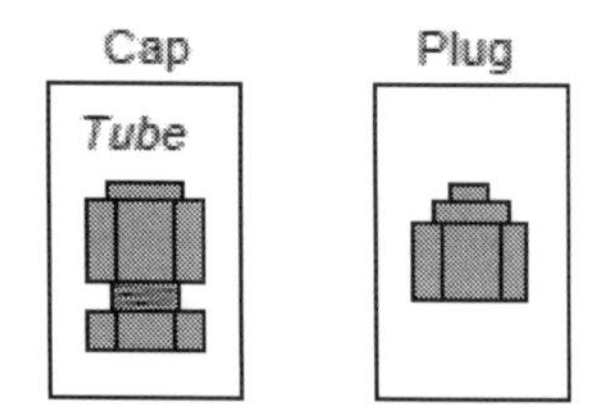

4.1.3 Bending instrument tubing

Tube bending is something of an art, especially when done with stainless steel tubing. It is truly magnificent to see a professionally-crafted array of stainless steel instrument tubes, all bends perfectly made, all terminations square, all tubes parallel when laid side by side and perfectly perpendicular when crossing.

If possible, a goal in tube bending is to eliminate as many connections as possible. Connections invite leaks, and leaks are problematic. Long runs of instrument tubing made from standard 20 foot tube sections, however, require junctions be made somewhere, usually in the form of tube unions. When multiple tube unions must be placed in parallel tube runs, it is advisable to offset the unions so it is easier to get a wrench around the tube nuts to turn them. The philosophy here, as always, is to build the tubing system with future work in mind. A photograph of several tube junctions shows one way to do this:

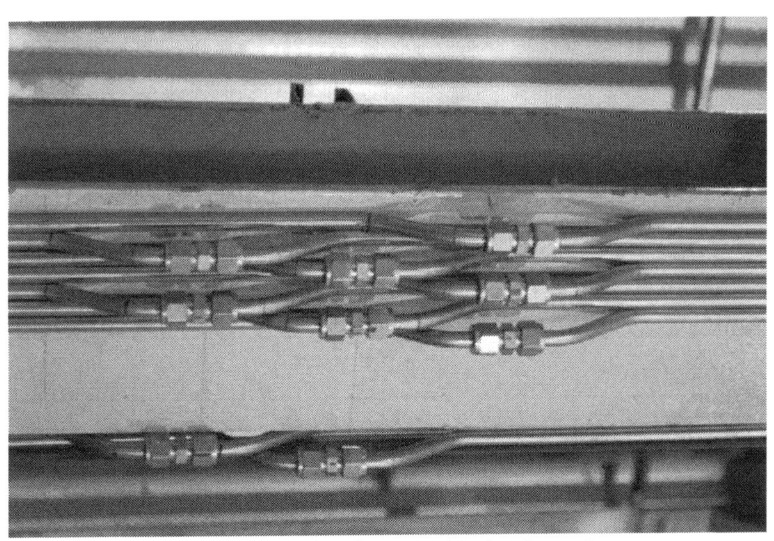

Figure: Example of Several Tube Junctions

If an instrument tube must connect between a stationary object and a vibrating object, a straight (square) run of tube is actually not desirable, since it will not have much flexibility to absorb the vibration. Instead, a vibration loop should be made in the tube, giving it the necessary elasticity to handle the vibrational stresses. An example of a vibration loop placed in the air supply tube going to a control valve appears in this photograph:

Figure: Example of Vibration Loop

When bending such a loop, it is helpful to use the circumference of a large pipe as a mandrel to form the tube rather than attempt to form a loop purely by hand.

4.1.4 Special tubing tools

A variety of specialized tools exist to help tubing installers work with compression-style tube fittings. One of these special devices is an electronic power tool manufactured by American Power Tool expressly for use with instrument tube fittings:

Figure: Electronic Power Tool Courtesy American Power Tool

The Aeroswage SX-1 has a microprocessor-controlled electric motor programmed to rotate a tube fitting's nut to a precise angular dimension, in order to properly swage the fitting. The tool comes complete with a holding jig to engage the body of the tube fitting, in order that all tightening torque is borne by the tool and not imposed on the person operating the tool:

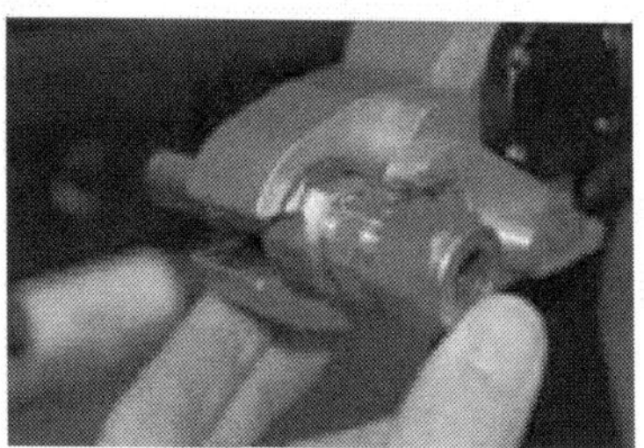

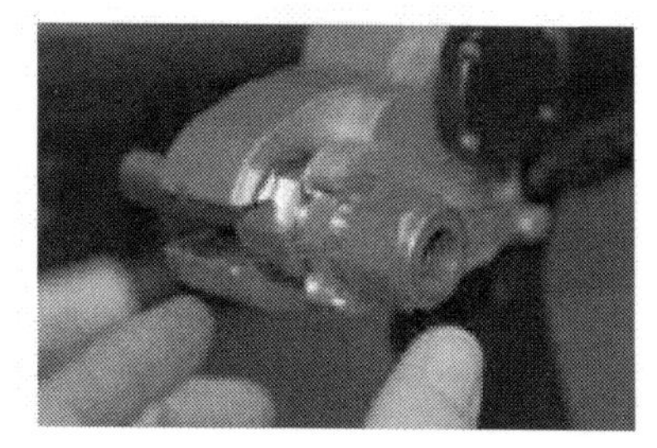

Not only does this feature reduce the amount of stress placed on the tube fitter's hand and wrist, but it also enables the tool to be used in the demanding environment of zero gravity, for example aboard a space station. In such an environment, torque applied to the tool operator could be disastrous, as the human operator has no weight to stabilize herself.

This next pair of photos shows how the tool is able to support itself on a piece of stiff (1/2 inch stainless steel) tubing, and indeed may even be operated hands-free:

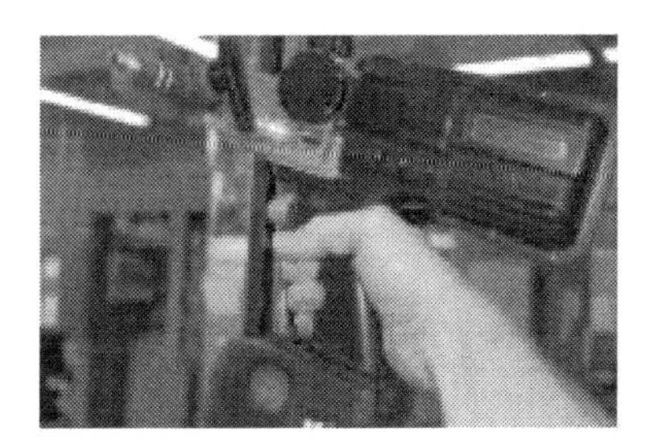

The amount of rotation is programmable, enabling the tool to be used with different kinds of fittings. For standard industrial Swagelok compression fitting sizes (1/4 inch, 3/8 inch, and 1/2 inch), the recommended swaging rotation of 1-1/4 turns may be entered into the tool as a tightening angle of 450 degrees:

Being a microprocessor-controlled device, the SX-1 has the ability to digitally record all actions. This is useful in high-reliability production environments (e.g. aerospace tube installation) where individual tube fitting data are archived for safety and quality control purposes. This data may be downloaded to a personal computer through a serial port connection on the side of the tool. Here you can see the tool's digital display showing the recorded action number, tightening angle, date, and time:

For large instrument compression fittings, hydraulic swaging tools are also available to provide the force necessary to properly compress the ferrule(s) onto the tube. Instrument tube manufacturers will provide specific recommendations for the installation of non-standard tube types, sizes, and materials, and also recommend particular swaging tools to use with their fittings.

5 Chapter: Basic Measurement and Control Concepts

The basic set of units used on this course is the SI unit system. This can be summarized in the following table:

Quantity	Unit	Abbreviation
Length	metre	m
Mass	kilogram	kg
Time	second	s
Current	ampere	A
Temperature	degree Kelvin	°K
Voltage	volt	V
Resistance	ohm	Ω
Capacitance	farad	F
Inductance	henry	H
Energy	joule	J
Power	watt	W
Frequence	hertz	Hz
Charge	coulomb	C
Force	newton	N
Magnetic Flux	weber	Wb
Magnetic Flux Density	webers/metre^2	Wb/m^2

Table : SI Units.

5.1 Basic Measurement Performance Terms and specifications

There are a number of criteria that must be satisfied when specifying process measurement equipment. Below is a list of the more important specifications.

5.1.1 Accuracy

The accuracy specified by a device is the amount of error that may occur when measurements are taken. It determines how precise or correct the measurements are to the actual value and is used to determine the suitability of the measuring equipment.

Accuracy can be expressed as any of the following:

- error in units of the measured value
- percent of span
- percent of upper range value
- percent of scale length
- percent of actual output value

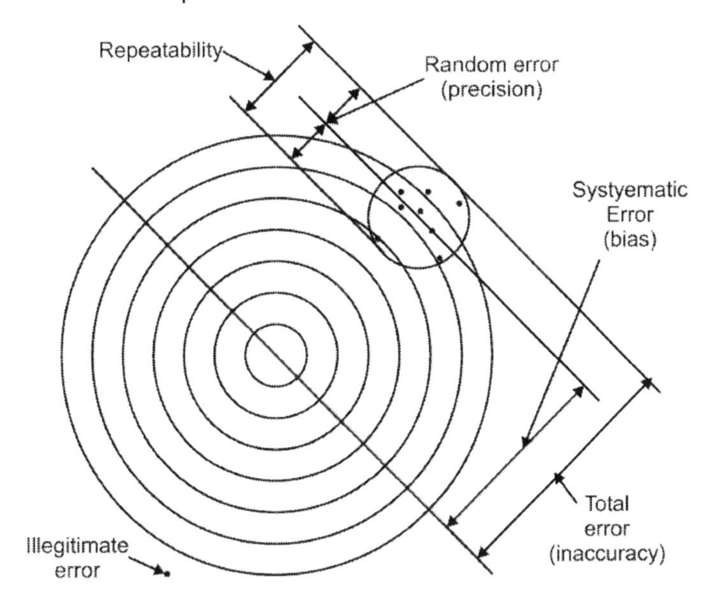

Figure: Accuracy Terminology

Accuracy generally contains the total error in the measurement and accounts for linearity, hysteresis and repeatability.

Reference accuracy is determined at reference conditions, ie. constant ambient temperature, static pressure, and supply voltage. There is also no allowance for drift over time.

5.1.2 Range of Operation

The range of operation defines the high and low operating limits between which the device will operate correctly, and at which the other specifications are guaranteed. Operation outside of this range can result in excessive errors, equipment malfunction and even permanent damage or failure.

5.1.3 Budget/Cost

Although not so much a specification, the cost of the equipment is certainly a selection consideration. This is generally dictated by the budget allocated for the application. Even if all the other specifications are met, this can prove an inhibiting factor.

5.2 Advanced Measurement Performance Terms and Specifications

More critical control applications may be affected by different response characteristics. In these circumstances the following may need to be considered:

5.2.1 Hysteresis

This is where the accuracy of the device is dependent on the previous value and the direction of variation. Hysteresis causes a device to show an inaccuracy from correct value, as it is affected by the previous measurement.

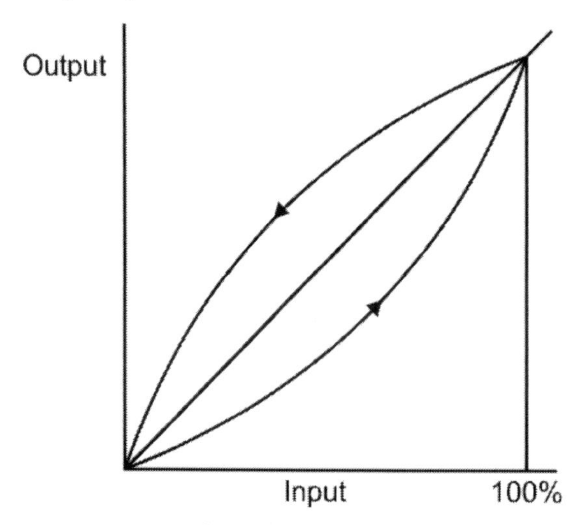

Figure: Hysteresis

5.2.2 Linearity

Linearity is how close a curve is to a straight line. The response of an instrument to changes in the measured medium can be graphed to give a response curve. Problems can arise if the response is not linear, especially for continuous control applications. Problems can also occur in point control as the resolution varies depending on the value being measured.

Linearity expresses the deviation of the actual reading from a straight line. For continuous control applications, the problems arise due to the changes in the rate the output differs from the instrument. The gain of a non-linear device changes as the change in output over input varies. In a closed loop system changes in gain affect the loop dynamics. In such an application, the linearity needs to be assessed. If a problem does exist, then the signal needs to be linearised.

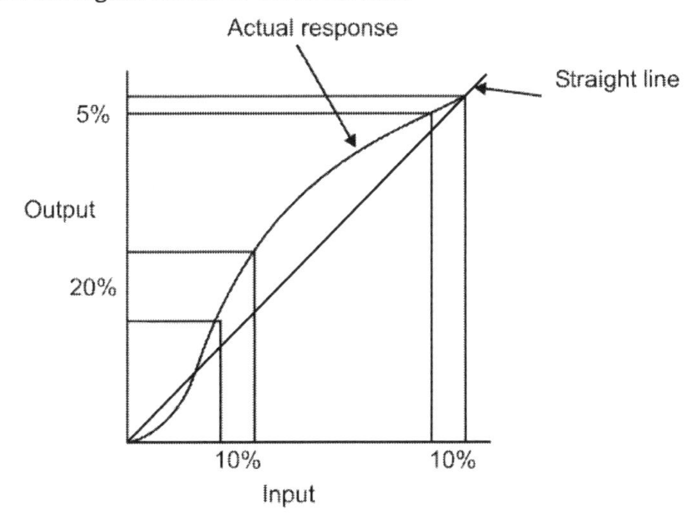

Figure: Linearity

5.2.3 Repeatability

Repeatability defines how close a second measurement is to the first under the same operating conditions, and for the same input. Repeatability is generally within the accuracy range of a device and is different from hysteresis in that the operating direction and conditions must be the same.

Continuous control applications can be affected by variations due to repeatability. When a control system sees a change in the parameter it is controlling, it will adjust its output accordingly. However if the change is due to the repeatability of the measuring device, then the controller will over-control. This problem can be overcome by using the deadband in the controller; however repeatability becomes a problem when an accuracy of say, 0.1% is required, and a repeatability of 0.5% is present.

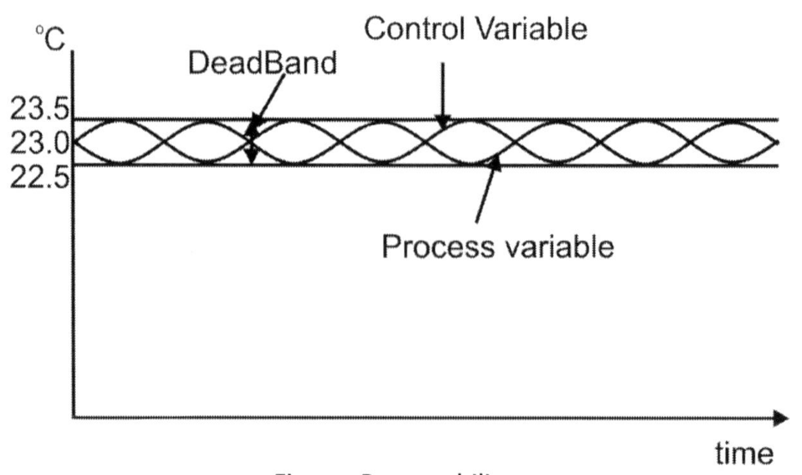

Figure: Repeatability

Ripples or small oscillations can occur due to overcontrolling. This needs to be accounted for in the initial specification of allowable values.

5.2.4 Response

When the output of a device is expressed as a function of time (due to an applied input) the time taken to respond can provide critical information about the suitability of the device. A slow responding device may not be suitable for an application. This typically applies to continuous control applications where the response of the device becomes a dynamic response characteristic of the overall control loop. However, in critical alarming applications where devices are used for point measurement, the response may be just as important.

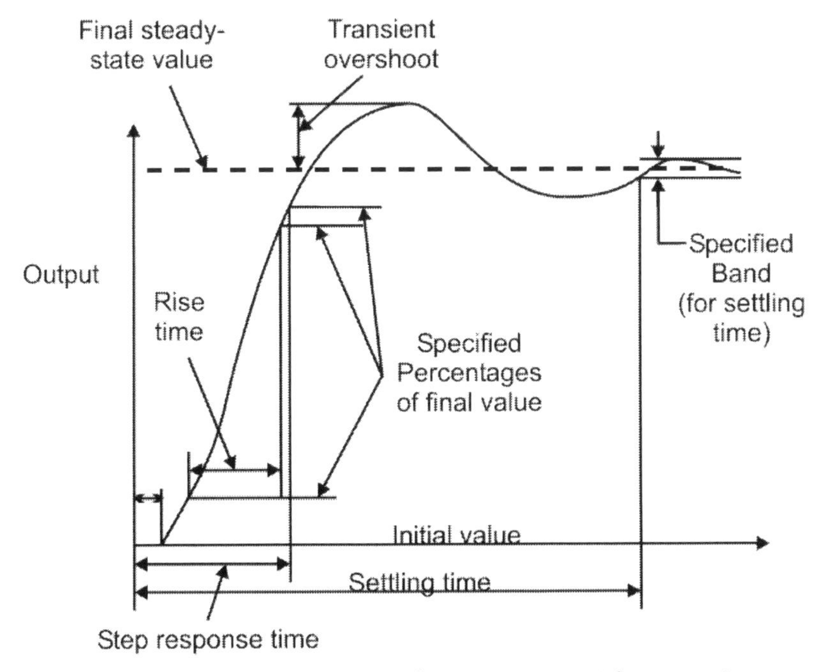

Figure: Typical time response for a system with a step input

5.3 Field Instrumentation Documentation

5.3.1 Process flow diagram (PFD)

"Drawing that shows the general process flow between major pieces of equipment of a plant and the expected operating conditions at the target production rate."

Since the purpose of the process flow diagram is to document the basic process design and assumptions, for example, the operating pressure and temperature of a reactor at normal production rates, it does not include many details concerning piping and field instrumentation. In some cases, however, the process engineer may include in the PFD an overview of key measurements and control loops that are needed to achieve and maintain the design operating conditions.

During the design process, the process engineer will typically use high-fidelity process simulation tools to verify and refine the process design. The values for operating pressures, temperatures, and flows that are included in the PFD may have been determined using these design tools. A PFD does not show minor components, piping systems, piping ratings and designations.

A PFD should include:

- Process Piping
- Major equipment symbols, names and identification numbers
- Control, valves and valves that affect operation of the system

- Interconnection with other systems
- Major bypass and recirculation lines
- System ratings and operational values as minimum, normal and maximum flow, temperature and pressure
- Composition of fluids

An example of a PFD is shown in below Figure. In this example, the design conditions are included in the lower portion of the drawing.

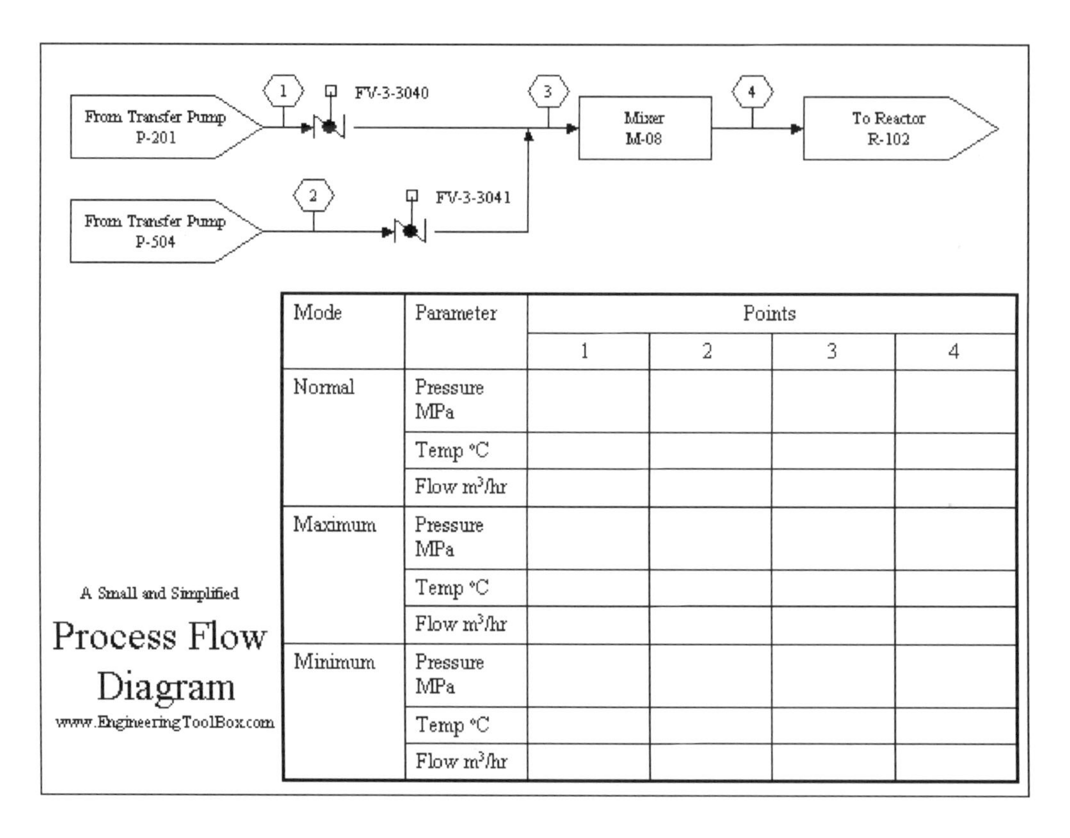

Mode	Parameter	Points			
		1	2	3	4
Normal	Pressure MPa				
	Temp °C				
	Flow m³/hr				
Maximum	Pressure MPa				
	Temp °C				
	Flow m³/hr				
Minimum	Pressure MPa				
	Temp °C				
	Flow m³/hr				

5.3.2 Piping and Instrumentation Diagram

"Drawing that shows the instrumentation and piping details for plant equipment."

The instrumentation department of an engineering firm is responsible for the selection of field devices that best matches the process design requirements. This includes the selection of the transmitters that fit the operating conditions, the type and sizing of valves, and other implementation details.

The decisions that are made concerning field instrumentation, the assignment of device tags, and piping details are documented using a piping and instrumentation diagram (P&ID). A piping and instrumentation diagram is similar to a process flow diagram in that it includes an illustration of the major equipment. However, the P&ID

includes much more detail about the piping associated with the process, to include manually operated blocking valves. It shows the field instrumentation that will be wired into the control system, as well as local pressure, temperature, or level gauges that may be viewed in the field but are not brought into the control system.

The Instrumentation, Systems, and Automation Society (ISA) is one of the leading process control trade and standards organizations. The ISA has developed a set of symbols for use in engineering drawings and designs of control loops (ISA S5.1 instrumentation symbol specification). You should be familiar with ISA symbology so that you can demonstrate possible process control loop solutions on paper to your customer. Figure shows a control loop using ISA symbology. Drawings of this kind are known as *piping and instrumentation drawings* (P&ID).

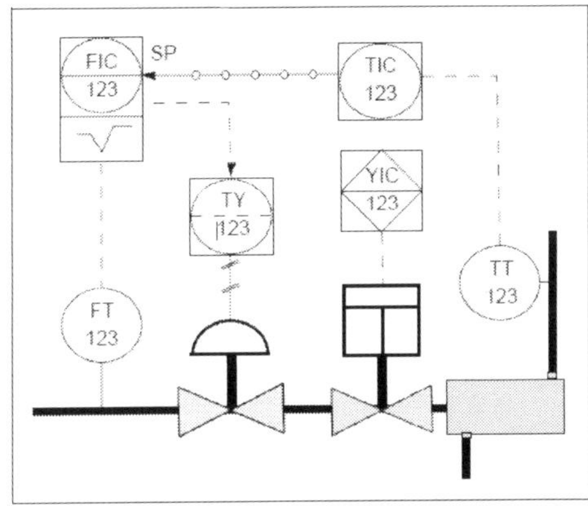

Piping and Instrumentation Drawing
(P&ID)

Some of the typical symbols used are indicated in the figures below.

Instrument Representation on Flow Diagrams

	Central Control Room		Auxiliary Location		
	Accessible to Operator	Behind the panel or otherwise inaccessible to Operator	Accessible to Operator	Behind the panel or otherwise inaccessible to Operator	Feild Mounted Instrument
Discrete Instruments					
Shared Hardware Shared Display, Shared Control					
Software Computer Function					
Shared Logic Programmable Logic Control					

Figure Instrument representation on flow diagrams (a).

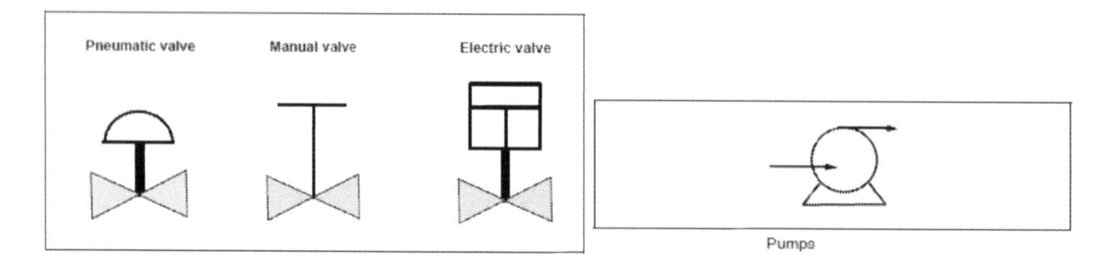

Piping and Connections

Piping and connections are represented with several different symbols (Figure):

- A heavy solid line represents piping
- A thin solid line represents process connections to instruments (e.g., impulse piping)
- A dashed line represents electrical signals (e.g., 4–20 mA connections)
- A slashed line represents pneumatic signal tubes
- A line with circles on it represents data links

Other connection symbols include capillary tubing for filled systems (e.g., remote diaphragm seals), hydraulic signal lines, and guided electromagnetic or sonic signals.

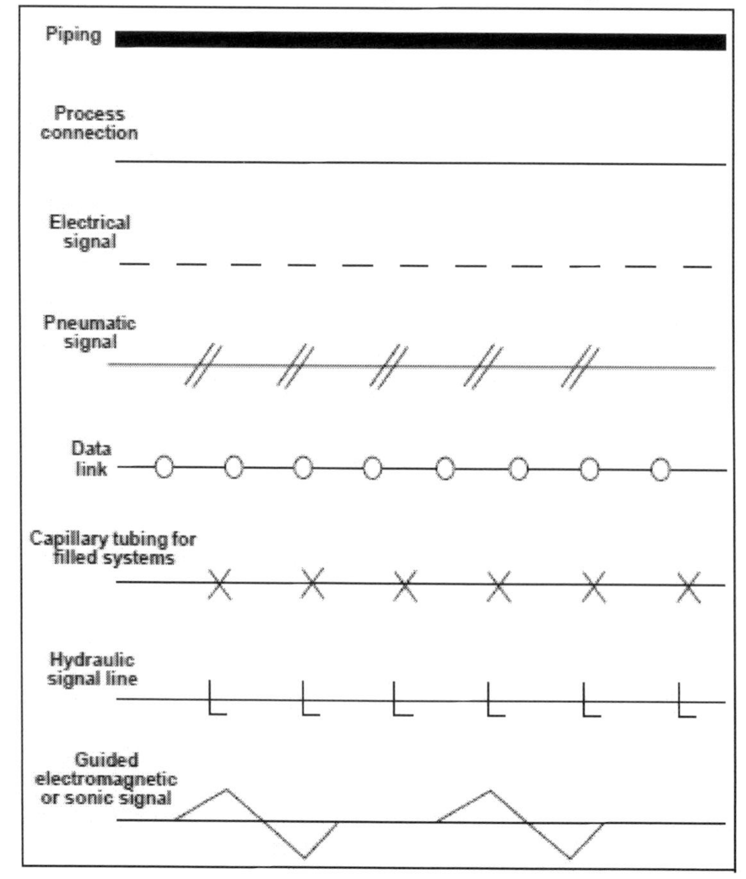

Figure: Piping and Connection Symbols

Identification Letters Activities

Identification letters on the ISA symbols (e.g., TT for temperature transmitter) indicate:

- The variable being measured (e.g., flow, pressure, temperature)
- The device's function (e.g., transmitter, switch, valve, sensor, indicator)
- Some modifiers (e.g., high, low, multifunction)

Table on next page shows the ISA identification letter designations. The initial letter indicates the measured variable. The second letter indicates a modifier, readout, or device function. The third letter usually indicates either a device function or a modifier. For example, "FIC" on an instrument tag represents a flow indicating controller. "PT" represents a pressure transmitter. You can find identification letter symbology information on the ISA Web site at http://www.isa.org.

Tag Numbers

Numbers on P&ID symbols represent instrument tag numbers. Often these numbers are associated with a particular control loop (e.g., flow transmitter 123). See Figure

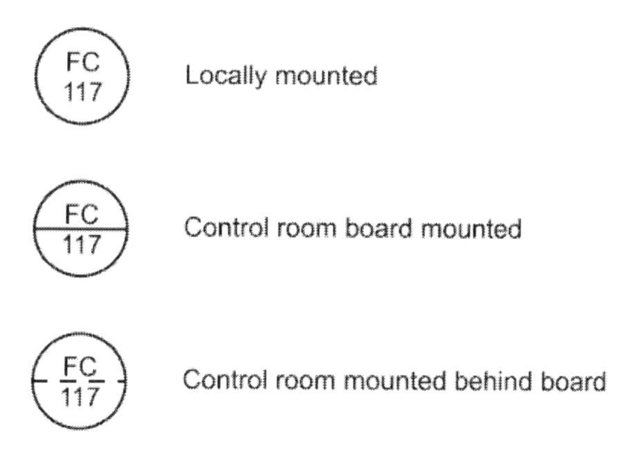

Figure: Letter codes and balloon symbols

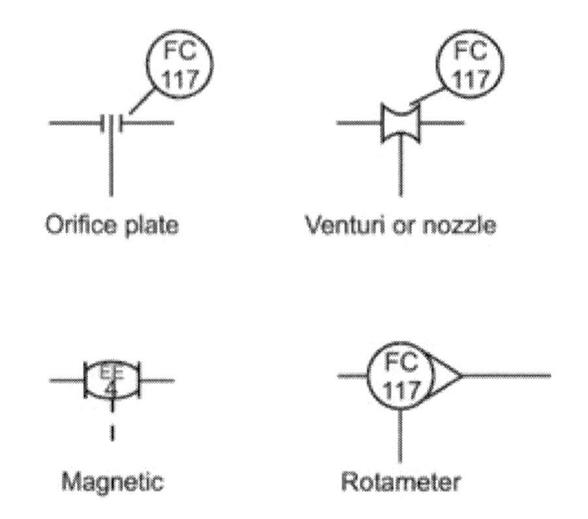

Figure: P& ID symbols for transducers and other elements

	Measured Variable	Modifier	Readout	Device Function	Modifier
A	Analysis		Alarm		
B	Burner, combustion		User's choice	User's choice	User's choice
C	User's choice			Control	
D	User's choice	Differential			
E	Voltage		Sensor (primary element)		
F	Flow rate	Ration (fraction)			
G	User's choice		Glass, viewing device		
H	Hand				High
I	Electrical Current		Indication		
J	Power	Scan			
K	Time, time schedule	Time rate of change		Control station	
L	Level		Light		Low
M	User's choice	Momentary			Middle, intermediate
N	User's choice		User's choice	User's choice	User's choice
O	User's choice		Orifice, restriction		
P	Pressure, vacuum		Point, test connection		
Q	Quantity	Integrate, totalizer			
R	Radiation		Record		
S	Speed, frequency	Safety		Switch	
T	Temperature			Transmit	
U	Multivariable		Multifunction	Multifunction	Multifunction
V	Vibration, mechanical analysis			Valve, damper, louver	
W	Weight, force		Well		
X	Unclassified	X axis	Unclassified	Unclassified	Unclassified
Y	Event, state, or presence	Y axis		Relay, compute, convert	
Z	Position, dimension	Z axis		Driver, actuator	

ISA Identification Letters

The P&ID acts as a directory to all field instrumentation and control that will be installed on a process and thus is a key document to the control engineer. Since the instrument tag (tag number) assigned to field devices is shown on this document, the instrument tag associated with, for example, a measurement device or actuator of interest may be quickly found. Also, based on the instrument tags, it is possible to quickly identify the instrumentation and control associated with a piece of equipment. For example, a plant operator may report to Maintenance that a valve on a piece of equipment is not functioning correctly. By going to the P&ID the maintenance person can quickly identify the tag assigned to the valve and also learn how the valve is used in the control of the process. Thus, the P&ID plays an important role in the design, installation and day to day maintenance of the control system. It is a key piece of information in terms of understanding what is currently being used in the plant for process control. An example of a P&ID is shown in below Figure

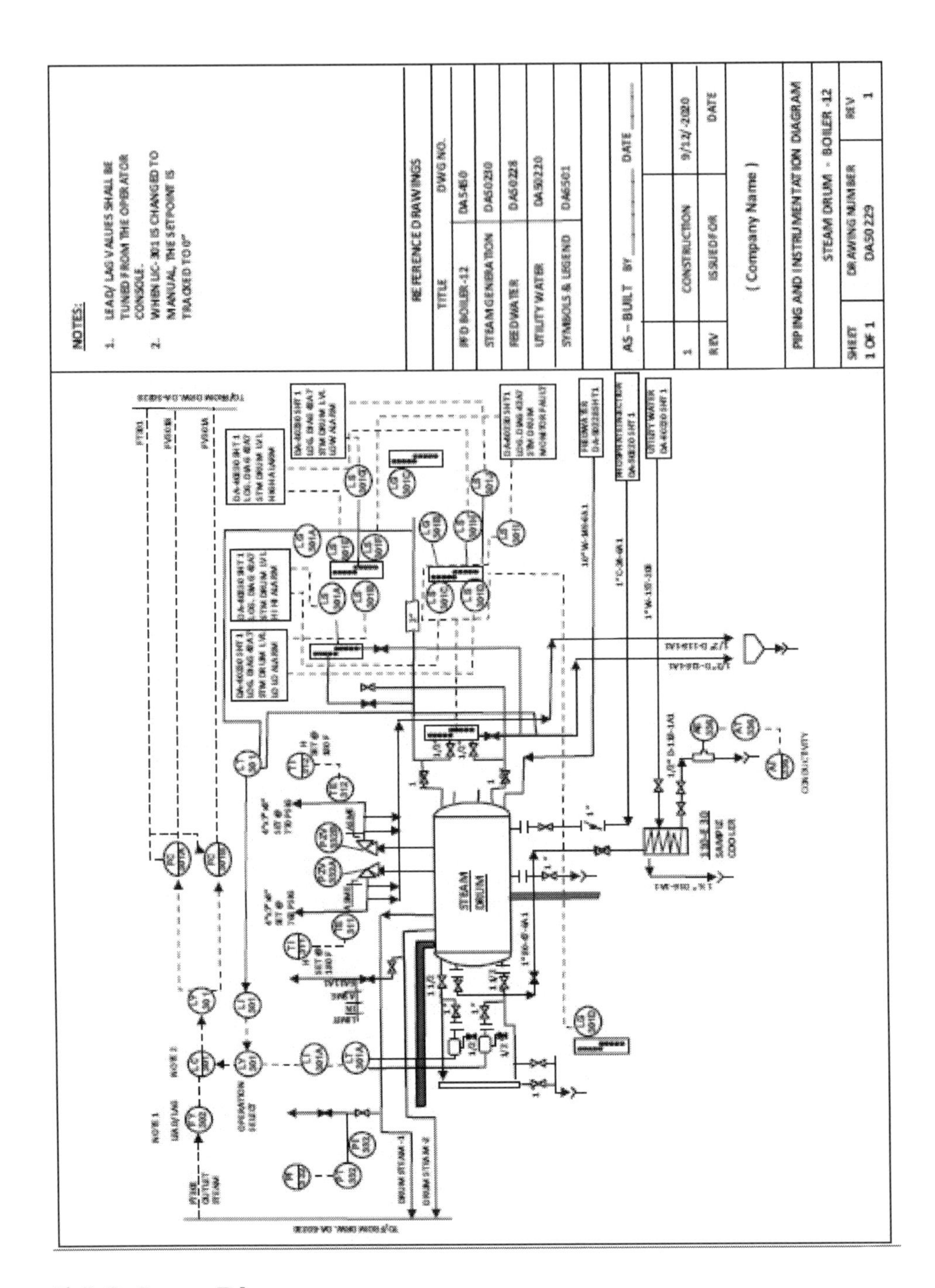

5.3.3 Loop Diagram

"Drawing that shows field device installation details including wiring and the junction box (if one is used) that connects the field device to the control system."

The piping and instrumentation diagram identifies, but does not describe in detail, the field instrumentation that is used by the process control system, as well as field devices such as manual blocking valves that are needed in plant operations. Many of the

installation details associated with field instrumentation, such as the field devices, measurement elements, wiring, junction block termination, and other installation details are documented using a loop diagram. ISA has defined the ISA-5.4 standard for Instrument Loop Diagrams.
Loop diagram, also commonly known as a loop sheet, is created for each field device that has been given a unique tag number. The loop diagrams for a process area are normally bound into a book and are used to install and support checkout of newly installed field devices. After plant commissioning, the loop Diagrams provide the wiring details that a maintenance person needs to find and troubleshoot wiring to the control system.

The loop diagram typically contains a significant amount of detail. For example, if a junction box is used as an intermediate wiring point, the loop diagram will contain information on the wiring junctions from the field device to the control system. An example of a loop diagram is shown in below Figure.

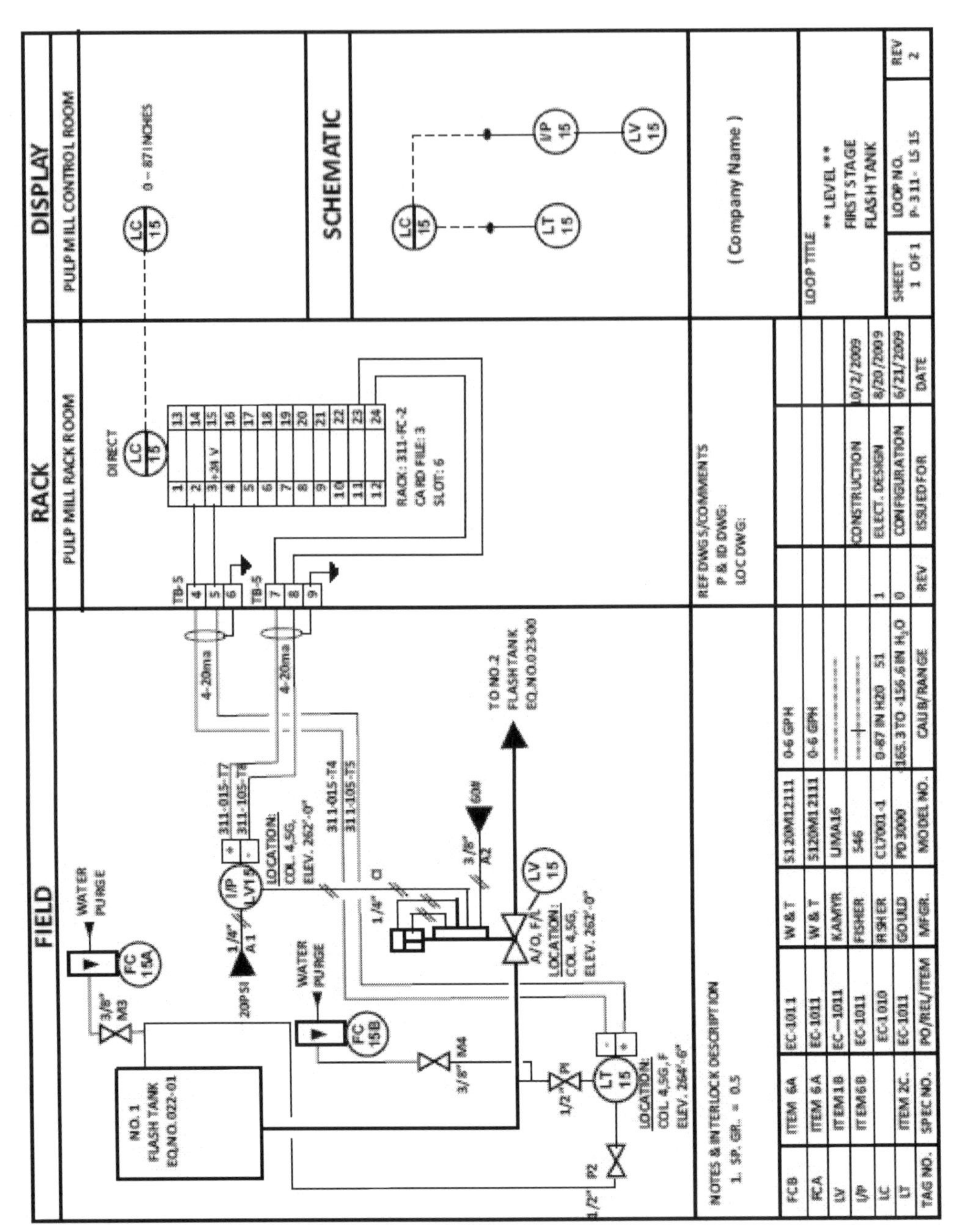

As is illustrated in this example, junction box connections are shown on the line that shows the division between the field and the rack room. The loop diagram shows the termination numbers used in the junction box and the field device and for wiring to the control system input and output cards. Also, the Display and Schematic portions of the loop diagram provide information on how the field input and output are used in the control system.

Figure shows installation details for a two-wire level transmitter that is powered through the control system analog input card. Also, connections are shown between the control system analog output card and an I/P transducer and pneumatic valve actuator. Details such as the 20 psi air supply to the I/P and the 60 psi air supply to the actuator are shown on this drawing. Based on information provided by the loop diagram, we know that the I/P will be calibrated to provide a 3–15 psi signal to the valve actuator. In addition, specific details are provided on the level measurement installation. Since the installation shows sensing lines to the top and bottom of the tank, it becomes clear that the tank is pressurized, and that level will be sensed based on the differential pressure.

In this particular installation, the instrumentation engineer has included purge water to keep the sensing line from becoming plugged by material in the tank. Even fine details such as the manual valve to regulate the flow of purge water are included in the loop diagram to guide the installation and maintenance of the measurement device.

In this example, the loop diagram shows the installation of a rotameter. A rotameter consists of a movable float inserted in a vertical tube and may be used to provide an inexpensive mechanical means of measuring volumetric
flow rate in the field.

5.3.4 Documentation Example

The example included in this section is designed to illustrate many of the recommendations of the ISA-5.1 standard. By studying the control system and field instrumentation documentation, a lot can be learned about the process and the way the control system is designed to work. Such background information can be extremely helpful when you are working with or troubleshooting the control system for a new process.

Batch Reactor Control System

Batch processing plays an important role in some industries. In a batch process, a vessel is charged with feed material that is processed in the vessel through mechanical or chemical means (or both). At the end of the process, the product—which may be a finished product or an intermediate product for use in another process—is removed from the vessel. The batch reactor example illustrated in Figure is known as "continuous feed batch," in which feed material is continuously added to the vessel throughout part or all of the batch processing. The reaction takes place, and finally the product is pumped from the vessel. In comparing the basic components to those seen in the previous two examples, the only new element introduced by this example is an eductor (a device that produces vacuum by means of the Venturi effect) that is used to remove gases created by the batch reaction.

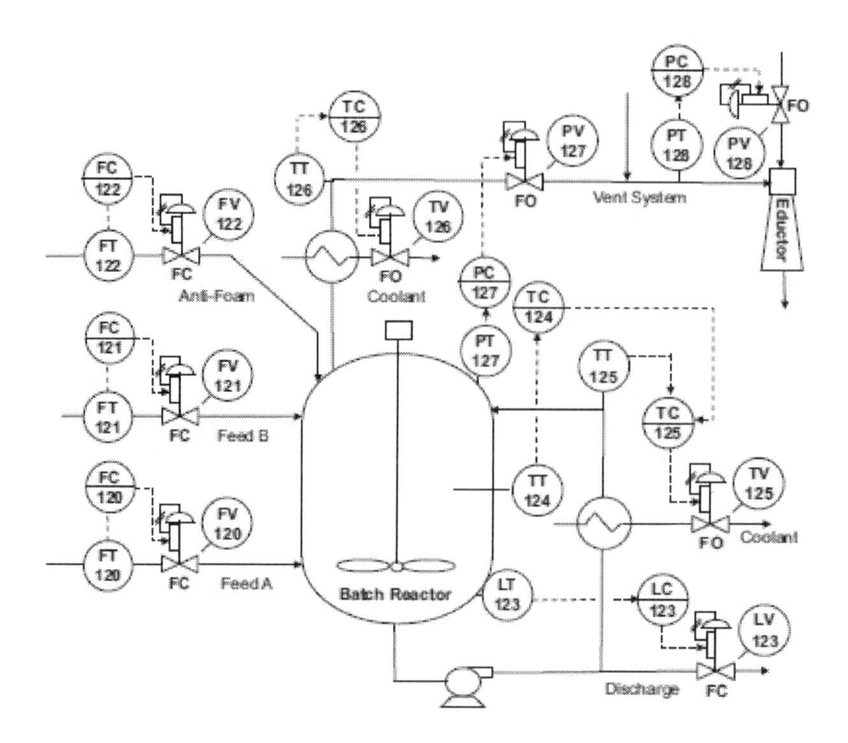

5.4 *Effects of Selection Criteria*

5.4.1 Advantages

Wide operating range

The range of operation not only determines the suitability of the device for a particular application but can be chosen for a range of applications. This can reduce the inventory in a plant as the number of sensors and models decrease. This also increases system reliability as sensing equipment can be interchanged as the need arises.

An increased operating range also gives greater over and under-range protection, should the process perform outside of specifications.

Widening the operating range of the sensing equipment may be at the expense of resolution. Precautions also need to be made when changing the range of existing equipment. In the case of control systems, the dynamics of the control loop can be affected.

Fast Response

With a fast response, delays are not added into the system. In the case of continuous control, lags can accumulate with the various control components and result in poor or slow control of the process. In a point or alarming application, a fast speed of response can assist in triggering safety or shutdown procedures that can reduce the amount of equipment failure or product lost.

Often a fast response is achieved by sacrificing the mechanical protection of the transducer element.

Good Sensitivity

Improved sensitivity of a device means that more accurate measurements are possible. The sensitivity also defines the magnitude of change that occurs. High sensitivity in the measuring equipment means that the signal is easily read by a controller or other equipment.

High Accuracy

This is probably one of the most important selection criteria. The accuracy determines the suitability of the measuring equipment to the application and is often a trade off with cost.

High accuracy means reduced errors in measurement; this also can improve the integrity and performance of a system.

High Overrange Protection

This is more a physical limitation on the protection of the equipment. In applications where the operating conditions are uncertain or prone to failure, it is good practice to 'build-in' suitable protection for the measuring equipment.

High overrange protection is different to having a wide operating range in that it does not measure when out of range. The range is kept small to allow sufficient resolution, with the overrange protection ensuring a longer operating life.

Simple Design and Maintenance

A simple design means that there are less "bits that can break". More robust designs are generally of simple manufacture.

Maintenance is reduced with less pieces to wear, replace or assemble. There are also savings in the time it takes to service, repair and replace, with the associated procedures being simplified.

Cost

Any application that requires a control solution or the interrogation of process information is driven by a budget. It therefore is no surprise that cost is an important selection criterion when choosing measurement equipment.

The cost of a device is generally increased by improvements in the following specifications:
- Accuracy
- Range of operation
- Operating environment (high temperature, pressure etc.)

The technology used and materials of construction do affect the cost, but are generally chosen based on the improvement of the other selection criteria (typically those listed above).

Repeatability

Good repeatability ensures measurements vary according to process changes and not due to the limitations of the sensing equipment. An error can still exist in the measurement, which is defined by the accuracy. However tighter control is still possible as the variations are minimised and the error can be overcome with a deadband.

Size

This mainly applies to applications requiring specifically sized devices and has a bearing on the cost.
Small devices have the added advantage of:

- Can be placed in tight spaces
- Limited obstruction to the process
- Very accurate location of the measurement required (point measurement)

Large devices have the added advantage of:

- Area measurements

Stable

If a device drifts or loses calibration over time then it is considered to be unstable. Drifting can occur over time, or on repeated operation of the device. In the case of thermocouples, it has been proven that drift is more extreme when the thermocouple is varied over a wide range quite often, typically in furnaces that are repeatedly heated to high temperatures from the ambient temperature.

Even though a device can be recalibrated, there are a number of factor that make it undesirable:

- Labour required
- Possible shutdown of process for access
- Accessibility
-

Resolution

Whereas the accuracy defines how close the measurement is to the actual value, the resolution is the smallest measurable difference between two consecutive measurements.

The resolution defines how much detail is in the measured value. The control or alarming is limited by the resolution.

Robust

This has the obvious advantage of being able to handle adverse conditions. However this can have the added limitation of bulk.

Self Generated Signal

This eliminates the need for supplying power to the device.

Most sensing devices are quite sensitive to electrical power variations, and therefore if power is required it generally needs to be conditioned.

Temperature Corrected

Ambient temperature variations often affect measuring devices. Temperature correction eliminates the problems associated with these changes.

Intrinsic Safety

Required for specific service applications. This requirement is typically used in environments where electrical or thermal energy can ignite the atmospheric mixture.

Simple to Adjust

This relates to the accessibility of the device. Helpful if the application is not proven and constant adjustments and alterations are required.

A typical application may be the transducer for ultrasonic level measurement. It is not uncommon to weld in brackets for mounting, only to find the transducer needs to be relocated.

Suitable for Various Materials

Selecting a device that is suitable for various materials not only ensures the suitability of the device for a particular application, but can it to be used for a range of applications. This can reduce the inventory in a plant as the number of sensors and models are decreased. This also increases system reliability as sensing equipment can be interchanged as the need arises.

Non Contact

This is usually a requirement based on the type of material being sensed. Noncontact sensing is used in applications where the material causes build-up on the probe or sensing devices. Other applications are where the conditions are hazardous to the operation of the equipment. Such conditions may be high temperature, pressure or acidity.

Reliable Performance

This is an obvious advantage with any sensing device, but generally is at the expense of cost for very reliable and proven equipment. More expensive and reliable devices need to be weighed up against the cost of repair or replacement, and also the cost of loss of production should the device fail. The costs incurred should a device fail, are not only the loss of production (if applicable), but also the labour required to replace the equipment. This also may include travel costs or appropriately certified personnel for hazardous equipment or areas.

Unaffected by Density
Many applications measure process materials that may have variations in density. Large variations in the density can cause measurement problems unless accounted for. Measuring equipment that is unaffected by density provides a higher accuracy and is more versatile

Unaffected by Moisture Content
Applies primarily to applications where the moisture content can vary, and where precautions with sensing equipment are required. It is quite common for sensing equipment, especially electrical and capacitance, to be affected by moisture in the material.

The effect of moisture content can cause problems in both cases, ie. when a product goes from a dry state to wet, or when drying out from a wet state.

Unaffected by Conductivity
The conductivity of a process material can change due to a number of factors, and if not checked can cause erroneous measurements. Some of the factors affecting conductivity are:
- pH
- salinity
- temperature

Mounting External to the Vessel
This has the same advantages as non-contact sensing. However it is also possible to sense through the container housing, allowing for pressurized sensing. This permits maintenance and installation without affecting the operation of the process.

Another useful advantage with this form of measurement is that the detection obstructions in chutes or product in boxes can be performed unintrusively.

High Pressure Applications
Equipment that can be used in high pressure applications generally reduces error by not requiring any further transducer devices to retransmit the signal. However the cost is usually greater than an average sensor due to the higher pressure rating.

This is more a criterion that determines the suitability of the device for the application.

High Temperature Applications
This is very similar to the advantages of high pressure applications, and also determines the suitability of the device for the application.

Dual Point Control
This mainly applies to point control devices. With one device measuring two or even three process points, ON-OFF control can be performed simply with the one device.

This is quite common in level control. This type of sensing also limits the number of tapping points required into the process.

Polarity Insensitive
Sensing equipment that is polarity insensitive generally protects against failure from incorrect installation.

Small Spot or Area Sensing
Selecting instrumentation for the specific purpose reduces the problems and errors in averaging multiple sensors over an area, or deducing the spot measurement from a crude reading.

Generally, spot sensing is done with smaller transducers, with area or average sensing being performed with large transducers.

Remote Sensing
Sensing from afar has the advantage of being non-intrusive and allowing higher temperature and pressure ratings. It can also avoid the problem of mounting and accessibility by locating sensing equipment at a more convenient location.

Well Understood and Proven
This, more than anything, reduces the stress involved when installing new equipment, both for its reliability and suitability.

No Calibration Required
Pre-calibrated equipment reduces the labour costs associated with installing new equipment and also the need for expensive calibration equipment.

No Moving Parts
The advantages are:
Long operating life
- Reliable operation with no wear or blockages

If the instrument does not have any moving or wearing components, then this provides improved reliability and reduced maintenance.
Maintenance can be further reduced if there are no valves or manifolds to cause leakage problems. The absence of manifolds and valves results in a particularly safe installation; an important consideration when the process fluid is hazardous or toxic.

Complete Unit Consisting of Probe and Mounting
An integrated unit provides easy mounting and lowers the installation costs, although the cost of the equipment may be slightly higher.

FLOW APPLICATIONS

Low Pressure Drop

A device that has a low pressure drop presents less restriction to flow and also has less friction. Friction generates heat, which is to be avoided. Erosion (due to cavitation and flashing) is more likely in high pressure drop applications.

Less Unrecoverable Pressure Drop

If there are applications that require sufficient pressure downstream of the measuring and control devices, then the pressure drops across these devices needs to be taken into account to determine a suitable head pressure. If the pressure drops are significant, then it may require higher pressures. Equipment of higher pressure ratings (and higher cost) are then required.

Selecting equipment with low pressure losses results in safer operating pressures with a lower operating cost.

High Velocity Applications

It is possible in high velocity applications to increase the diameter of the section which gives the same quantity of flow, but at a reduced velocity. In these applications, because of the expanding and reducing sections, suitable straight pipe runs need to be arranged for suitable laminar flow.

Operate in Higher Turbulence

Devices that can operate with a higher level of turbulence are typically suited to applications where there are limited sections of straight length pipe.

Fluids Containing Suspended Solids

These devices are not prone to mechanical damage due to the solids in suspension, and can also account for the density variations.

Require Less Straight Pipe Up and Downstream

This is generally a requirement applied to equipment that can accommodate a higher level of turbulence. However the device may contain straightening vanes which assist in providing laminar flow.

Price does not Increase Dramatically with Size

This consideration applies when selecting suitable equipment, and selecting a larger instrument sized for a higher range of operation.

Good Rangeability

In cases where the process has considerable variations (in flow for example), and accuracy is important across the entire range of operation, the selecting of equipment with good rangeability is vital.

Suitable for Very Low Flow Rates

Very low flow rates provide very little energy (or force) and as such can be a problem with many flow devices. Detection of low flow rates requires particular consideration.

Unaffected by Viscosity
The viscosity generally changes with temperature, and even though the equipment may be rated for the range of temperature, problems may occur with the fluidity of the process material.

No Obstructions
This primarily means no pressure loss. It is also a useful criteria when avoiding equipment that requires maintenance due to wear, or when using abrasive process fluids.

Installed on Existing Installations
This can reduce installation costs, but more importantly can avoid the requirement of having the plant shutdown for the purpose or duration of the installation.

Suitable for Large Diameter Pipes
Various technologies do have limitations on pipe diameter, or the cost increases rapidly as the diameter increases.

5.4.2 Disadvantages

The disadvantages are obviously the opposite of the advantages listed previously. The following is a discussion of effects of the disadvantages and reasons for the associated limitations.

Hysteresis
Hysteresis can cause significant errors. The errors are dependent on the magnitude of change and the direction of variation in the measurement.
One common cause of hysteresis is thermoelastic strain.

Linearity
This affects the resolution over the range of operation. For a unit change in the process conditions, there may be a 2% change at one end of the scale, with a 10% change at the other end of the scale. This change is effectively a change in the sensitivity or gain of the measuring device.

In point measuring applications this can affect the resolution and accuracy over the range. In continuous control applications where the device is included in the control loop, it can affect the dynamic performance of the system.

Indication Only
Devices that only perform indication are not suited for automated control systems as the information is not readily accessible. Errors are also more likely and less predictable as they are subject to operator interpretation.
These devices are also generally limited to localised measurement only and are isolated from other control and recording equipment.

Sensitive to Temperature Variations
Problems occur when equipment that is temperature sensitive is used in applications where the ambient temperature varies continuously. Although temperature compensation is generally available, these devices should be avoided with such applications.

Shock and Vibration
These effects not only cause errors but can reduce the working life of equipment, and cause premature failure.

Transducer Work Hardened
The physical movement and operation of a device may cause it to become harder to move. This particularly applies to pressure bellows, but some other devices do have similar problems.

If it is unavoidable to use such equipment, then periodic calibration needs to be considered as a maintenance requirement.

Poor Overrange Protection
Care needs to be taken to ensure that the process conditions do not exceed the operating specifications of the measuring equipment. Protection may need to be supplied with additional equipment.

Poor overrange protection in the device may not be a problem if the process is physically incapable of exceeding the operating conditions, even under extreme fault conditions.

Unstable
This generally relates to the accuracy of the device over time. However the accuracy can also change due to large variations in the operation of the device due to the process variations. Subsequently, unstable devices require repeated calibration over time or when operated frequently.

Size
Often the bulkiness of the equipment is a limitation. In applications requiring area or average measurements then too small a sensing device can be a disadvantage in that it does not “see” the full process value.

Dynamic Sensing Only
This mainly applies to shock and acceleration devices where the impact force is significant. Typical applications would involve piezoelectric devices.

Special Cabling
Measurement equipment requiring special cabling bears directly on the cost of the application. Another concern with cabling is that of noise and cable routing. Special

conditions may also apply to the location of the cable in reference to high voltage, high current, high temperature, and other low power or signal cabling.

Signal Conditioning
Primarily used when transmitting signals over longer distances, particularly when the transducer signal requires amplification. This is also a requirement in noisy environments. As with cabling, this bears directly on the cost and also may require extra space for mounting.

Stray Capacitance Problems
This mainly applies to capacitive devices where special mounting equipment may be required, depending on the application and process environment.

Maintenance
High maintenance equipment increases the labour which become a periodic expense. Some typical maintenance requirements may include the following:

- Cleaning
- Removal/replacement
- Calibration

If the equipment is fragile then there is the risk of it being easily damaged due to repeated handling.

Sampled Measurement Only
Measurement equipment that requires periodic sampling of the process (as opposed to continual) generally relies on statistical probability for the accuracy. More pertinent in selecting such devices is the longer response and update times incurred in using such equipment.
Sampled measurement equipment is mainly used for quality control applications where specific samples are required and the quality does not change rapidly.

Pressure Applications
This applies to applications where the measuring equipment is mounted in a pressurised environment and accessibility is impaired. There are obvious limitations in installing and servicing such equipment. In addition are the procedures and experience required for personnel working in such environments.

Access
Access to the process and measuring equipment needs to be assessed for the purpose of:

- The initial installation
- Routine maintenance

The initial mounting of the measuring equipment may be remote from the final installation; as such the accessibility of the final location also needs to be considered. This may also have a bearing on the orientation required when mounting equipment.

Requires Compressed Air
Pneumatic equipment requires compressed air. It is quite common in plants with numerous demands for instrument air to have a common compressor with pneumatic hose supplying the devices.

The cost of the installation is greatly increased if no compressed air is available for such a purpose. More common is the requirement to tap into the existing supply, but this still requires the installation of air lines.

Material Build-up
Material build-up is primarily related to the type of process material being measured. This can cause significant errors, or degrade the operating efficiency of a device over time. There are a number of ways to avoid or rectify the problems associated with material build-up:

- Regular maintenance
- Location (or relocation) of sensing equipment
- Automated or self cleaning (water sprays)

Constant Relative Density
Measurement equipment that relies on a constant density of process material is limited in applications where the density varies. Variations in the density will not affect the continued operation of the equipment, but will cause increased errors in the measurement. A typical example would be level measurement using hydrostatic pressure.

Radiation
The use of radioactive materials such as Cobalt or Cesium often gives accurate measurements. However, problems arise from the hazards of using radioactive materials which require special safety measures. Precautions are required when housing such equipment, to ensure that it is suitably enclosed and installation safety requirements are also required for personal safety.

Licensing requirements may also apply with such material.

Electrolytic Corrosion
The application of a voltage to measuring equipment can cause chemical corrosion to the sensing transducer, typically a probe. Matching of the process materials and metals used for the housing and sensor can limit the effects; however in extreme mismatches, corrosion is quite rapid.

Susceptible to Electrical Noise
In selecting equipment, this should be seen as an extra cost and possibly more equipment or configuration time is required to eliminate noise problems.

More Expensive to Test and Diagnose
More difficult and expensive equipment can also require costly test and diagnosis equipment. For 'one-off' applications, this may prove an inhibiting factor. The added expense and availability of specialised services should also be considered.

Not Easily Interchangeable
In the event of failure or for inventory purposes, having interchangeable equipment can reduce costs and increase system availability. Any new equipment that is not easily replaced by anything already existing, could require an extra as a spare.

High Resistance
Devices that have a high resistance can pick up noise quite easily. Generally high resistance devices require good practice in terms of cable selection and grounding to minimise noise pickup.

Accuracy Based on Technical Data
The accuracy of a device can also be dependent on how well the technical data is obtained from the installation and data sheets. Applications requiring such calculations are often subject to interpretation.

Requires Clean Liquid
Measuring equipment requiring a clean fluid do so for a number of reasons:

- Constant density of process fluid
- Sensing equipment with holes can become easily clogged
- Solids cause interference with sensing technology

Orientation Dependent
Depending on the technology used, requirements may be imposed on the orientation when mounting the sensing transducer. This may involve extra work, labour and materials in the initial installation. A typical application for mounting an instrument vertically would be a variable area flowmeter.

Uni-Directional Measurement Only
This is mainly a disadvantage with flow measurement devices where flow can only be measured in the one direction. Although this may seem like a major limitation, few applications use bi-directional flows.

Not Suitable with Partial Phase Change
Phase change is where a fluid, due to pressure changes, reverts partly to a gas. This can cause major errors in measurements, as it is effectively a very large change in density. For those technologies that sense through the process material, the phase change can result in reflections and possibly make the application unmeasurable.

Viscosity Must be Known
The viscosity of a fluid is gauged by the Reynolds number and does vary with temperature. In applications requiring the swirling of fluids and pressure changes there is usually an operating range of which the fluids viscosity is required to be within.

Limited Life Due to Wear
Non-critical service applications can afford measuring equipment with a limited operating life, or time to repair. In selecting such devices, consideration needs to be given to the accuracy of the measurement over time.

Mechanical Failure
Failure of mechanical equipment cannot be avoided; however the effects and consequences can be assessed in determining the suitable technology for the application. Flow is probably the best example of illustrating the problems caused if a measurement transducer should fail. If the device fails, and it is of such a construction that debris may block the line or a valve downstream, then this can make the process inoperative until shutdown and repaired.

Filters
There are two main disadvantages with filters:

- Maintenance and cleaning
- Pressure loss across filter

The pressure loss can be a process limitation, but from a control point of view can indicate that the filter is in need of cleaning or replacement.

Flow Profile
The flow profile may need to be of a significant form for selected measuring equipment. Note that the flow profile is dependent on viscosity and turbulence.

Acoustically Transparent
Measuring transducers requiring the reflection of acoustic energy are not suitable where the process material is acoustically transparent. These applications would generally require some contact means of measurement.

5.5 Overall Control System

Below is a diagram outlining how instrument and control valves fit into the overall control system structure. The topic of controllers and tuning forms part of a separate workshop.

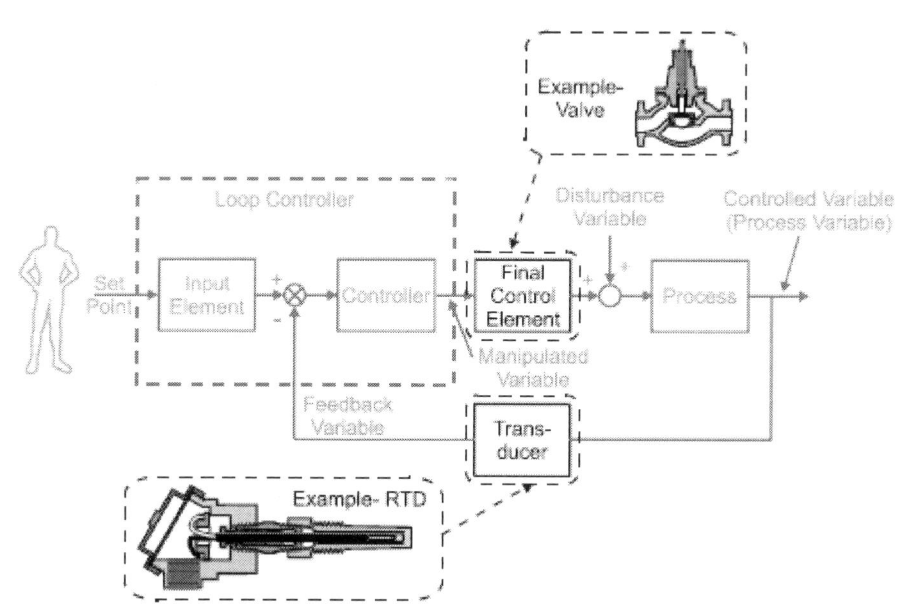

Figure: Instruments and control valves in the overall control system.

5.6 Typical Applications

Some typical applications are listed below.

HVAC (Heating, ventilation and air conditioning) Applications

- Heat transfer
- Billing
- Axial fans
- Climate control
- Hot and chilled water flows
- Forced air
- Fumehoods
- System balancing
- Pump operation and efficiency

Petrochemical Applications

- Co-generation
- Light oils
- Petroleum products
- Steam
- Hydrocarbon vapours
- Flare lines, stacks

Natural Gas

- Gas leak detection
- Compressor efficiency
- Fuel gas systems
- Bi-directional flows

- Mainline measurement
- Distribution lines measurement
- Jacket water systems
- Station yard piping

Power Industry

- Feed water
- Circulating water
- High pressure heaters
- Fuel oil
- Stacks
- Auxiliary steam lines
- Cooling tower measurement
- Low pressure heaters
- Reheat lines
- Combustion air

Emissions Monitoring

- Chemical incinerators
- Trash incinerators
- Refineries
- Stacks and rectangular ducts
- Flare lines

6 Chapter: Integration of the System

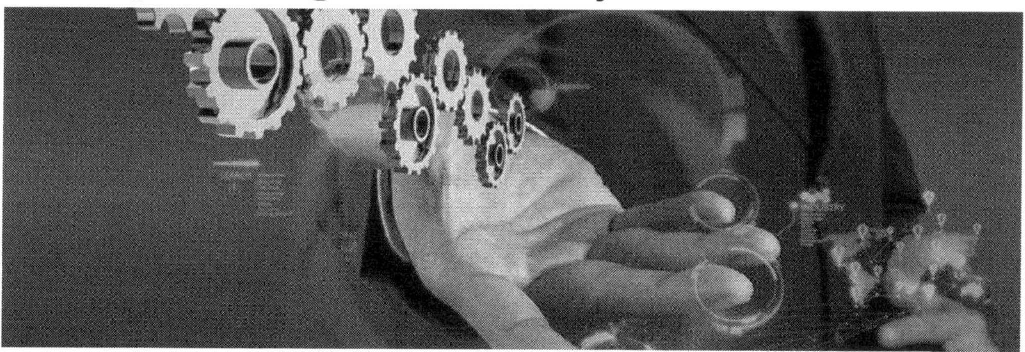

6.1 Calculation of Individual Instruments and Total Error for the System

The accuracy specified for an instrument (eg. 1%) is the error or inaccuracy of any measurement performed with that device. This is assuming that the device is operating within its specifications.

Error calculations become more complex when looking at multiple instruments, or systems with more than one component, or even devices that perform calculations on process measurements.

Throughout this section we will use a signal of 0 - 10V to represent 0 - 100%, this will simplify calculations.

6.1.1 Linear Devices in Series

When combining devices in series, the errors are multiplied. In multiplying any inaccuracy, it is a common mistake to simply multiply out the inaccuracies. The total error takes into account the total measurement with the variation.

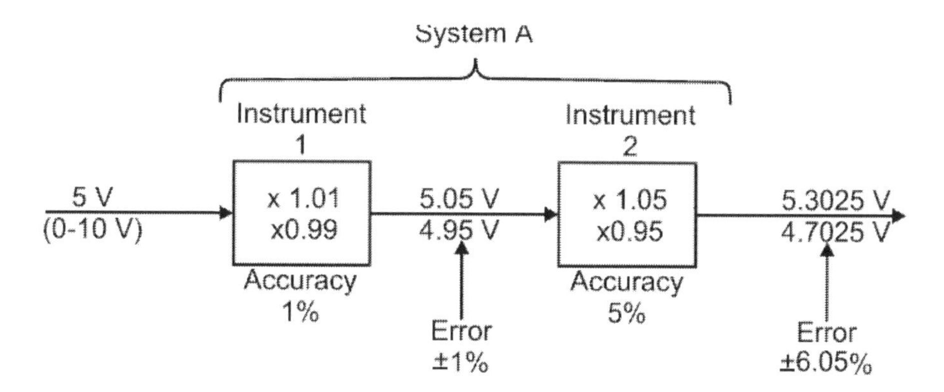

Figure: Multiple Instruments connected in Series.

If Instrument 1 (as shown) has an incoming signal of 5V, with an accuracy of 1%, then the error in the output will give a signal from 4.95V to 5.05V.

Note that: 5V x 1.01 = 5.05V maximum measurement due to error
5V x 0.99 = 4.95V minimum measurement due to error

When Instrument 2 receives 5.05V, with its accuracy of 5%, then the error in the output will give an upper value of 5.05V x 1.05 = 5.3025V. The minimum measurement due to error will be 4.95V x 0.95 = 4.7025.

Note that: 5.05V x 1.05 = 5.3025V maximum measurement due to error
4.95V x 0.95 = 4.7025V minimum measurement due to error

So the maximum output value due to the errors is:
5V x (Instrument 1 error) x (Instrument 2 error)
= 5V x ***1.01 x 1.05***
= 5V x ***1.0605***
= 5.3025V
The error in this system, System A, is ***6.05***%.

6.1.2 Non Linear Devices

One of the most common non linear devices used for process measurement is the flowmeter with differential producers. The flow in these devices is calculated from the square root of the measured pressure in the primary element. Any errors in the differential pressure measurement affects the calculated flow.

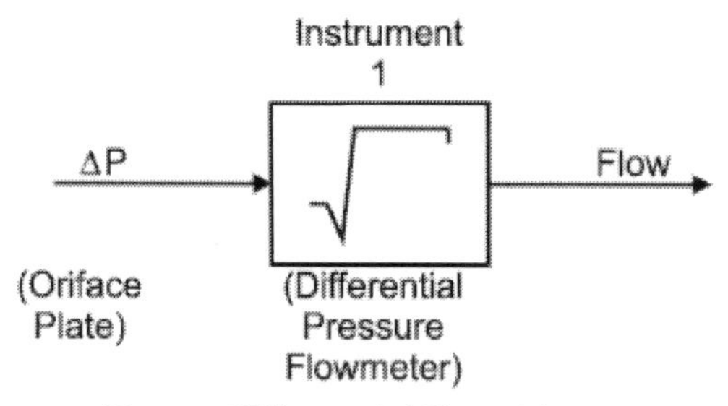

Figure: Differential Flow Meter

As the flow is deduced from the square root calculation, the magnitude of the errors are also affected by such a calculation.

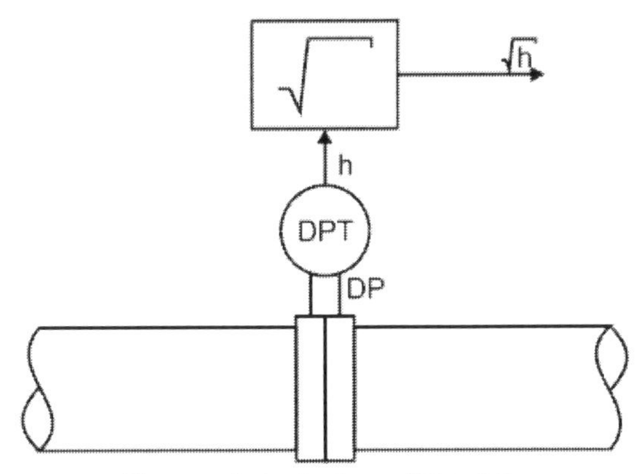

Figure: Calculation of Total Error

Actual flow, f (%)	Actual flow, df (0-1.0)	Change in DP, DDP	Change in transmitted signal, dh	Change in calculated flow, dSQRT(h)	Error in calculated flow
100	1.00	2.00	2.02	1.01	1%
80	0.80	1.28	1.30	0.82	2%
60	0.60	0.72	0.76	0.64	4%
40	0.40	0.32	0.41	0.51	11%
20	0.20	0.08	0.26	0.66	46%
10	0.10	0.02	0.25	1.25	115%
5	0.05	0.005	0.26	2.50	245%

Table: Calculation of error in non-linear device

Column 1: Actual flow, f(%)
This is the actual volumetric flow through the system.
Column 2: Actual flow, df (0-1.0)
This is the normalised flow, simply converted from percent for calculations.
Column 3: Change in DP, dDP
This is the change in DP detected for the specified flow rates.
This is calculated from the flow as follows:

$$DP = f^2$$

therefore, the change in DP is represented as,

$$dDP = 2f.\ df$$

Column 4: Change in transmitted signal, dh
The transmitted signal is the pressure measurement with the combined transmitter error. For this example an error of 0.25% is used.

$$dh = dDP$$

However, the error needs to be added in. This is done by root mean square as follows,

$$dh = SQRT(\ dDP^2 + e^2\ )$$

Column 5: Change in calculated flow, dSQRT(h)

dSQRT(h) = dh / 2f

Column 6: Error in calculated flow

Without an error in the transmitter (e = 0), then column 2 (Actual flow, df) should be equal to column 5 (Change in calculated flow). The differences are due to the error, e=0.25% in the transmitter and can be seen to be quite large at lower flowrates.

6.1.3 Other Non Linear Devices

Other such instruments with non linear calculations are:

- Flumes and rectangular weirs with flows calculated as follows,

$f = h^{3/2}$

- V-notch weirs use a separate formula,

$f = h^{5/2}$

6.2 Selection Considerations

A comprehensive list for selecting instruments would consider the following:

- Accuracy
- Reliability
- Purchase price
- Installed cost
- Cost of ownership
- Ease of use
- Process medium, liquid/ stem/ gas
- Degree of smartness
- Repeatability
- Intrusiveness
- Sizes available
- Maintenance
- Sensitivity to vibration

Particular requirements for flow would include:

- Capability of measuring liquid, steam and gas
- Rangeability
- Turndown
- Pressure drop
- Reynolds number
- Upstream and downstream piping requirements

A more systematic approach to selection process measurement equipment would cover the following steps:

Step 1 Application

This is the requirement and purpose of the measurement.

- Monitor
- Control
- Indicate
- Point or continuous

- Alarm

Step 2 Process Material Properties

Many process measuring devices are limited by the process material that they can measure.

- Solids, liquids, gas or steam
- Multiphase, liquid/gas ratio
- Viscosity
- Pressure
- Temperature

Step 3 Performance

This relates to the performance required in the application.

- Range of operation
- Accuracy
- Linearity (Accuracy may include linearity effects)
- Repeatability (Accuracy may include repeatability effects)
- Response time

Step 4 Installation

Mounting is one of the main concerns, but the installation does involve the access and other environmental concerns.

- Mounting
- Line size
- Vibration
- Access
- Submergence

Step 5 Economics

The associated costs determine whether the device is within the budget for the application.

- Purchase cost
- Installation cost
- Maintenance cost
- Reliability/ replacement cost

Step 6 Environmental And Safety

This relates to the performance of the equipment to maintain the operational specifications, and also failure and redundancy should be considered.

- Process emissions
- Hazardous waste disposal
- Leak potential
- Trigger system shutdown

Step 7 Measuring Device/Technology

At this stage the selection criteria is established and weighed up with readily available equipment. A typical example for flow is shown:

- DP - Orifice plate
- Variable area
- Velocity - Magmeter
- Vortex

- Turbine
- Propeller
- Positive displacement - Oval gear
- Rotary
- Mass flow - Coriolis
- Thermal

Step 8 Supply By Vendors

Limitations may be imposed, particularly with larger companies that have preferred suppliers, in which case the selections may be limited, or the procedure for purchasing new equipment may not warrant the time and effort for the application.

6.3 Testing and Commissioning of the Subsystems

Test procedures vary from one operation to the next and are generally dependent on how critical the equipment going into service is. In a non-critical application, where the instrument operation will not affect production, the preliminary checks can be performed on the device with it tested in working position.

More critical applications may involve the instrument to be tested on the bench before being placed in operation. Even after installation any outputs should be disabled until the correct operation of the device is proven.

Correct operation requires more than checking that the instrument works. To ensure that it is configured correctly, both for the process measurement and any alarming, trip or indication points should be tested. The interface to the Operator Interface Terminal (OIT) should also be checked.

Below are steps for a testing procedure for the worst case (most critical) application, with steps easily omitted for less critical cases as required.

- **Check correct installation of the instrument**
  This also includes grounding and isolation as required. Termination of the wire shields should be at one end only, unless used for an instrument ground path. Verify wire numbering and device tag number.
- **Check power supply**
  The reliability of the instrument is also dependent on the supply. Obviously if the main supply fails then that part of the process will not be able to proceed. However, from the instrument point of view, checks should be performed to verify the voltage rating and the proper allocation of breakers or fuses. This will also include the power supplied to I/O cards on such systems.
- **Before applying power**
  The field wiring should be isolated as well as possible from the digital system until loop commissioning or checkout is complete. This mainly applies to any output devices and can be done by removing fuses, unplugging terminal blocks or lifting isolation links.
- **Apply power**

Initial checks should be performed on the system sensibility.

- **Check: - Indicating lights on system modules**
  - Any alarms for validity
  - Temperature inputs for ambient readings
  - Pressure, level and flow for minimum readings
  - System communications
  - Smart instrument communications
- **Check loop voltages**
  - Loop voltages should be checked under load and monitored.
- **Check proper calibration of the instrument**

  The vendor usually does this to meet equipment specifications. If required this can be performed remotely with appropriated test equipment before installation into service.
- **Loop checkout**

  Each loop should be checked for proper operation from the instrument to the digital system. It often depends on the test equipment available and the process as to how much simulation is performed.
- **Simulations**

  A number of parts of the system can be simulated. This is not always a necessity, but by checking parts of the system faulty components can be eliminated. Such simulations are:
  - The transducer signal, or the input to the instrument
  - The instrument signal, or the input to the controller
  - The signal to the control device
- **Simulations can be performed at a number of values, but the more common being:**
  - The minimum or lower range value, 0%
  - The maximum or upper range value, 100%
  - The mid range value, 50%
  - Alarming points
  - Control points
- **Check trips, interlocks and shutdown procedure**

  This is vital before the system is placed into any form of automatic control operation. This is often performed by manually actuating switches and checking that interlocks function as required.
- **Automatic Control**

  PID controllers should be started in manual mode. Progression to automatic mode should be done under stable conditions and with one loop at a time. The actual procedure for achieving full automatic control of a plant is quite specific to the application. However, by activating automatic control with one loop at a time, it will be much easier to troubleshoot, test and tune the individual loops.

6.3.1 Prior considerations and budgeting requirements

This test procedure assumes that all associated personnel are familiar with instrumentation and digital systems. Allowances may be required should suitable experience not be available. This may require additional supervision or even the assistance of consulting support.

Loop checkout can be tedious and may require a substantial amount of organisation for large installations. Usually several stages of loop checkout occur before the final check. Sometimes, the application may require a more extensive test, such as running nitrogen, water, or a similar safe process medium through the facility and checking the control operation. The suitability of the medium used for testing must be checked, as it may not respond as the correct sensing material would, or may not work with the instruments engineered for the actual process.

The project budget and schedule will dictate the extent to which the system can be tested, but thorough testing increases familiarity of personnel involved and the chances for a smooth and safe start-up.

It is also good practice to involve the facility operators and maintenance personnel in the later stages of the installation and start-up of the process, this includes loop commissioning. This eases the handover to the facility personnel who will be using the system and are ultimately are responsible for production.

7 Chapter: Instrument Installation and Commissioning

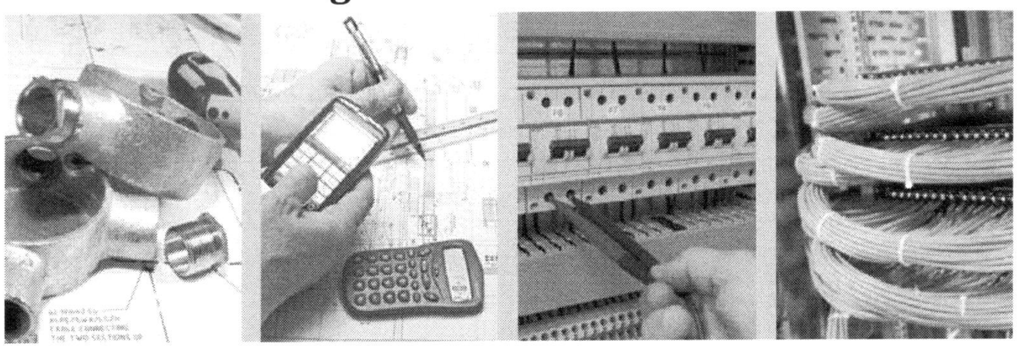

Plant safety and continuous effective plant operability are totally dependent upon correct installation and commissioning of the instrumentation systems. Process plants are increasingly becoming dependent upon automatic control systems, owing to the advanced control functions and monitoring facilities that can be provided in order to improve plant efficiency, product throughput, and product quality.

The instrumentation on a process plant represents a significant capital investment, and the importance of careful handling on site and the exactitude of the installation cannot be overstressed. Correct installation is also important in order to ensure long-term reliability and to obtain the best results from instruments which are capable of higher-order accuracies due to advances in technology. Quality control of the completed work is also an important function.

7.1 *General requirements*

Installation should be carried out using the best engineering practices by skilled personnel who are fully acquainted with the safety requirements and regulations governing a plant site. Prior to commencement of the work for a specific project, installation design details should be made available which define the scope of work and the extent of material supply and which give detailed installation information related to location, fixing, piping, and wiring. Such design details should have already taken account of established installation recommendations and measuring technology requirements. The details contained in this chapter are intended to give general installation guidelines.

7.2 *Storage and protection*

When instruments are received on a job site it is of the utmost importance that they are unpacked with care, examined for superficial damage, and then placed in a secure store which should be free from dust and suitably heated. In order to minimize handling, large items of equipment, such as control panels, should be programmed to go directly into their intended location, but temporary anti-condensation heaters should be installed if the intended air-conditioning systems have not been commissioned.

Throughout construction, instruments and equipment installed in the field should be fitted with suitable coverings to protect them from mechanical abuse such as paint spraying, etc. Preferably, after an installation has been fabricated, the instrument should be removed from the site and returned to the store for safe keeping until ready for precalibration and final loop checking. Again, when instruments are removed, care should be taken to seal the ends of piping, etc., to prevent ingress of foreign matter.

7.3 Mounting and accessibility

When instruments are mounted in their intended location, either on pipe stands, brackets, or directly connected to vessels, etc., they should be vertically plumbed and firmly secured. Instrument mountings should be vibration free and should be located so that they do not obstruct access ways which may be required for maintenance to other items of equipment. They should also be clear of obvious hazards such as hot surfaces or drainage points from process equipment.

Locations should also be selected to ensure that the instruments are accessible for observation and maintenance.
Where instruments are mounted at higher elevations, it must be ensured that they are accessible either by permanent or temporary means. Instruments should be located as close as possible to their process tapping points in order to minimize the length of impulse lines, but consideration should be paid to the possibility of expansion of piping or vessels which could take place under operating conditions and which could result in damage if not properly catered for. All brackets and supports should be adequately protected against corrosion by priming and painting.

When installing final control elements such as control valves, again, the requirement for maintenance access must be considered, and clearance should be allowed above and below the valve to facilitate servicing of the valve actuator and the valve internals.

7.4 Piping systems

All instrument piping or tubing runs should be routed to meet the following requirements:

1. They should be kept as short as possible.
2. They should not cause any obstruction that would prohibit personnel or traffic access.
3. They should not interfere with the accessibility for maintenance of other items of equipment.
4. They should avoid hot environments or potential fire-risk areas.
5. They should be located with sufficient clearance to permit lagging which may be required on adjacent pipework.

6. The number of joints should be kept to a minimum consistent with good practice.
7. All piping and tubing should be adequately supported along its entire length from supports attached to firm steelwork or structures (not handrails).

(*Note*: Tubing can be regarded as thin-walled seamless pipe that cannot be threaded and which is joined by compression fittings, as opposed to piping, which can be threaded or welded.)

7.4.1 Air Supplies

Air supplies to instruments should be clean, dry, and oil free. Air is normally distributed around a plant from a high-pressure
header (e.g., 6–7 bar g), ideally forming a ring main.
This header, usually of galvanized steel, should be sized to cope with the maximum demand of the instrument air users being serviced, and an allowance should be made for possible future expansion or modifications to its duty. Branch headers should be provided to supply individual instruments or groups of instruments. Again, adequate spare tappings should be allowed to cater for future expansion. Branch headers should be self draining and have adequate drainage/blow-off facilities. On small headers this may be achieved by the instrument air filter/regulators. Each instrument air user should have an individual filter regulator. Piping and fittings installed after filter regulators should be non-ferrous.

7.4.2 Pneumatic Signals

Pneumatic transmission signals are normally in the range of 0.2–1.0 bar (3–15psig), and for these signals copper tubing is most commonly used, preferably with a PVC outer sheath. Other materials are sometimes used, depending on environmental considerations (e.g., alloy tubing or stainless steel). Although expensive, stainless steel tubing is the most durable and will withstand the most arduous service conditions.
Plastic tubing should preferably only be used within control panels. There are several problems to be considered when using plastic tubes on a plant site, as they are very vulnerable to damage unless adequately protected, they generally cannot be installed at subzero temperatures, and they can be considerably weakened by exposure to hot surfaces. Also, it should be remembered that they can be totally lost in the event of a fire.
Pneumatic tubing should be run on a cable tray or similar supporting steelwork for its entire length and securely clipped at regular intervals. Where a number of pneumatic signals are to be routed to a remote control room they should be marshaled in a remote junction box and the signals conveyed to the control room via multitube bundles. Such junction boxes should be carefully positioned in the plant in order to minimize the lengths of the individually run tubes. (See Figure 4.1 for typical termination of pneumatic multitubes.)

7.4.3 Impulse Lines

These are the lines containing process fluid which run between the instrument impulse connection and the process tapping point, and are usually made up from piping and pipe fittings or tubing and compression fittings. Piping materials must be compatible with the process fluid.

Generally, tubing is easier to install and is capable of handling most service conditions provided that the correct fittings are used for terminating the tubing. Such fittings must be compatible with the tubing being run (i.e., of the same material).

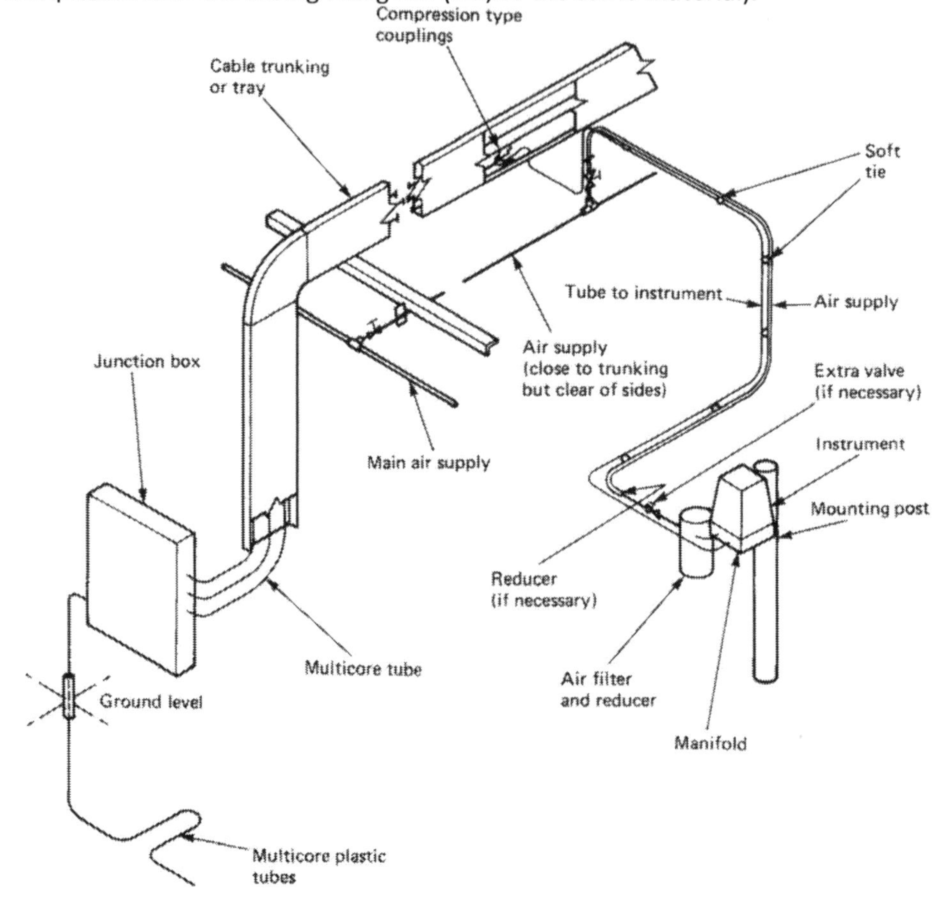

Figure :Typical field termination of pneumatic multitubes

Impulse lines should be designed to be as short as possible, and should be installed so that they are self-draining for liquids and self-venting for vapors or gases. If necessary, vent plugs or valves should be located at high points in liquid-filled lines and, similarly, drain plugs or valves should be fitted at low points in gas or vapor-filled lines. In any case, it should be ensured that there are provisions for isolation and depressurizing of instruments for maintenance purposes. Furthermore, filling plugs should be provided where lines are to be liquid scaled for chemical protection and, on services which are prone to plugging, rodding-out connections should be provided close to the tapping points.

7.5 Cabling

7.5.1 General Requirements

Instrument cabling is generally run in multicore cables from the control room to the plant area (either below or above ground) and then from field junction boxes in single pairs to the field measurement or actuating devices. For distributed microprocessor systems the inter-connection between the field and the control room is usually via duplicate data highways from remote located multiplexers or process interface units. Such duplicate highways would take totally independent routes from each other for plant security reasons.

Junction boxes must meet the hazardous area requirements applicable to their intended location and should be carefully positioned in order to minimize the lengths of individually run cables, always bearing in mind the potential hazards that could be created by fire.

Cable routes should be selected to meet the following requirements:
1. They should be kept as short as possible.
2. They should not cause any obstruction that would prohibit personnel or traffic access.
3. They should not interfere with the accessibility for maintenance of other items of equipment.
4. They should avoid hot environments or potential fire-risk areas.
5. They should avoid areas where spillage is liable to occur or where escaping vapors or gases could present a hazard.

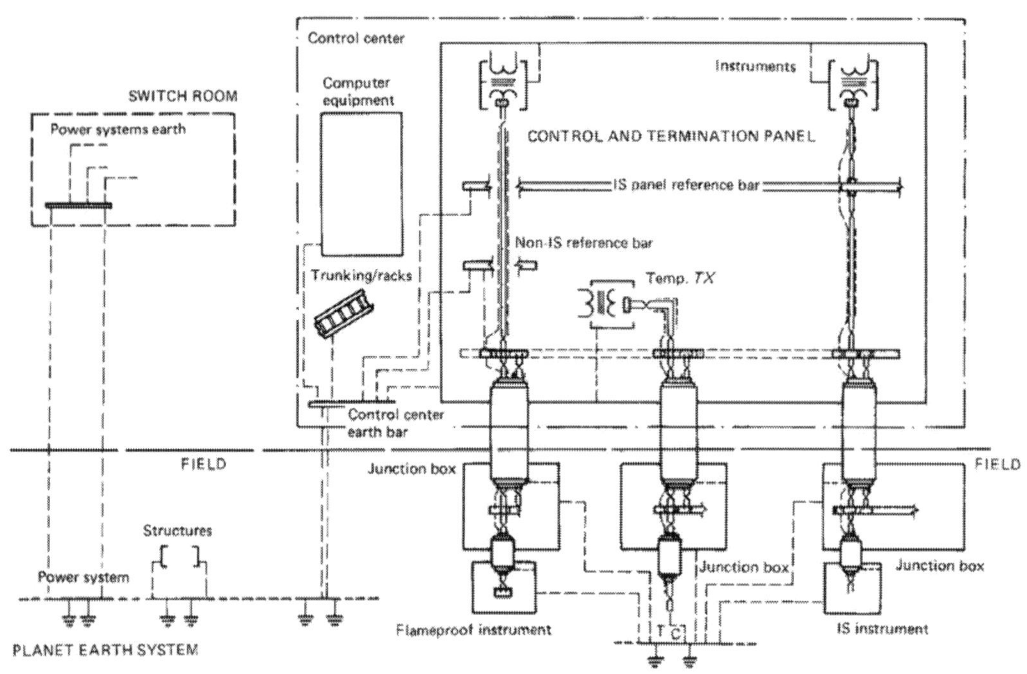

Figure: A typical control center grounding system

Cables should be supported for their whole run length by a cable tray or similar supporting steelwork. Cable trays should preferably be installed with their breadth in a vertical plane. The layout of cable trays on a plant should be carefully selected so that the minimum number of instruments in the immediate vicinity would be affected in the case of a local fire. Cable joints should be avoided other than in approved junction boxes or termination points. Cables entering junction boxes from below ground should be specially p protected by fire-resistant ducting or something similar test being made. Control valves should be tested *in situ* after the pipework fabrication has been finished and flushing operations completed. Control valves should be checked for correct stroking at 0, 50, and 100 percent open, and at the same time the valves should be checked for correct closure action.

7.5.2 Piping and Cable Testing

This is an essential operation prior to loop testing.

7.5.2.1 Pneumatic Lines

All air lines should be blown through with clean, dry air prior to final connection to instruments, and they should also be pressure tested for a timed interval to ensure that they are leak free. This should be in the form of a continuity test from the field end to its destination (e.g., the control room).

7.5.2.2 Process Piping

Impulse lines should also be flushed through and hydrostatically tested prior to connection of the instruments. All isolation valves or manifold valves should be checked for tight shutoff. On completion of hydrostatic tests, all piping should be drained and thoroughly dried out prior to reconnecting to any instruments.

7.5.2.3 Instrument Cables

All instrument cables should be checked for continuity and insulation resistance before connection to any instrument or apparatus. The resistance should be checked core to core and core to ground.
Cable screens must also be checked for continuity and insulation. Cable tests should comply with the requirements of Part 6 of the IEE Regulation for Electrical Installations (latest edition), or the rules and regulations with which the installation has to comply. Where cables are installed below ground, testing should be carried out before the trenches are back filled. Coaxial cables should be tested using sine-wave reflective testing techniques. As a prerequisite to cable testing it should be ensured that all cables and cable ends are properly identified.

7.5.2.4 Loop Testing

The purpose of loop testing is to ensure that all instrumentation components in a loop are in full operational order when interconnected and are in a state ready for plant commissioning.
Prior to loop testing, inspection of the whole installation, including piping, wiring, mounting, etc., should be carried out to ensure that the installation is complete and that the work has been carried out in a professional manner. The control room panels or display stations must also be in a fully functional state.

Loop testing is generally a two-person operation, one in the field and one in the control room who should be equipped with some form of communication, e.g., field telephones or radio transceivers. Simulation signals should be injected at the field end equivalent to 0, 50, and 100 percent of the instrument range, and the loop function should be checked for correct operation in both rising and falling modes. All results should be properly documented on calibration or loop check sheets. All ancillary components in the loop should be checked at the same time.

Alarm and shutdown systems must also be systematically tested, and all systems should be checked for "fail-safe" operation, including the checking of "burn-out" features on thermocouple installations. At the loop-checking stage all ancillary work should be completed, such as setting zeros, filling liquid seals, and fitting of accessories such as charts, ink, fuses, etc.

7.6 Plant commissioning

Commissioning is the bringing "on-stream" of a process plant and the tuning of all instruments and controls to suit the process operational requirements. A plant or section thereof is considered to be ready for commissioning when all instrument installations are mechanically complete and all testing, including loop testing, has been effected.

Before commissioning can be attempted it should be ensured that all air supplies are available and that all power supplies are fully functional, including any emergency standby supplies. It should also be ensured that all ancillary devices are operational, such as protective heating systems, air conditioning, etc. All control valve lubricators (when fitted) should be charged with the correct lubricant.

Commissioning is usually achieved by first commissioning the measuring system with any controller mode overridden. When a satisfactory measured variable is obtained, the responsiveness of a control system can be checked by varying the control valve position using the "manual" control function. Once the system is seen to respond correctly and the required process variable reading is obtained, it is then possible to switch to "auto" in order to bring the controller function into action. The controller responses should then be adjusted to obtain optimum settings to suit the automatic operation of plant.

Alarm and shutdown systems should also be systematically brought into operation, but it is necessary to obtain the strict agreement of the plant operation supervisor before any overriding of trip systems is attempted or shutdown features are operated.

Finally, all instrumentation and control systems would need to be demonstrated to work satisfactorily before formal acceptance by the plant owner.

8 Chapter: Calibration

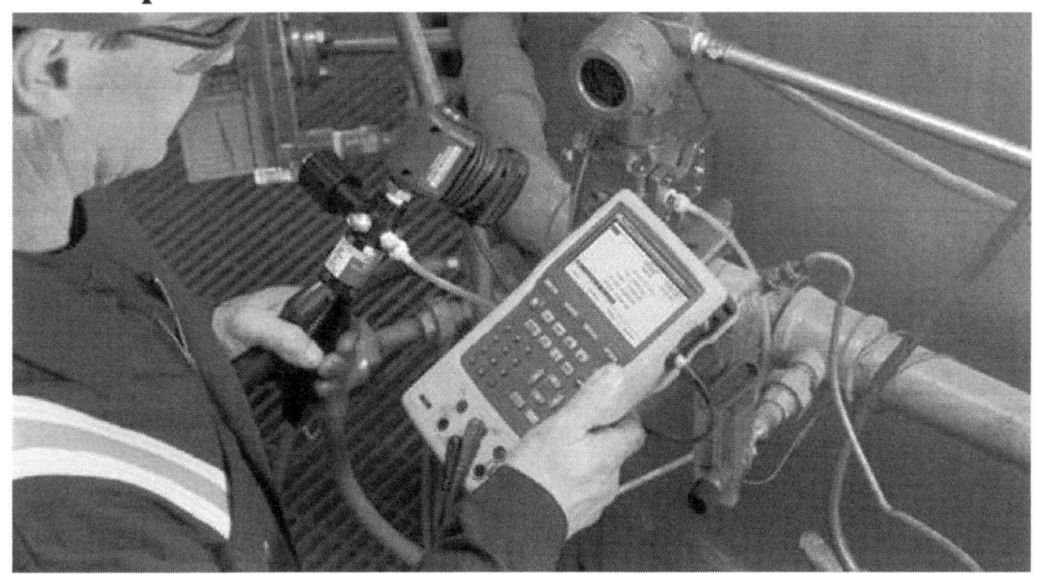

Definition: The process of comparing the response of an instrument to agree with a standard instrument over the measurement range.
Objective: To determine the deviation between measuring values and corresponding true values (an industrial calibrator is thus a "true value").

Calibration consists of comparing the output of the instrument or sensor under test against the output of an instrument of known accuracy when the same input (the measured quantity) is applied to both instruments. This procedure is carried out for a range of inputs covering the whole measurement range of the instrument or sensor.
Calibration ensures that the measuring accuracy of all instruments and sensors used in a measurement system is known over the whole measurement range, provided that the calibrated instruments and sensors are used in environmental conditions that are the same as those under which they were calibrated. For example, an electronic pressure transmitter may have a nameplate instrument range of 0–750 pounds per square inch, gauge (psig) and output of 4-to-20 milliamps (mA). However, the engineer has determined the instrument will be calibrated for 0-to-300 psig = 4-to-20 mA. Therefore, the calibration range would be specified as 0-to-300 psig = 4-to-20 mA. In this example, the zero input value is 0 psig and zero output value is 4 mA. The input span is 300 psig and the output span is 16 mA. Different terms may be used at your facility. Just be careful not to confuse the range the instrument is capable of with the range for which the instrument has been calibrated.

8.1 Instrument Errors

Instrument error refers to the combined accuracy and precision of a measuring instrument, or the difference between the actual value and the value indicated by the instrument (error). Measuring instruments are usually calibrated on some regular frequency against a standard. The most rigorous standard is one maintained by a standards organization such as NIST in the United States, or the ISO in European countries.

Any given instrument is prone to errors either due to aging or due to manufacturing tolerances. Here are some of the common terms used when describing the performance of an instrument.

Range
The range of an instrument is usually regarded as the difference between the maximum and minimum reading. For example a thermometer that has a scale from 20 to 100oC has a range of 80oC. This is also called the FULL SCALE DEFLECTION (f.s.d.).

Accuracy
The accuracy of an instrument is often stated as a % of the range or full scale deflection. For example a pressure gauge with a range 0 to 500 kPa and an accuracy of plus or minus 2% f.s.d. could have an error of plus or minus 10 kPa. When the gauge is indicating 10 kPa the correct reading could be anywhere between 0 and 20 kPa and the actual error in the reading could be 100%. When the gauge indicates 500 kPa the error could be 2% of the indicated reading.

Repeatability
If an accurate signal is applied and removed repeatedly to the system and it is found that the indicated reading is different each time, the instrument has poor repeatability. This is often caused by friction or some other erratic fault in the system.

Stability
Instability is most likely to occur in instruments involving electronic processing with a high degree of amplification. A common cause of this is adverse environment factors such as temperature and vibration. For example, a rise in temperature may cause a transistor to increase the flow of current which in turn makes it hotter and so the effect grows and the displayed reading DRIFTS. In extreme cases the displayed value may jump about. This, for example, may be caused by a poor electrical connection affected by vibration.

Time Lag Error
In any instrument system, it must take time for a change in the input to show up on the indicated output. This time may be very small or very large depending upon the system. This is known as the response time of the system. If the indicated output is incorrect because it has not yet responded to the change, then we have time lag error.
A good example of time lag error is an ordinary glass thermometer. If you plunge it into hot water, it will take some time before the mercury reaches the correct level. If you

read the thermometer before it settled down, then you would have time lag error. A thermocouple can respond much more quickly than a glass thermometer but even this may be too slow for some applications.

When a signal changes a lot and quite quickly, (speedometer for example), the person reading the dial would have great difficulty determining the correct value as the dial may be still going up when in reality the signal is going down again.

Reliability
Most forms of equipment have a predicted life span. The more reliable it is, the less chance it has of going wrong during its expected life span. The reliability is hence a probability ranging from zero (it will definitely fail) to 1.0 (it will definitely not fail).

Drift
This occurs when the input to the system is constant but the output tends to change slowly. For example when switched on, the system may drift due to the temperature change as it warms up.

8.2 Instrument Calibration

Most instruments contain a facility for making two adjustments. These are

- The RANGE adjustment.
- The ZERO adjustment.

In order to calibrate the instrument an accurate gauge is required. This is likely to be a SECONDARY STANDARD. Instruments calibrated as a secondary standard have themselves been calibrated against a PRIMARY STANDARD.

8.2.1 Procedure

An input representing the minimum gauge setting should be applied. The output should be adjusted to be correct. Next the maximum signal is applied. The range is then adjusted to give the required output. This should be repeated until the gauge is correct at the minimum and maximum values.

8.2.2 Calibration Errors

It makes sense that calibration is required for a new instrument. We want to make sure the instrument is providing accurate indication or output signal when it is installed. But why can't we just leave it alone as long as the instrument is operating properly and continues to provide the indication we expect?

Instrument error can occur due to a variety of factors: drift, environment, electrical supply, addition of components to the output loop, process changes, etc. Since a calibration is performed by comparing or applying a known signal to the instrument under test, errors are detected by performing a calibration. An error is the algebraic difference between the indication and the actual value of the measured variable. Typical errors that occur include:

Range and Zero Error

After obtaining correct zero and range for the instrument, a calibration graph should be produced. This involves plotting the indicated reading against the correct reading from the standard gauge. This should be done in about ten steps with increasing signals and then with reducing signals. Several forms of error could show up. If the zero or range is still incorrect the error will appear as shown.

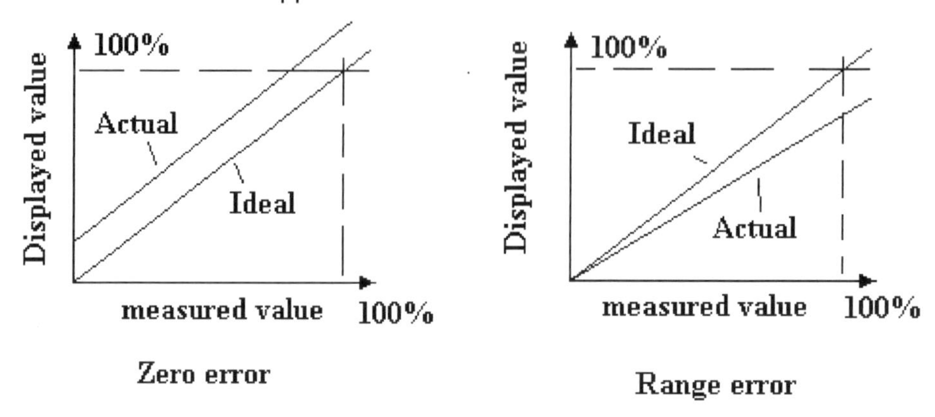

Hysteresis and Non Linear Errors

Hysteresis is produced when the displayed values are too small for increasing signals and too large for decreasing signals. This is commonly caused in mechanical instruments by loose gears and linkages and friction. It occurs widely with things involving magnetisation and demagnetisation.

The calibration may be correct at the maximum and minimum values of the range but the graph joining them may not be a straight line (when it ought to be). This is a non linear error. The instrument may have some adjustments for this and it may be possible to make it correct at mid range as shown.

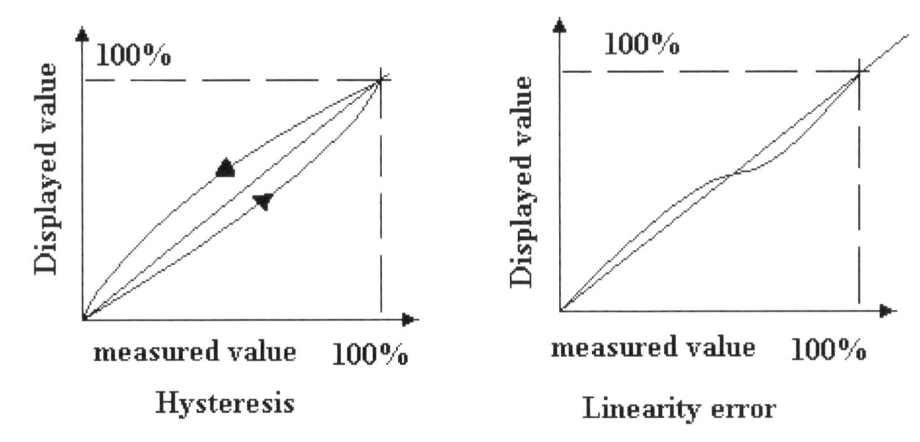

To detect and correct instrument error, periodic calibrations are performed. Even if a periodic calibration reveals the instrument is perfect and no adjustment is required, we would not have known that unless we performed the calibration. And even if adjustments are not required for several consecutive calibrations, we will still perform the calibration check at the next scheduled due date. Periodic calibrations to specified tolerances using approved procedures are an important element of any quality system.

8.3 Characteristics Of a Calibration

Calibration Tolerance: Every calibration should be performed to a specified tolerance. The terms tolerance and accuracy are often used incorrectly. In ISA's The Automation, Systems, and Instrumentation Dictionary , the definitions for each are as follows:

Accuracy: The ratio of the error to the full scale output or the ratio of the error to the output, expressed in percent span or percent reading, respectively.

Tolerance: Permissible deviation from a specified value; may be expressed in measurement units, percent of span, or percent of reading.

As you can see from the definition s, there are subtle differences between the terms. It is recommended that the tolerance, specified in measurement units, is used for the calibration requirements performed at your facility. By specifying an actual value, mistakes caused by calculating percentages of span or reading are eliminated. Also, tolerances should be specified in the units measured for the calibration.

For example, you are assigned to perform the calibration of the previously mentioned 0-to-300 psig pressure transmitter with a specified calibration tolerance of ±2 psig. The output tolerance would be:

$$\begin{array}{r} 2 \text{ psig} \\ \div\ 300 \text{ psig} \\ \times\ \ 16 \text{ mA} \\ \hline 0.1067 \text{ mA} \end{array}$$

The calculated tolerance is rounded down to 0.10 mA, because rounding to 0.11 mA would exceed the calculated tolerance. It is recommended that both ±2 psig and ±0.10 mA tolerances appear on the calibration data sheet if the remote indications and output milliamp signal are recorded.

Note the manufacturer's specified accuracy for this instrument may be 0.25% full scale (FS). Calibration tolerances should not be assigned based on the manufacturer's specification only. Calibration tolerances should be determined from a combination of factors. These factors include:

- Requirements of the process
- Capability of available test equipment
- Consistency with similar instruments at your facility
- Manufacturer's specified tolerance

Example: The process requires ±5°C; available test equipment is capable of ±0.25°C; and manufacturer's stated accuracy is ±0.25°C. The specified calibration tolerance must be between the process requirement and manufacturer's specified tolerance. Additionally the test equipment must be capable of the tolerance needed. A calibration tolerance of ±1°C might be assigned for consistency with similar instruments and to meet the recommended accuracy ratio of 4:1.

Accuracy Ratio: This term was used in the past to describe the relationship between the accuracy of the test standard and the accuracy of the instrument under test. The term is still used by those that do not understand uncertainty calculations (uncertainty is described below). A good rule of thumb is to ensure an accuracy ratio of 4:1 when performing calibrations. This means the instrument or standard used should be four times more accurate than the instrument being checked. Therefore, the test equipment (such as a field standard) used to calibrate the process instrument should be four times more accurate than the process instrument, the laboratory standard used to calibrate the field standard should be four times more accurate than the field standard, and so on.

With today's technology, an accuracy ratio of 4:1 is becoming more difficult to achieve. Why is a 4:1 ratio recommended? Ensuring a 4:1 ratio will minimize the effect of the accuracy of the standard on the overall calibration accuracy. If a higher level standard is found to be out of tolerance by a factor of two, for example, the calibrations performed using that standard are less likely to be compromised.

Suppose we use our previous example of the test equipment with a tolerance of ±0.25°C and it is found to be 0.5°C out of tolerance during a scheduled calibration. Since we took in to consideration an accuracy ratio of 4:1 and assigned a calibration tolerance of ±1°C to the process instrument, it is less likely that our calibration performed using that standard is compromised.

The out-of-tolerance standard still needs to be investigated by reverse traceability of all calibrations performed using the test standard. However, our assurance is high that the process instrument is within tolerance. If we had arbitrarily assigned a calibration tolerance of ±0.25°C to the process instrument, or used test equipment with a calibration tolerance of ±1°C, we would not have the assurance that our process instrument is within calibration tolerance. This leads us to traceability.

Traceability: All calibrations should be performed traceable to a nationally or internationally recognized standard . For example, in the United States, the National Institute of Standards and Technology (NIST), formerly National Bureau of Standards (NBS) maintains the nationally recognized standards. *Traceability* is defined by ANSI/NCSL Z540-1-1994 (which replaced MIL-STD-45662A) as "the property of a result of a measurement whereby it can be related to appropriate standards, generally national or international standards, through an unbroken chain of comparisons."

Note this does not mean a calibration shop needs to have its standards calibrated with a primary standard. It means that the calibrations performed are traceable to NIST through all the standards used to calibrate the standards, no matter how many levels exist between the shop and NIST.

Traceability is accomplished by ensuring the test standards we use are routinely calibrated by "higher level" reference standards. Typically the standards we use from the shop are sent out periodically to a standards lab which has more accurate test equipment. The standards from the calibration lab are periodically checked for calibration by "higher level" standards, and so on until eventually the standards are tested against Primary Standards maintained by NIST or another internationally recognized standard.

The calibration technician's role in maintaining traceability is to ensure the test standard is within it's calibration interval and the unique identifier is recorded on the applicable calibration data sheet when the instrument calibration is performed. Additionally, when test standards are calibrated, the calibration documentation must be reviewed for accuracy and to ensure it was performed using NIST traceable equipment.

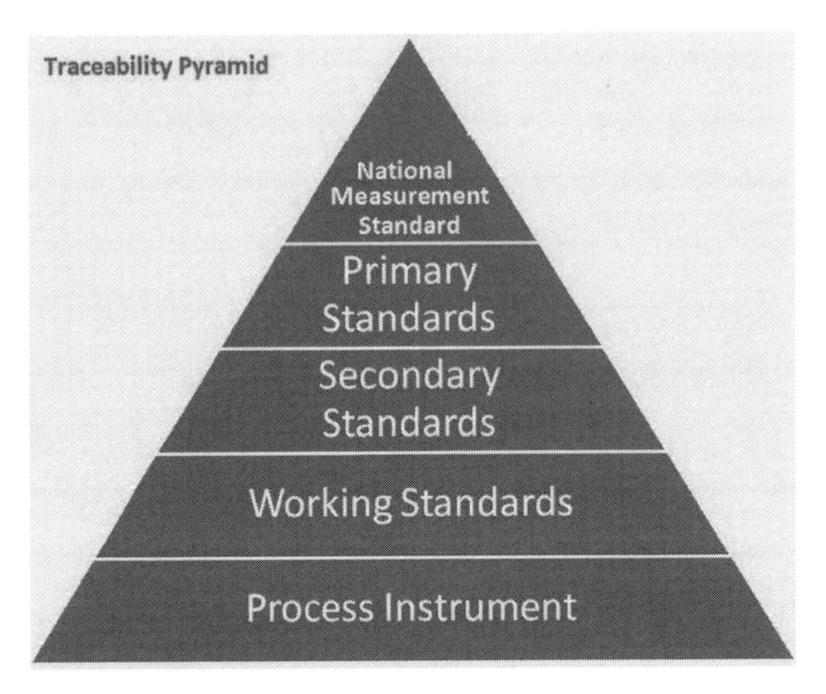

Figure : Traceability Pyramid

The calibration facilities provided within the instrumentation department of a company provide the first link in the calibration chain. Instruments used for calibration at this level are known as ***working standards***. As such working standard instruments are kept by the instrumentation department of a company solely for calibration duties, and for no other purpose, then it can be assumed that they will maintain their accuracy over a reasonable period of time because use-related deterioration in accuracy is largely eliminated. However, over the longer term, the characteristics of even such standard instruments will drift, mainly due to ageing effects in components within them. Therefore, over this longer term, a programme must be instituted for calibrating working standard instruments at appropriate intervals of time against instruments of yet higher accuracy. The instrument used for calibrating working standard instruments is known as a ***secondary reference standard***. This must obviously be a very well-engineered instrument that gives high accuracy and is stabilized against drift in its performance with time. This implies that it will be an expensive instrument to buy. It also requires that the environmental conditions in which it is used be carefully controlled in respect of ambient temperature, humidity etc.

Uncertainty: Parameter, associated with the result of a measurement that characterizes the dispersion of the values that could reasonably be attributed to the measurand. Uncertainty analysis is required for calibration labs conforming to ISO

17025 requirements. Uncertainty analysis is performed to evaluate and identify factors associated with the calibration equipment and process instrument that affect the calibration accuracy. Calibration technicians should be aware of basic uncertainty analysis factors, such as environmental effects and how to combine multiple calibration equipment accuracies to arrive at a single calibration equipment accuracy. Combining multiple calibration equipment or process instrument accuracies is done by calculating the square root of the sum of the squares, illustrated below:

Calibration equipment combined accuracy

$$\sqrt{(\text{calibrator1 error})^2 + (\text{calibrator2 error})^2 + (\text{etc. error})^2}$$

Process instrument combined accuracy

$$\sqrt{(\text{sensor error})^2 + (\text{transmitter error})^2 + (\text{indicator error})^2 + (\text{etc. error})^2}$$

8.4 Temperature Calibration

In order to maintain consistent quality of manufactured products, it is necessary to perform calibrations on process sensors and instruments. There are several philosophies for calibration of the measurement and control circuits. The basis of the chosen method should always be to include the temperature sensor.

It doesn't make sense only to calibrate and adjust the electronic part of the loop. A rule of thumb says that only 10% of the total error is in the electronics, the other 90% is in the sensing element. So it is essential that the temperature sensor is tested, meaning physically exposed to the desired temperature. A dry-block calibrator is an easy method to create the "process" temperature.

The output from the sensor can be taken from anywhere in the loop. And the rest of loop might be tested electronically.

8.4.1 Principle of Dry Block Temperature Calibrator

"Heating up a metal block and keeping the temperature stable"
This is the very basic principle of a dry-block calibrator. The design gives the user a lot of advantages compared to the more traditional liquid baths.

- Heat up and cool down much faster
- No hazardous hot liquid
- Much wider temperature ranges
- Physically smaller and lighter
- Designed for industrial applications
- Models with completely integrated calibration solutions

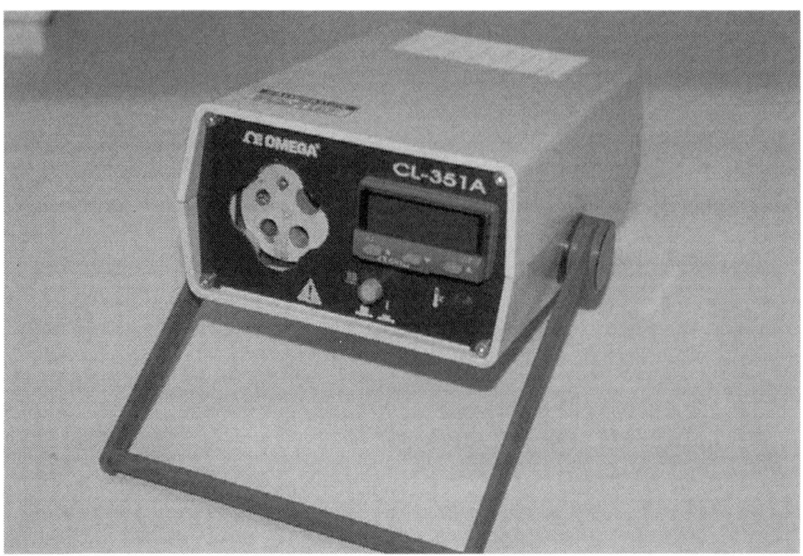

A dry-block temperature calibrator intended for bench-top service is shown in this photograph

Combined Uncertainity
A calibration is a matter of qualifying the sensor-under-test. Only be knowing the limitations of the sensor it is possible to maximize the process loop. In other words, a temperature reading is only valid if the uncertainty value can be accounted for, eg 60°C ±2°C.

In this case the uncertainty is ±2°C. This means that a temperature reading of 60°C could have any value between 58°C and 62°C. The lower the uncertainty the more accurate is the measurements.

8.5 Pressure Calibration

There are many different factors that need to be considered when selecting a pressure standard. It is important to evaluate the requirements of the task being performed and

- The overall accuracy of the pressure standard for the application
- Specific application considerations including test fluid, pressure range and task being performed.
- Cost of the inaccurate pressure measurement either on the purchase or sale of product
- Adequacy of the pressure standard to test or adjust safety related systems.
- Requirements for routine maintenance inspections

As a general guideline, the calibration standard used must be four times (4x) more accurate than the device being calibrated.

8.5.1 Expression of Pressure Measurement

Low Pressure		Vacuum		Pressure	
Inches H_2O (20°C)	= 0.036063 PSI	Inches Hg (0°C)	= 0.49114999 PSI	Pascal = 1Newton/m²	= 0.0001450377 PSI
Inches H_2O (60°C)	= 0.036092 PSI	Torr = 1mm Hg (0°C)	= 0.019718 PSI	Bar = 100 Kilopascal	= 14.50377 PSI
Inches H_2O (4°C)	= 0.036126 PSI	Millimeters Hg (20°C)	= 0.019266878 PSI	Kg/cm²	= 14.22334 PSI
Millimeters H_2O (20°C)	= 0.0014198 PSI				
Millimeters H_2O (4°C)	= 0.00142228 PSI				

8.5.2 Pneumatic Deadweight testers

These self-regulating testers offer a high level of accuracy independent of the operator and have earned their reputation as the "standard of primary standards". They are available for pressure as low as 4 in H2O (10 cmH2O) or as high as 1500 psi (100 bar).

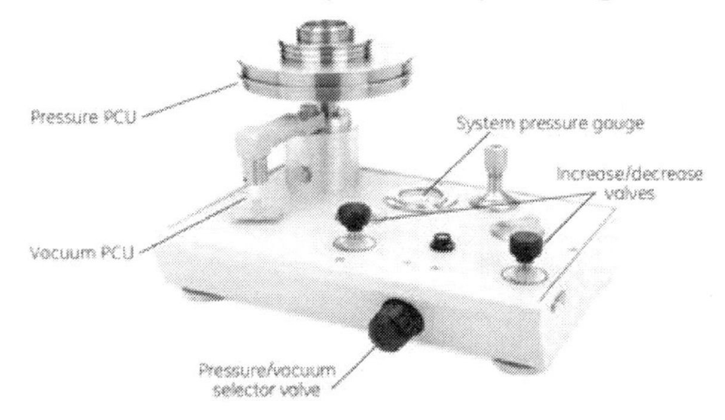

Figure: AMETEK pneumatic tester

Applications

- The ideal primary standard
- Calibrate instruments used in critical process control loops and extend the time between calibrations due to better accuracy and uncertainty.
- Calibrate instruments on custody transfer stations and reduce the uncertainty of the measurement
- Reduce maintenance costs by calibrating your portable digital calibrators and achieve the highest possible accuracy
- Ideal for hazardous locations

Principle of Operation

Pneumatic testers are self-regulating, primary type pressure standards. An accurate calibrating pressure is produced by bringing into equilibrium the pneumatic pressure on the underside of a ball of known area by weights of known mass on the top.

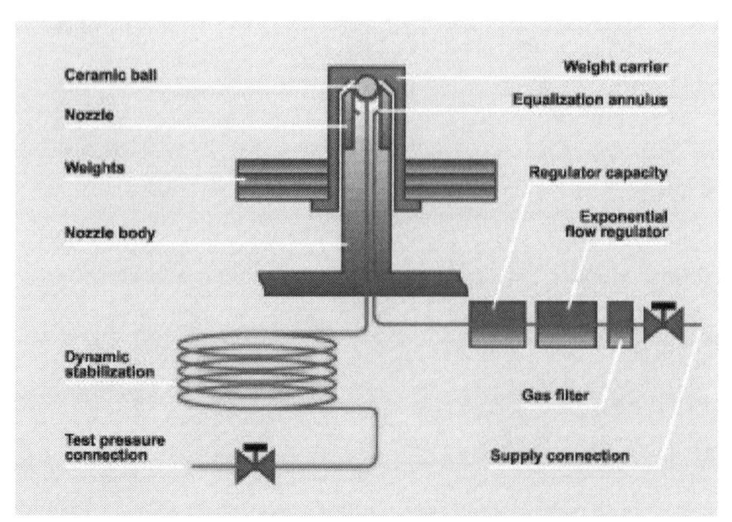

Figure: Pneumatic Deadweight tester Operation

The precision ceramic ball is floated within a tapered stainless steel nozzle. A flow regulator introduces pressure under the ball, lifting it in the tapered annulus until equilibrium is reached. At this point the ball is floating and the vented flow is equal to the fixed flow from the supply regulator. The pressure, which is also the output pressure, is proportional to the weight load. During the operation the ball is centred by a dynamic film of air, eliminating physical contact between the ball and the nozzle.

8.6 Hydraulic Deadweight Testers

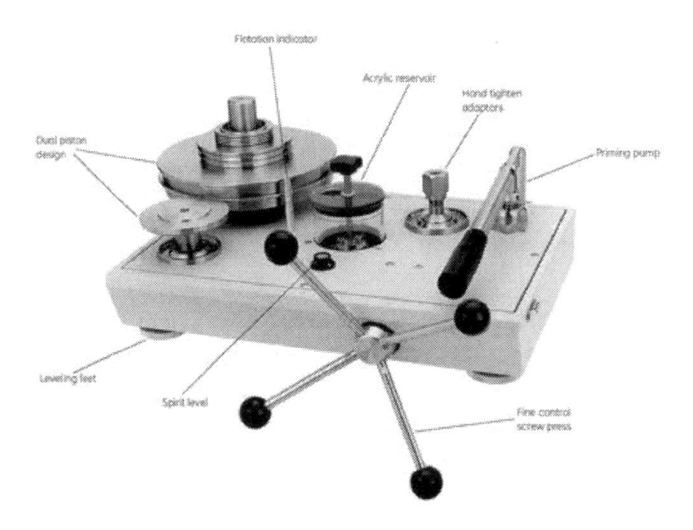

Figure: Hydraulic Deadweight Tester

- Repeatability better than ±0.005% of reading
- Accuracy up to ±0.025% of reading and even better using a DADT converter

- Dual volume control valve delivers high volume to rapidly build pressure
- Vernier adjustment for fine adjustment of pump pressure
- Re-entrant type of piston and cylinder assembly maintain accuracy as test pressure increases and improve the spinning/float time.
- Single and dual column versions.
- Enclosed piston/cylinder assembly for safe operation
- Available engineering units: psi, bar, kPa, kg/cm^2

8.7 Dual Volume Hand Pump

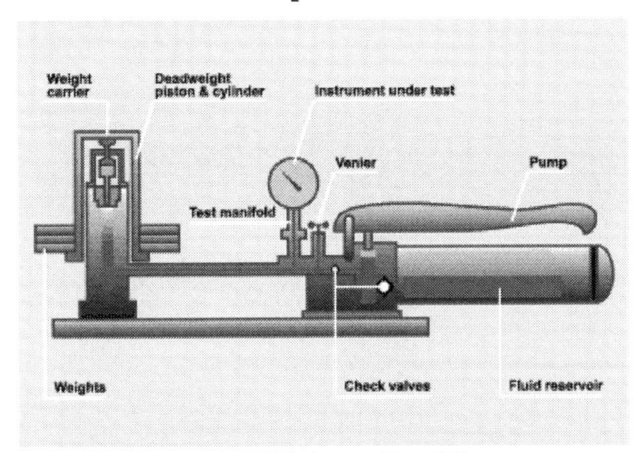

Figure: Dual Volume Hand Pump

The lever action hand pump incorporates a dual volume control valve. This allows the pump to deliver a high volume in order to rapidly fill systems and build pressure. The low volume permits easy pumping at high pressures and a more gradual approach to the point of pressure calibration. A vernier adjustment is provided for fine adjustment to the desired pressure.

8.8 Overhung Weight Carriers

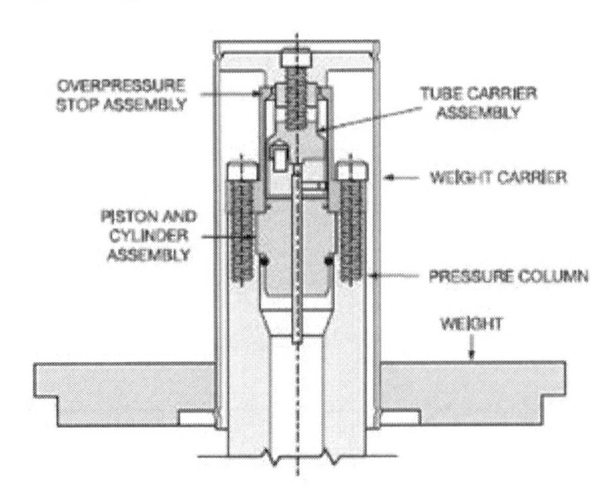

Figure: Overhung Weight Carriers

The weights are placed on a weight tube that is suspended from the piston assembly. Suspended weights have a lowers center of gravity that minimizes side thrust on the measuring piston and cylinder assembly resulting in improved measurement accuracy, and minimized wear on the piston and cylinder assembly.

8.9 Re-Entrant Piston/Cylinder

Within the re-entrant piston and cylinder design, the test fluid is applied to a chamber on the outside of the cylinder as well as to the inside of the cylinder. The area of the outside of the cylinder is larger than the inside giving a reduced clearance between the piston and the cylinder at higher pressures.

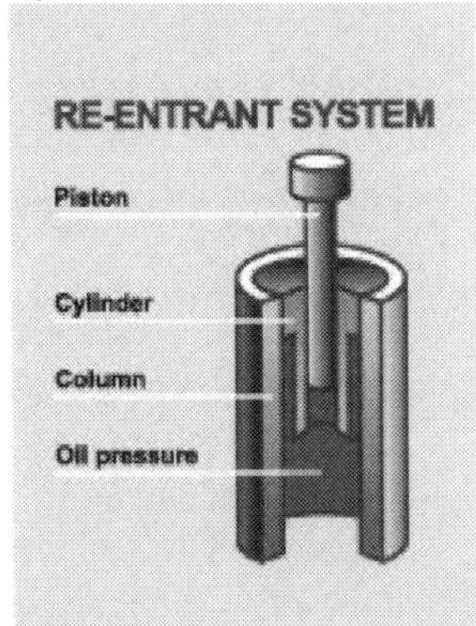

Figure: Re-entrant Piston /Cylinder

This design reduces the rate of fluid leakage, thus increasing the time available for calibration of instruments prior to pumping to restore fluid loss.

8.10 U-Tube Manometers

U-tube manometers for relative and absolute pressures, vacuum and differential pressures.

These devices are primary standards and are void of faults within their physical tolerances. They are designed to measure pressure and are suitable for calibrating high precision pressure sensors.

- Very low maintenance cost. No need for recalibration
- No electronic parts. Can be used anywhere
- Ideal for calibration of transmitters in clean-room applications
- Differential pressure for flow applications
- BAROSCOPE for absolute pressure in metrology laboratories and aerospace applications
- Calibration of handheld instruments used in HVAC applications

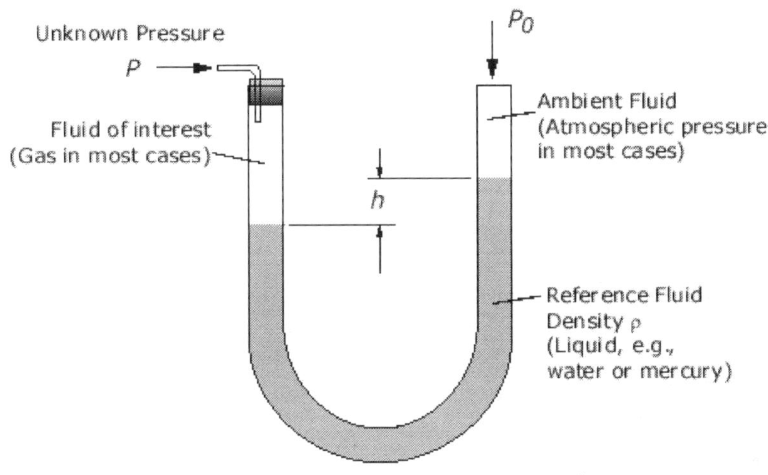

Gage Pressure $\Delta P = P - P_0 = \rho g h$

Figure : U-tube Manometer

Basic Knowledge

The measuring principle of these liquid manometers is based on the following relationship:

$\Delta p = \Delta h \times pm \times g$

The pressure (p) to be measured is to be compared with the height (h) of a liquid column. If the pressure exerted on the two surfaces of the so-called confined liquid is not the same, there is a deflection and consequently a difference in height. The confined liquid continues to rise until the effect of the force of the pressure differentials and weight of the liquid columns are identical. In accordance with the law of physics, the effect of the liquid column on the pressure in the liquid is, in essence, only dependent on height (h) of the liquid column and on density (pm) of the liquid. Further influences are relatively low and know. For highly precise measurements, correction calculations can be made. Recalibration is not necessary.

8.11 Flowmeter calibration methods

There are various methods available for the calibration of flowmeters, and the requirement can be split into two distinct categories: *in situ* and laboratory. Calibration of liquid flowmeters is generally somewhat more straightforward than that of gas flowmeters since liquids can be stored in open vessels and water can often be utilized as the calibrating liquid.

8.11.1 Flowmeter Calibration Methods for Liquids

The main principles used for liquid flowmeter calibration are *in situ:* insertion-point velocity and dilution gauging/tracer method; laboratory: master meter, volumetric, gravimetric, and pipe prover.

8.11.1.1 In situ Calibration Methods

Insertion-Point Velocity One of the simpler methods of *in situ* flowmeter calibration utilizes point-velocity measuring devices where the calibration device chosen is positioned in the flowstream adjacent to the flowmeter being calibrated and such that mean flow velocity can be measured. In difficult situations a flow traverse can be carried out to determine flow profile and mean flow velocity.

Dilution Gauging/Tracer Method This technique can be applied to closed-pipe and open-channel flowmeter calibration. A suitable tracer (chemical or radioactive) is injected at an accurately measured constant rate, and samples are taken from the flowstream at a point downstream of the injection point, where complete mixing of the injected tracer will have taken place. By measuring the tracer concentration in the samples, the tracer dilution can be established, and from this dilution and the injection rate the volumetric flow can be calculated. This principle is illustrated in Figure 5.11.1. Alternatively, a pulse of tracer material may be added to the flowstream, and the time taken for the tracer to travel a known distance and reach a maximum concentration is a measure of the flow velocity.

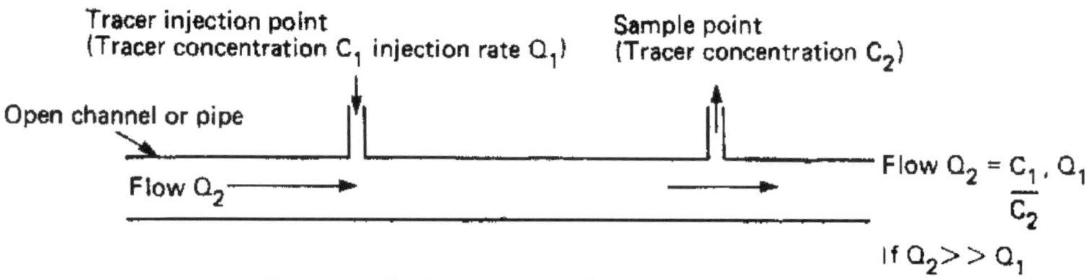

Figure : Dilution Gauging by tracer injection

8.11.1.2 Laboratory Calibration Methods

Master Meter For this technique a meter of known accuracy is used as a calibration standard. The meter to be calibrated and the master meter are connected in series and are therefore subject to the same flow regime. It must be borne in mind that to ensure consistent accurate calibration, the master meter itself must be subject to periodic recalibration.

Volumetric Method I n this technique, flow of liquid through the meter being calibrated is diverted into a tank of known volume. When full, this known volume can be compared with the integrated quantity registered by the flowmeter being calibrated.

Gravimetric Method Where the flow of liquid through the meter being calibrated is diverted into a vessel that can be weighed either continuously or after a predetermined time, the weight of the liquid is compared with the registered reading of the flowmeter being calibrated (see Figure below).

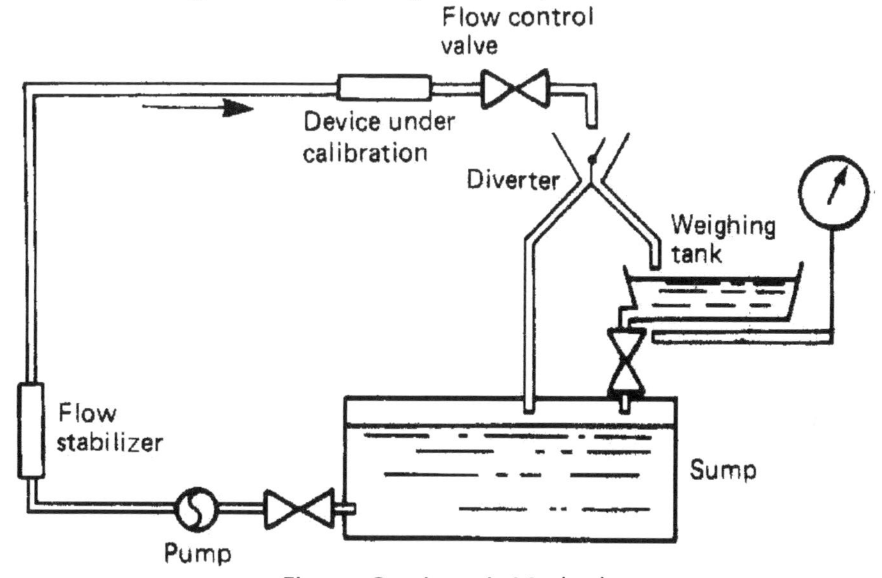

Figure: Gravimetric Method

Pipe Prover This device, sometimes known as a *meter prover*, consists of a U-shaped length of pipe and a piston or elastic sphere. The flowmeter to be calibrated is installed on the inlet to the prover, and the sphere is forced to travel the length of the pipe by the flowing liquid. Switches are inserted near both ends of the pipe and operate when the sphere passes them. The swept volume of the pipe between the two switches is determined by initial calibration, and this known volume is compared with that registered by the flowmeter during calibration. A typical pipe-prover loop is shown in the below Figure.

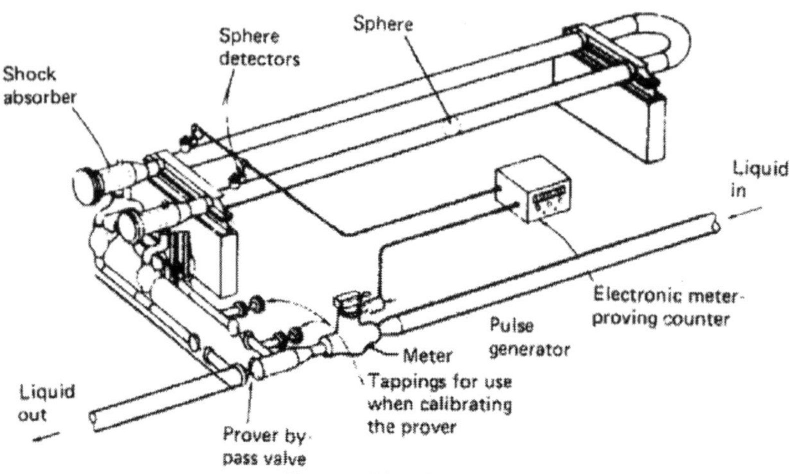

Figure: Pipe Prover

8.11.2 Flowmeter Calibration Methods for Gases

Methods suitable for gas flowmeter calibration are in situ: as for liquids; and laboratory: soap-film burette, water displacement method, and gravimetric.

8.11.2.1 Laboratory Calibration Methods

Soap-Film Burette This method is used to calibrate measurement systems with gas flows in the range of 10e-7 to 10e-4 m3/s. Gas flow from the meter on test is passed through a burette mounted in the vertical plane. As the gas enters the burette, a soap film is formed across the tube and travels up it at the same velocity as the gas. By measuring the time of transit of the soap film between graduations of the burette, it is possible to determine flow rate. A typical calibration system is illustrated in the below Figure.

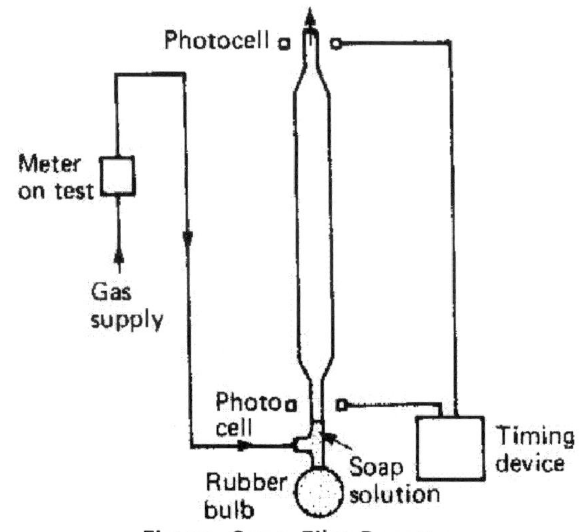

Figure: Soap-Film Burette

Water-Displacement Method I n this method a cylinder closed at one end is inverted over a water bath, as shown in below Figure . As the cylinder is lowered into the bath, a trapped volume of gas is developed. This gas can escape via a pipe connected to the cylinder out through the flowmeter being calibrated. The time of the fall of the cylinder combined with the knowledge of the volume/length relationship leads to a determination of the amount of gas displaced, which can be compared with that measured by the flowmeter under calibration.

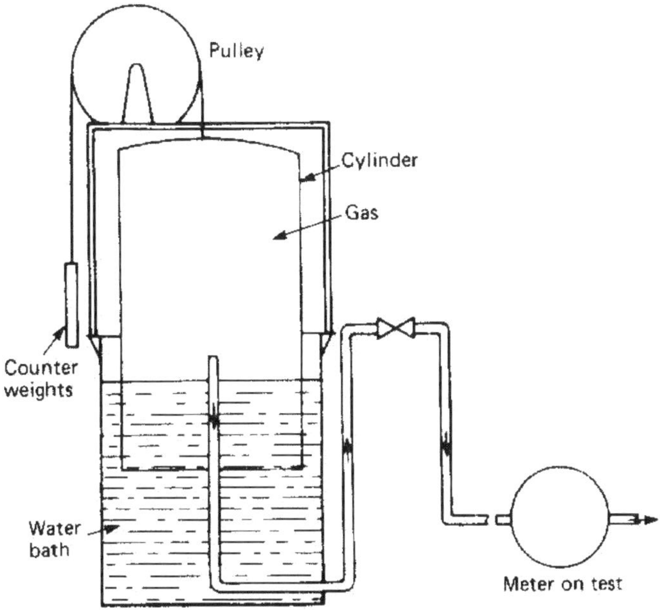

Figure: Water-Displacement Method

Gravimetric Method Here gas is diverted via the meter under test into a gas-collecting vessel over a measured period of time. When the collecting vessel is weighed before diversion and again after diversion, the difference will be due to the enclosed gas, and flow can be determined. This flow can then be compared with that measured by the flowmeter.

It should be noted that the cost of developing laboratory flow calibration systems as outlined can be quite prohibitive; it may be somewhat more cost-effective to have systems calibrated by the various national standards laboratories (such as NIST, NEL, and SIRA) or by manufacturers rather than committing capital to what may be an infrequently used system.

8.11.3 Calibration of level measuring systems

Contents that are traded for money, such as petrochemicals, foods, milk, and alcohol, must be measured to standards set by the relevant Weights and Measures authority. Official approval of the measuring system and its procedures of use and calibration is required. In such cases the intrinsic value of the materials will decide the accuracy of such measurements, and this often means that the system and calibrations must comply to very strict codes and be of the highest accuracy possible.

Elevation & Suppression

If the d/p cell is not located at an elevation that corresponds to 0% level in the tank, it must be calibrated to account for the difference in elevation. This calibration adjustment is called zero elevation when the cell is located above the lower tap, and is called zero suppression or zero depression when the cell is located below the lower tap. Most d/p cells are available with elevation and suppression ranges of 600% and 500% of calibrated span, respectively, as long as the calibrated span does not exceed 100% of the upper range limit of the cell.

For example, assume that an electronic d/p cell can be calibrated for spans between 0-10 psid (which is its lower range limit, LRL) and 0-100 psid (which is its upper range limit, URL). The cell is to be used on a 45-ft tall closed water tank, which requires a hydrostatic range of 0-20 psid. The cell is located about 11 feet (5 psid) above the lower tap of the tank; therefore, a zero elevation of 5 psid is needed. The d/p cell can handle this application, because the calibrated span is 20% of the URL and the elevation is 25% of the calibrated span.

Dry and Wet Calibration

When calibrating a d/p transmitter, there are two approaches to this. One is to isolate the transmitter from the process, drain and vent the transmitter to remove all of the process liquid, and then apply pressure to the high side of the transmitter using a dry gas such as air or Nitrogen (recommended for any transmitter containing oil or other volatile substance. This is a "dry" calibration.

A wet calibration involves the use of a "head chamber", which consists of two cylindrical chambers mounted between two metal plates. The plates are grooved, with O-rings inserted in the grooves. The cylindrical chambers are sealed into the grooves in the plates using threaded rods which hold the assembly together. Holes are drilled in the plates and tapped to allow fittings to be installed in the holes. The chambers are filled approximately half way with water. The fittings on the bottom plate have valves which allow the water to stay in the chamber. There is also an equalizing valve between the fittings on the bottom to allow the water level in the two chambers to be equalized.

After tubing from the valves is connected to the transmitter, the water can be forced from one side of the head chamber, down through the transmitter valve block and out the transmitter vents. This process is then repeated on the other chamber so that all air is removed from the transmitter and it is solid with water. At that point the pressure is removed from the head chamber and the water levels are allowed to equalize.

Once that is done, the equalizing valves for both the transmitter and the head chamber are closed. Air pressure is then applied to the chamber connected to the high side of the transmitter. The water levels, being equal apply equal pressure to both sides of the transmitter, cancelling each other out. The only pressure then, which affects the output of the transmitter is the air pressure applied to the water in the chamber. (Pascal's Law applies here).

8.12 Signal Calibration

8.12.1 Complete Loop Calibration

In order to maintain consistent quality of manufactured product, it is necessary to perform periodic calibrations on process sensors and instruments. There are several philosophies for calibration of the measurement and control circuit. The most obvious and representative method is to calibrate the complete measurement circuit, ie from the sensor through the transmitter to the indicator or controller.

8.12.2 Split Loop Calibration

Each component of the process measurement and control system is calibrated as an independent item. The transmitter can be calibrated using a precision signal calibrator. The indicator or controller can be calibrated either as described for the transmitter or by using the mAcal loop calibrator which simulates an output of 4-20mA proportional to the temperature (or pressure). The sensor part can be brought to the workshop for a time saving automatic calibration.

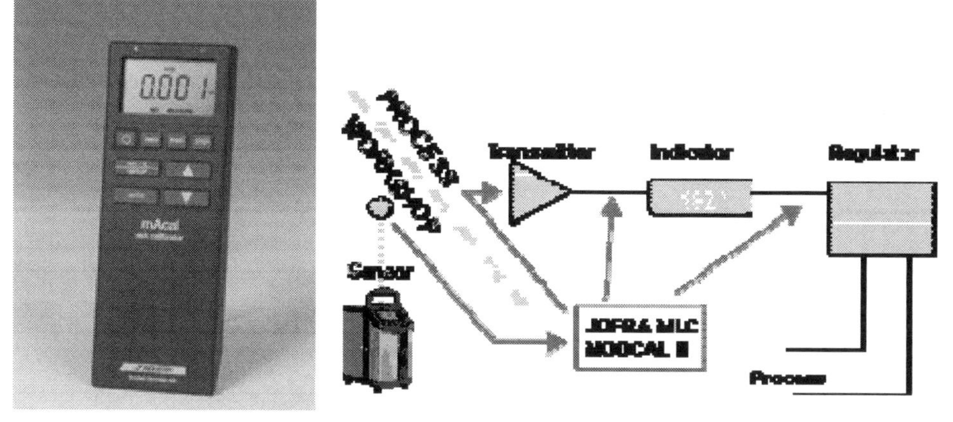

9 Chapter: Automation

Automation is the use of control systems and information technologies to reduce the need for human work in the production of goods and services. In the scope of industrialization, automation is a step beyond mechanization. Whereas mechanization provided human operators with machinery to assist them with the muscular requirements of work, automation greatly decreases the need for human sensory and mental requirements as well. Automation plays an increasingly important role in the world economy and in daily experience.

Automation, much like mechanization, depends on machines to execute functions many of which were first performed manually. Mechanization can be seen as the stepping stone between manual labor and automation—it eliminates the need for physical labor, but operators are still needed to oversee machine operations and provide maintenance and feedback. Automation systems, however, eliminate the need for an operator by including feedback and sensory programs. The result is highly independent machine systems that can carry out a task from start to finish, without human assistance.

Automated machines have been seamlessly integrated into countless industries, from carrying out manufacturing tasks to handling telephone switchboards. In quotidian life, we encounter automated systems each time we use an ATM. The level of human dependence is high, as is the functions we entrust them with—managing our finances, our phone calls, our computers. With such an array of functions, it's not surprising that not all automated systems are the same. Depending on the exact function, one of several different tools may be responsible for an automated system: an artificial neural network, distributed control system, human machine interface, supervisory control and data acquisition, or a programmable logic controller.

The main advantages of automation are:

- Replacing human operators in tasks that involve hard physical or monotonous work.
- Replacing humans in tasks done in dangerous environments (i.e. fire, space, volcanoes, nuclear facilities, underwater, etc.)
- Performing tasks that are beyond human capabilities of size, weight, speed, endurance, etc.
- Economy improvement: Automation may improve in economy of enterprises, society or most of humanity. For example, when an enterprise invests in automation, technology recovers its investment; or when a state or country increases its income due to automation like Germany or Japan in the 20th Century.
- Reduces operation time and work handling time significantly.
- Frees up workers to take on other roles.
- Provides higher level jobs in the development, deployment, maintenance and running of the automated processes.

Different types of automation tools exist:

- ANN - Artificial neural network
- DCS - Distributed Control System
- HMI - Human Machine Interface
- SCADA - Supervisory Control and Data Acquisition
- PLC - Programmable Logic Controller
- PAC - Programmable automation controller
- Instrumentation
- Motion control AND Robotics

Artificial Neural Network (ANN)

An artificial neural network is a mathematical or computational model whose rhythms mimic those of biological neurons. The structure of the network is adaptive, meaning it can change based on the external or internal exchange of information throughout the network. Artificial neural networks are used to identify patterns in pools of data and to classify relationships (such as sequence recognition). Applications include e-mail spam filtering, system control (such as in a car), pattern recognition in systems (such as radars), pattern recognition in speech, movement, and text, and financial automated trading systems.

Distributed Control System (DCS)

A distributed control system is one in which there are separate controls throughout the system. The controls are not centrally located, but tend to be spread out depending on which region of the system needs monitoring—each control is connected to the others in a communication network. These kinds of systems are typically used in manufacturing processes, especially when the action or production is continuous. The controllers can be specified for a given process, and manipulated to enhance or monitor machine performance. Traffic lights are usually controlled by distributed control systems, and they can also be applied in oil refining and central station power generation.

Human Machine Interface (HMI)

Commonly referred to as a user interface, a human machine interface system depends on human interaction with the system in order to function. A user will provide input, and the system in turn will produce output that coincides with the user's intent. In order for this to work, users must have access to the system and a means by which to manipulate it. ATMs, for example, are designed so users can easily dictate what the system is supposed to do while enabling it to easily respond and provide the appropriate results. Buttons that read *withdrawal* or *make a deposit* provide the user with any easy way to trigger a chain of commands within the internal system. The desired result, either the intake of a deposit or the ejection of cash, can then be achieved.

Supervisory Control and Data Acquisition (SCADA)

A supervisory control and data acquisition system (SCADA) is a larger, industrial control network that is often comprised of smaller sub-systems, including human machine interface systems connected to remote terminal units, which work to translate sensor signals into comprehensible data. These systems can work together to control an entire manufacturing site, or even an entire region by connecting several different manufacturing plants. SCADA systems bear a high resemblance to distributed control systems, and at times it may be difficult to differentiate between the two. The key difference lies in what they ultimately do—SCADA systems do not control each process in real time, rather they coordinate processes. Generally speaking, however, the two systems are highly similar and are often used in identical applications.

Programmable Logic Controllers (PLC)

Programmable logic controllers are real time systems, meaning there is a set deadline and time frame in which the desired result must be achieved. The PLC system is essentially a computer that controls manufacturing machines in an industrial production line, so it has multiple capabilities, such as varied temperature ranges and input and output settings, as well as the ability to weather dust and other unfavorable conditions. Programmable logic controllers can be used to program a variety of day-to-day applications, such as amusement park rides.

Industrial Automation

Industrial automation or numerical control is the use of control systems such as computers to control industrial machinery and processes, replacing human operators. Currently, for manufacturing companies, the purpose of automation has shifted from increasing productivity and reducing costs, to broader issues, such as increasing quality and flexibility in the manufacturing process.

Advantages of Industrial Automation:

- High quality,
- Repeatability,
- Reduced manufacturing lead time,
- Increase in production,
- Labor cost is reduced.

10 Chapter: PLC (Programmable Logic Controller)

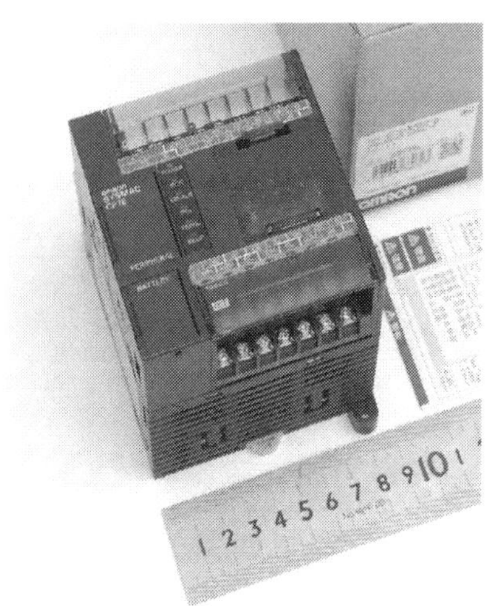

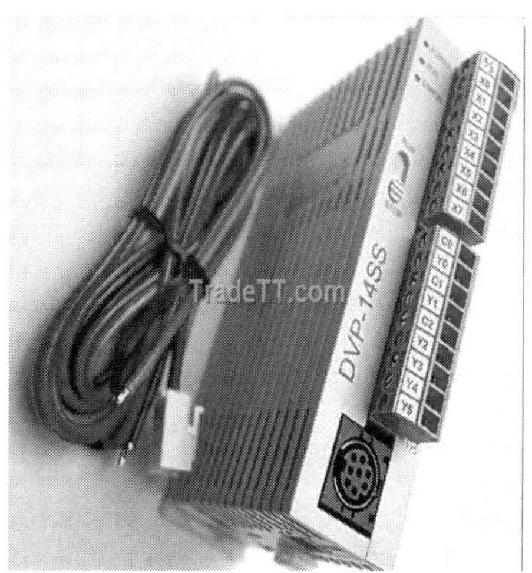

Introduction to PLC

PLC (Programmable Logic Controller) is an electronic device, previously called "sequence controller". In 1978, NEMA (National Electrical Manufacture Association) in the United States officially named it as "programmable logic controller". PLC reads the status of the external input devices, e.g. keypad, sensor, switch and pulses, and execute by the microprocessor logic, sequential, timing, counting and arithmetic operations according the status of the input signals as well as the pre-written program stored in the PLC. The generated output signals are sent to output devices as the switch of a relay, electromagnetic valve, motor drive, control of a machine or operation of a procedure for the purpose of machine automation or processing procedure. The peripheral devices (e.g. personal computer/handheld programming panel) can easily edit or modify the program and monitor the device and conduct on-site program maintenance and adjustment. The widely used language in designing a PLC program is the ladder diagram.

With the development of the electronic technology and wider applications of PLC in the industry, for example in position control and the network function of PLC, the input/output signals of PLC include DI (digital input), AI (analog input), PI (pulse input), NI (numeric input), DO (digital output), AO (analog output), and PO (pulse output). Therefore, PLC will still stand important in the industrial automation field in the future.

What does 'PLC' mean?

A PLC (Programmable Logic Controllers) is an industrial computer used to monitor inputs, and depending upon their state make decisions based on its program or logic, to control (turn on/off) its outputs to automate a machine or a process.

NEMA defines a PROGRAMMABLE LOGIC CONTROLLER as:

"A digitally operating electronic apparatus which uses a programmable memory for the internal storage of instructions by implementing specific functions such as logic sequencing, timing, counting, and arithmetic to control, through digital or analog input/output modules, various types of machines or processes".

Traditional PLC Applications

*In automated system, PLC controller is usually the central part of a process control system.

*To run more complex processes it is possible to connect more PLC controllers to a central computer.

Disadvantages of PLC control

- Too much work required in connecting wires.
- Difficulty with changes or replacements.
- Difficulty in finding errors; requiring skillful work force.
- When a problem occurs, hold-up time is indefinite, usually long.

Advantages of PLC control

* Rugged and designed to withstand vibrations, temperature, humidity, and noise.
* Have interfacing for inputs and outputs already inside the controller.
* Easily programmed and have an easily understood programming language.

Major Types of Industrial Control Systems

Industrial control system or ICS comprise of different types of control systems that are currently in operation in various industries. These control systems include PLC, SCADA and DCS and various others:

PLC

They are based on the Boolean logic operations whereas some models use timers and some have continuous control. These devices are computer based and are used to control various process and equipments within a facility. PLCs control the components in the DCS and SCADA systems but they are primary components in smaller control configurations.

DCS

Distributed Control Systems consists of decentralized elements and all the processes are controlled by these elements. Human interaction is minimized so the labor costs and injuries can be reduced.

Embedded Control

In this control system, small components are attached to the industrial computer system with the help of a network and control is exercised.

SCADA

Supervisory Control and Data Acquisition refers to a centralized system and this system is composed of various subsystems like Remote Telemetry Units, Human Machine Interface, Programmable Logic Controller or PLC and Communications

10.1 Basic PLC Operation:

PLCs consist of input modules or points, a Central Processing Unit (CPU), and output modules or points. An input accepts a variety of digital or analog signals from various field devices (Sensors) and converts them into a logic signal that can be used by the CPU. The CPU makes decisions and executes control instructions based on program instructions in memory. Output modules convert control instructions from the CPU into a digital or analog signal that can be used to control various field devices (Actuators). A programming device is used to input the desired instructions. These instructions determine what the PLC will do for a specific input. An operator interface device allows process information to be displayed and new control parameters to be entered.

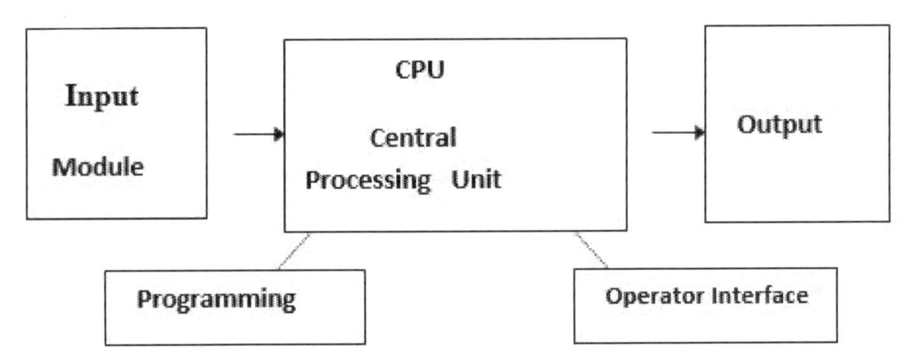

Figure: Basic PLC Operation

Pushbuttons (sensors), in this simple example, connected to PLC inputs, can be used to start and stop a motor connected to a PLC through a motor starter (actuator).

The Working concept of PLC as the name suggest is as follow:

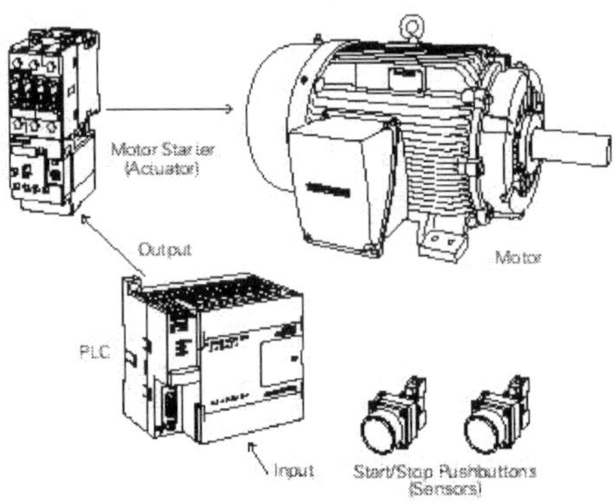

Figure: PLC Working

Programmable is showing its ability that can be easily changed according to the program made. Logic is showing its ability in processing the input of arithmetic (ALU)

which compares the operation add, multiply, divide, subtract and negation. Controller is showing its ability to control and manage the process to produce the desired output.

10.1.1 Hardware Components of a PLC System

Central Processing Unit (CPU)

CPU – Microprocessor based, may allow arithmetic operations, logic operators, block memory moves, computer interface, local area network, functions, etc. CPU makes a great number of check-ups of the PLC controller itself so eventual errors would be discovered early.

System Busses

The internal paths along which the digital signals flow within the PLC are called busses. The system has four busses:

- The CPU uses the data bus for sending data between the different elements,
- The address bus to send the addresses of locations for accessing stored data,
- The control bus for signals relating to internal control actions,
- The system bus is used for communications between the I/O ports and the I/O unit.

Memory

System (ROM) to give permanent storage for the operating system and the fixed data used by the CPU. RAM for data. This is where information is stored on the status of input and output devices and the values of timers and counters and other internal devices. EPROM for ROM's that can be programmed and then the program made permanent.

I/O Sections

Inputs monitor field devices, such as switches and sensors. Outputs control other devices, such as motors, pumps, solenoid valves, and lights.

Power Supply

Most PLC controllers work either at 24 VDC or 220 VAC. Some PLC controllers have electrical supply as a separate module, while small and medium series already contain the supply module.

Programming Device

The programming device is used to enter the required program into the memory of the processor. The program is developed in the programming device and then transferred to the memory unit of the PLC.

The PLC consists of:

Central Processing Unit (CPU) which contains the application program memory (a RAM and ROM) PLC. CPU is much variation depending on the brand and its type. Interface module Input /Output (I/O) is connected directly to the input (switches, sensors) and output (motor, solenoid, lamps).

Digital Input Point
- DC 24 V input
- DC 5 V input / TTL (Transistors Logic)
- AC / DC 24 V input
- AC 110 V input
- AC 220 V input

Analog input
Analog input Point Linear
- 0 to 10 V DC
- DC 10 V - 10 V DC
- 4 to 20 mA DC

Output unit
Digital Output: Digital Output Point 1
- Relay Output
- 110 V AC output
- 220 V AC output
- DC 24 V output, the type of PNP and NPN types.

Analog Output: Output Point Linear
- 0 to 1 V DC
- 10 V DC - 10 V DC
- 4 to 20 mA DC

Besides that PLC also has additional peripheral called programming console, it is to transfer the PLC program. PLC programming can also be done with special software. Programming console is a panel that contains RAM (Random Access Memory) that functions as a semi-permanent storage in a program that was created or modified.

PLC programming is written into the console must be in the mnemonic form. This device can be connected directly to the CPU by using extention cable, it can be installed or removed at any time. If the execution of the program has gone through one cycle the programming console can be lifted and moved to another CPU. While the first CPU can still run the program, but it must be in RUN or MONITOR mode

10.1.2 Types of Programmable Logic Controllers

Programmable logic controllers (PLCs) have several types. Based on the size of the module, type PLC divided into several types:

1. Micro PLC or Small PLC
It is the simplest PLC with the power supply module, CPU, I / O modules and communication ports in a single chassis. This PLC types are usually limited to a few I / O

discrete and can be expanded. There are various micro PLCs on the market today. The vast majority offer analog I / O. with just about any micro PLC, or for that matter PLC in general, when the application requires the monitoring of various analog signals, a separate module is required for each signal (voltage, current, temperature). Examples of this type are CP1H Omron, Siemens S7-200, Fuji Electric SPB.

2. Medium PLC

It is PLC which has CPU module, I / O or communication port are separately. Each module is connected by connector or backplane. It has the capacity more than 2000 I / O. Examples of this type are Omron CS1, Siemens S7-300.

3. Large PLC

This kind of PLC is nearly equal to the medium one but it has large I/O capacity and more able to be connected with the higher control systems. Examples of this type are CVM1 Omron, Siemens S7-400.

Based on the type of input PLC / output PLC can be divided into:

1. Discrete I / O, It is a logic shaped digital input and output as high level with 24VDC, or low level with 0V or as a relay contact output that can be irrigated up to 240VAC.

2. Special I / O, it is I / O that has special functions such as.
a. Analog Input Modules
b. Modules temperature PT100 module or thermocouple (low-level analog inputs)
c. High Speed Counter Module is a frequency logic with generally high level 5V, 12V or 24V.
d. Fuzzy Logic Module
e. PID Module
f. Servo Module
g. Communication protocol module form made by each manufacturer, for example Field bus, Mod bus, Profi bus, Ethernet, Sysmac way, Device Net, Control Net.

Types Of Memory and its description

RAM

Random Access Memory (RAM) is memory where data can be directly accessed at any address. Data can be written to and read from RAM. RAM is used as a temporary storage area. RAM is volatile, meaning that the data stored in RAM will be lost if power is lost. A battery backup is required to avoid losing data in the event of a power loss.

ROM

Read Only Memory (ROM) is a type of memory that data can be read from but not written to. This type of memory is used to protect data or programs from accidental erasure. ROM memory is nonvolatile. This means a user program will not lose data during a loss of electrical power. ROM is normally used to store the programs that define the capabilities of PLC.

EPROM

Erasable Programmable Read Only Memory (EPROM) provides some level of security against unauthorized or unwanted changes in a program. EPROMs are designed so that data stored in them can be read, but not easily altered. Changing EPROM data requires a special effort. UVEPROMs (ultraviolet erasable programmable read only memory) can only be erased with an ultraviolet light. EEPROM (electronically erasable programmable read only memory), can only be erased electronically.

Memory Size

Kilo, abbreviated K, normally refers to 1000 units. When talking about computer or PLC memory, however, 1K means 1024. This is because of the binary number system (210=1024). This can be 1024 bits, 1024 bytes, or 1024 words, depending on memory type.

10.1.3 PLC Scan

The PLC program is executed as part of a repetitive process referred to as a scan. A PLC scan starts with the CPU reading the status of inputs. The application program is executed using the status of the inputs. Once the program is completed, the CPU performs internal diagnostics and communication tasks. The scan cycle ends by updating the outputs, then starts over. The cycle time depends on the size of the program, the number of I/Os, and the amount of communication required.

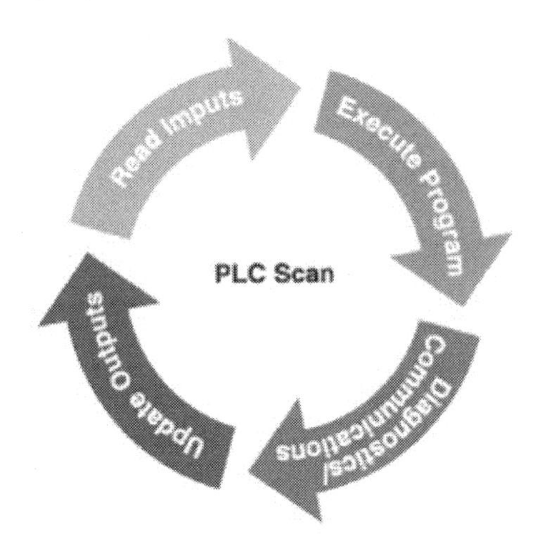

Figure: PLC Scan

10.1.4 Number Systems

Since a PLC is a computer, it stores information in the form of On or Off conditions (1 or 0), referred to as binary digits (bits). Sometimes binary digits are used individually and sometimes they are used to represent numerical values.

10.1.4.1 Decimal System

Various number systems are used by PLCs. All number systems have the same three characteristics: digits, base, weight. The Decimal system, which is commonly used in everyday life, has the following characteristics:
Ten digits 0, 1, 2, 3, 4, 5, 6, 7, 8, 9
Base 10
Weights 1, 10, 100, 1000, ...

10.1.4.2 Binary System:

The binary system is used by programmable controllers. The binary system has the following characteristics:

Two digits 0, 1
Base 2
Weights Powers of base 2 (1, 2, 4, 8, 16, ...)

Bits, Bytes, and Words:

Each binary piece of data is a bit. Eight bits make up one byte. Two bytes, or 16 bits, make up one word.

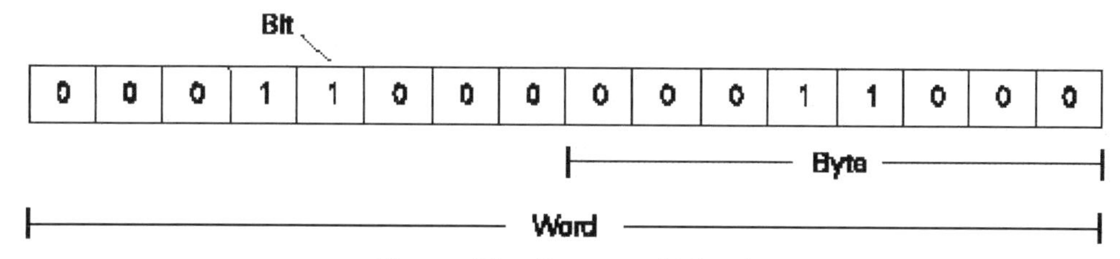

Figure: Bits, Bytes and Words

Logic 0, Logic 1: Programmable controllers can only understand a signal that is On or Off (present or not present).

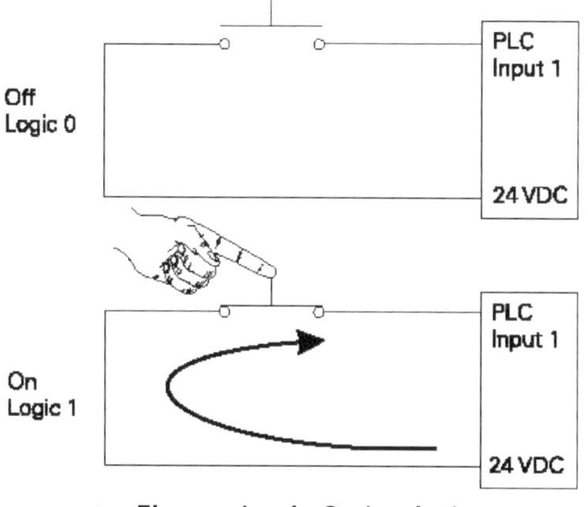

Figure: Logic 0 , Logic 1

The binary system is a system in which there are only two numbers, 1 and 0. Binary 1 indicates that a signal is present, or the switch is On. Binary 0 indicates that the signal is not present, or the switch is Off.

10.1.4.3 A List of Programming Languages

- Ladder diagram (LD)
- Sequential Function Charts (SFC)
- Function Block Diagram (FBD)
- Structured Text (ST)
- Instruction List (IL)

Ladder Logic
Ladder logic is the main programming method used for PLCs. As mentioned before, ladder logic has been developed to mimic relay logic. The decision to use the relay logic diagrams was a strategic one. By selecting ladder logic as the main programming method, the amount of retraining needed for engineers and trades people was greatly reduced.
The first PLC was programmed with a technique that was based on relay logic wiring schematics. This eliminated the need to teach the electricians, technicians and engineers how to program - so this programming method has stuck and it is the most common technique for programming in today's PLC.

Instruction list
There are other methods to program PLCs. One of the earliest techniques involved mnemonic instructions. These instructions can be derived directly from the ladder logic diagrams and entered into the PLC through a simple programming terminal.

Sequential Function Charts (SFC)
SFC have been developed to accommodate the programming of more advanced systems. These are similar to flowcharts, but much more powerful. This method is much different from flowcharts because it does not have to follow a single path through the flowchart.

Structured Text (ST)
Programming has been developed as a more modern programming language. It is quite similar to languages such as BASIC and Pascal.
Structured Text (ST) is a high level textual language that is a Pascal like language. It is very flexible and intuitive for writing control algorithms.

Function Block Diagram (FBD)
FBD is another graphical programming language. The main concept is the data flow that starts from inputs and passes in block(s) and generates the output.

Data Files

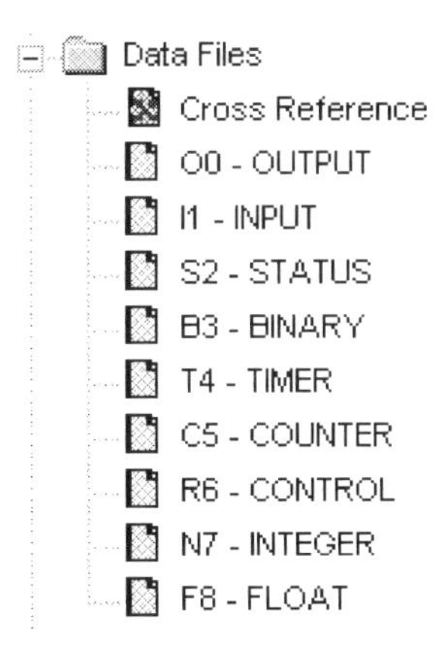

Data File Descriptions

File #	Type	Description
O0	Output	This file stores the state of output terminals for the controller.
I1	Input	This file stores the state of input terminals for the controller.
S2	Status	This file stores controller operation information useful for troubleshooting controller and program operation.
B3	Bit	This file stores internal relay logic.
T4	Timer	This file stores the timer accumulator and preset values and status bits.
C5	Counter	This file stores the counter accumulator and preset values and status bits.
R6	Control	This file stores the length, pointer position, and status bits for control instructions such as shift registers and sequencers.
N7	Integer	This file is used to store bit information or numeric values with a range of -32767 to 32768.
F8	Floating Point	This file stores a # with a range of 1.1754944e-38 to 3.40282347e+38.

10.1.4.4 Allen-Bradley Driver Addressing

The table below identifies the value ranges for Allen-Bradley PLCs. Bitwise addressing for binary and integer formats is accomplished by following the element with "/bit-number".

Allen-Bradley Addressing	Value Range
Output Image	O:0 – O:30
Input Image	I:0 – I:30
Status	S:0 – S:n
Binary	B3:0 – B3:255 (bitwise - B3:1/15 16th bit in element number 1)
Timer	T4:0 – T4:255
Counter	C5:0 – C5:255
Control	R6:0 – R6:255
Integer	N7:0 – N7:255 (bitwise - N7:1/5 6th bit)
Floating Point	F8:0 – F8:255

BASIC INSTRUCATION OF PLC

Examine if Closed (XIC)

Use the XIC instruction in your ladder program to determine if a bit is On . When the instruction is executed, if the bit addressed is on (1), then the instruction is evaluated as true. When the instruction is executed, if the bit addressed is off (0), then the instruction is evaluated.

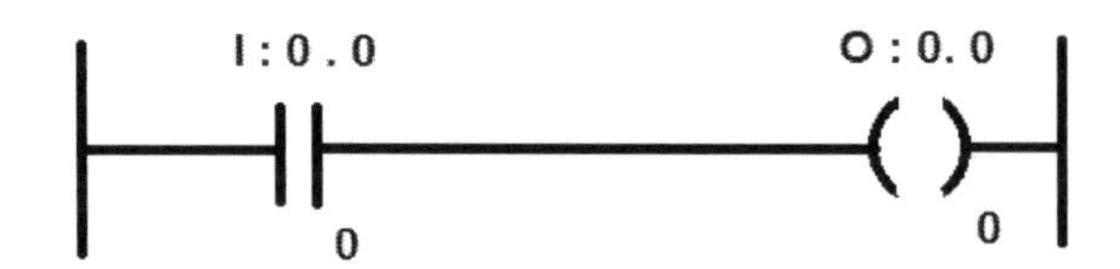

Bit Address State XIC Instruction 0 False 1 True

Examine if Open (XIO)

Use the XIO instruction in your ladder program to determine if a bit is Off . When the instruction is executed, if the bit addressed is off (0), then the instruction is evaluated as true. When the instruction is executed, if the bit addressed is on (1), Then the instruction is evaluated as false.

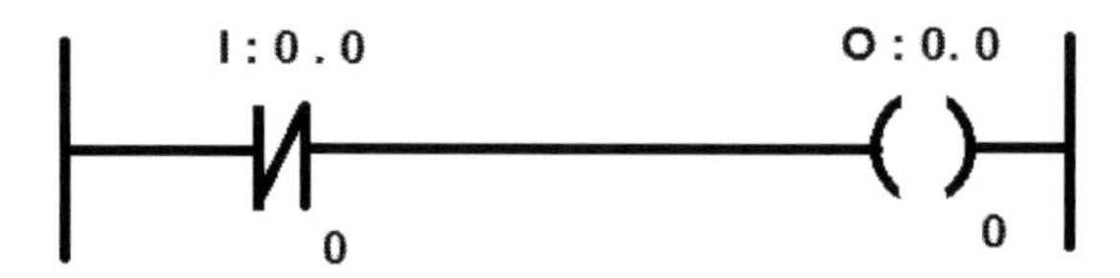

Bit Address State XIO Instruction 0 True 1 False

Output Energize (OTE)

Use the OTE instruction in your ladder program to turn on a bit when rung conditions are evaluated as true.

O : 0. 0

( )

0

A bit that is set within a subroutine using an OTE instruction remains set until the subroutine is scanned again.

Output Latch (OTL) and Output Unlatch (OTU)

OTL and OTU are retentive output instructions. OTL can only turn on a bit, while OTU can only turn off a bit. These instructions are usually used in pairs, with both instructions addressing the same bit.

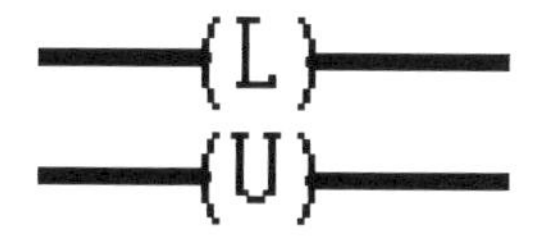

10.2 Sourcing and Sinking Concept

“Sinking” and “Sourcing” terms are very important in connecting a PLC correctly with external environment. These terms are applied only for DC modules.

The most brief definition of these two concepts would be:

SINKING = Common GND line (-)

SOURCING = Common VCC line (+)

Sink - A wiring arrangement in which the I/O device provides current to the I/O module

Source - A wiring arrangement in which the I/O module provides current to the I/O device

Most commonly used DC module options in PLCs are:

*Sinking input and

*Sourcing output module

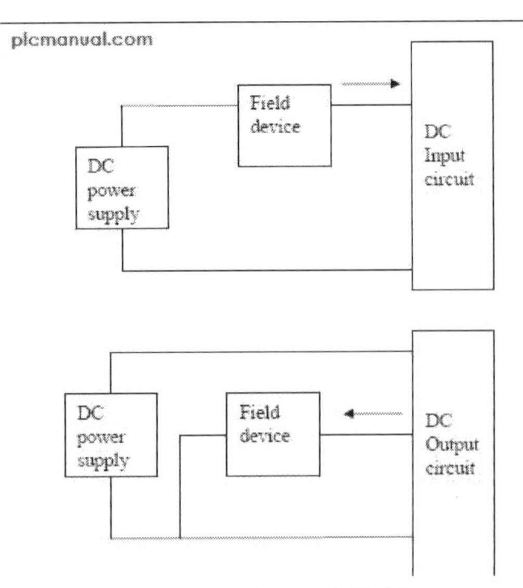

Figure: Sourcing and Sinking

- Sinking I/O circuits on the I/O modules receive (sink) current from sourcing field devices. Sinking output modules used for interfacing with electronic equipment.
- Sourcing I/O: Sourcing output modules used for interfacing with solenoids.

PLC AC I/O circuits accommodate either sinking or sourcing field devices. Solid-state DC I/O circuits require that they used in a specific sinking or sourcing circuit depending on the internal circuitry.

PLC contact (relay) output circuits AC or DC accommodate either sinking or sourcing field devices.

10.3 Different START / STOP Logics

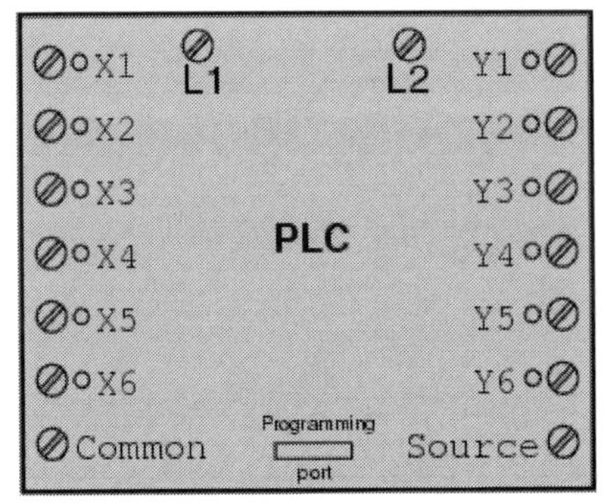

Inside the PLC housing, connected between each input terminal and the Common terminal, is an opto-coupler device (Light-Emitting Diode) that provides an electrically isolated "high" logic signal to the computer's circuitry (a photo-transistor interprets the LED's light) when there is AC power applied between the respective input terminal and the Common terminal. An indicating LED on the front panel of the PLC gives visual indication of an "energized" input: Output signals are generated by the PLC's computer circuitry activating a switching device (transistor, TRIAC, or even an electromechanical relay), connecting the "Source" terminal to any of the "Y-" labeled output terminals. The "Source" terminal, correspondingly, is usually connected to the L1 side of the AC power source. As with each input, an indicating LED on the front panel of the PLC gives visual indication of an "energized" output:

There are the four different combinations of start/stop logics.

Pushbutton in h/w	START	STOP
1.	NO	NO
2.	NO	NC
3.	NC	NO
4.	NC	NC

START/NO & STOP/NO

The pushbutton switch connected to input X1 serves as the "Start" switch, while the switch connected to input X2 serves as the "Stop." Another contact in the program, named Y1, uses the output coil status as a seal-in contact, directly, so that the motor contactor will continue to be energized after the "Start" pushbutton switch is released. You can see the normally-closed contact X2 appear in a colored block, showing that it is in a closed ("electrically conducting") state

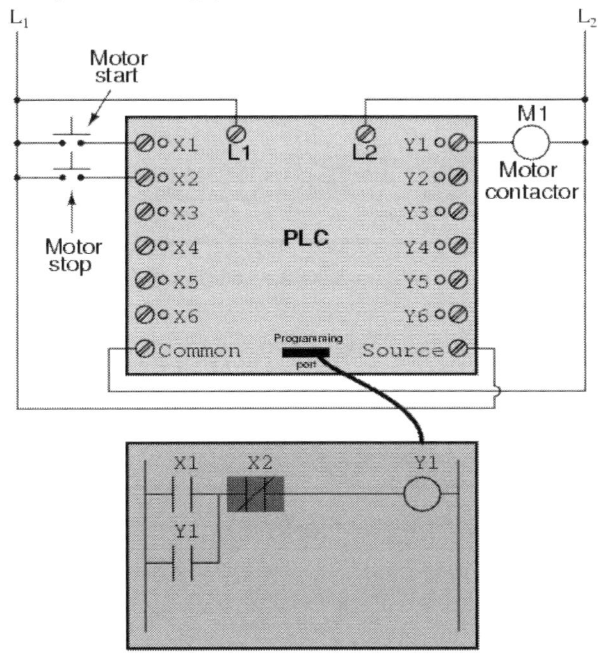

If we were to press the "Start" button, input X1 would energize, thus "closing" the X1 contact in the program, sending "power" to the Y1 "coil," energizing the Y1 output and

applying 120 volt AC power to the real motor contactor coil. The parallel Y1 contact will also "close," thus latching the "circuit" in an energized state:

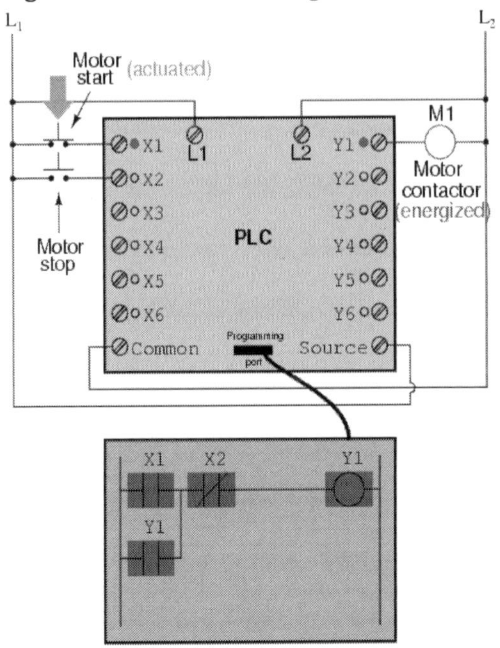

Now, if we release the "Start" pushbutton, the normally-open X1 "contact" will return to its "open" state, but the motor will continue to run because the Y1 seal-in "contact" continues to provide "continuity" to "power" coil Y1, thus keeping the Y1 output energized:

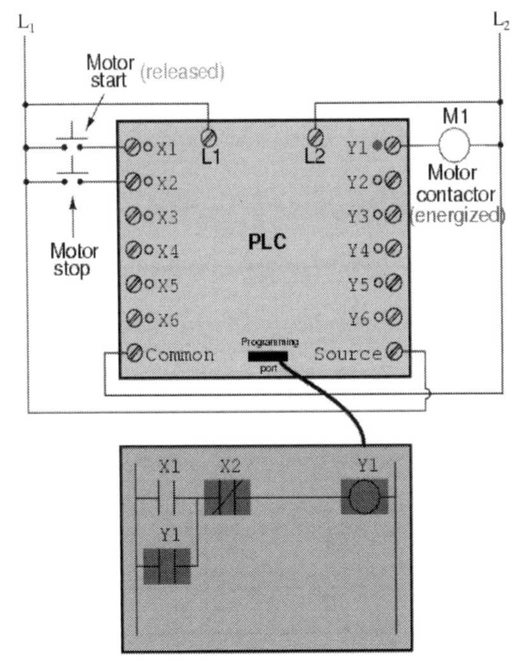

To stop the motor, we must momentarily press the "Stop" pushbutton, which will energize the X2 input and "open" the normally-closed "contact," breaking continuity to the Y1 "coil:"

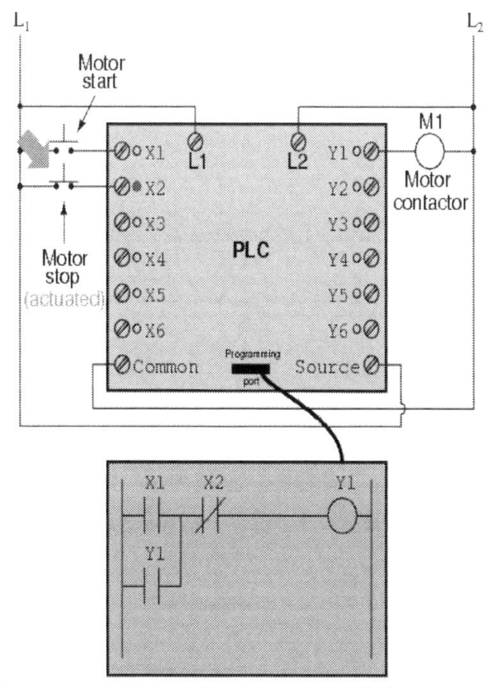

When the "Stop" pushbutton is released, input X2 will de-energize, returning "contact" X2 to its normal, "closed" state. The motor, however, will not start again until the "Start" pushbutton is actuated, because the "seal-in" of Y1 has been lost:

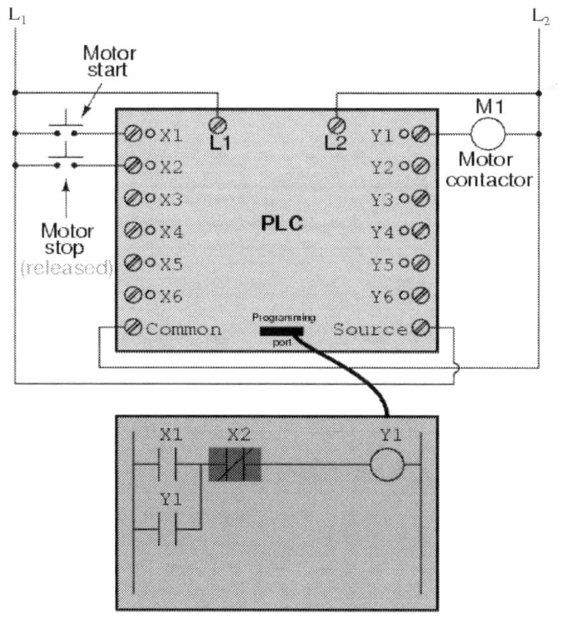

An important point to make here is that fail-safe design is just as important in PLC-controlled systems as it is in electromechanical relay-controlled systems. One should always consider the effects of failed (open) wiring on the device or devices being controlled. In this motor control circuit example, we have a problem: if the input wiring for X2 (the "Stop" switch) were to fail open, there would be no way to stop the motor!
The solution to this problem is a reversal of logic between the X2 "contact" inside the PLC program and the actual "Stop" pushbutton switch:

START/NO & STOP/NC

When the normally-closed "Stop" pushbutton switch is unactuated (not pressed), the PLC's X2 input will be energized, thus "closing" the X2 "contact" inside the program. This allows the motor to be started when input X1 is energized, and allows it to continue to run when the "Start" pushbutton is no longer pressed. When the "Stop" pushbutton is actuated, input X2 will de-energize, thus "opening" the X2 "contact" inside the PLC program and shutting off the motor. So, we see there is no operational difference between this new design and the previous design.
However, if the input wiring on input X2 were to fail open, X2 input would de-energize in the same manner as when the "Stop" pushbutton is pressed. The result, then, for a wiring failure on the X2 input is that the motor will immediately shut off. This is a safer design than the one previously shown, where a "Stop" switch wiring failure would have resulted in an inability to turn off the motor.

START/NC & STOP/NO

When the normally-open "Stop" pushbutton switch is unactuated (not pressed), the PLC's X2 input will be energized, thus "closing" the X2 "contact" inside the program.

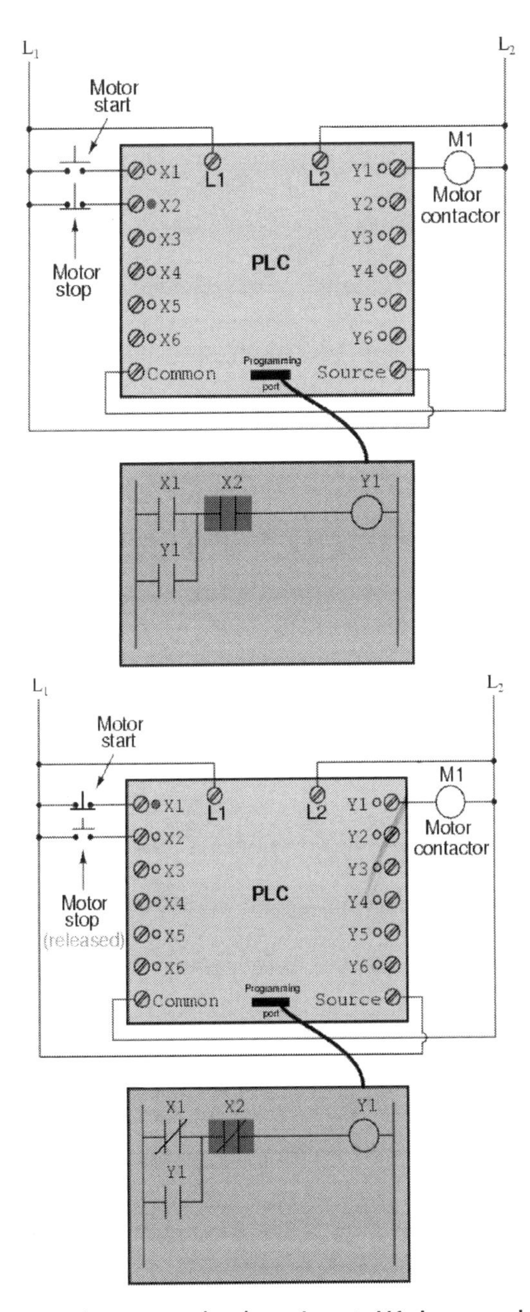

This allows the motor to be started when input X1 is energized, and allows it to continue to run when the "Start" pushbutton is no longer pressed. When the "Stop" pushbutton is actuated, input X2 will de-energize, thus "opening" the X2 "contact" inside the PLC program and shutting off the motor. So, we see there is no operational difference between this new design and the previous design.

START/NC & STOP/NC

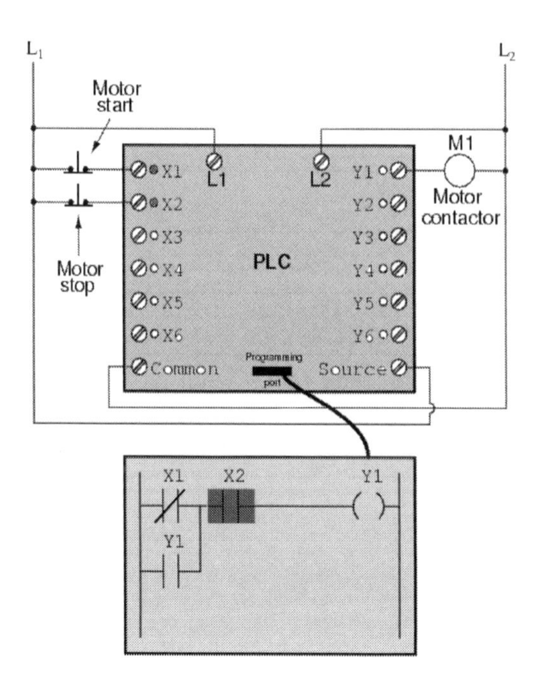

Logic Gates In Ladder Language

Ladder logic is a programming language that represents a program by a graphical diagram based on the circuit diagrams of relay logic hardware. It is primarily used to develop software for programmable logic controllers (PLCs) used in industrial control applications. The name is based on the observation that programs in this language resemble ladders, with two vertical rails and a series of horizontal rungs between them

Logic AND

The truth table of logic AND is as follows,

A	B	O
0	0	0
0	1	0
1	0	0
1	1	1

A B O

Conversion to Ladder

A B O

Logic OR

The truth table of logic OR is as follows,

A	B	O
0	0	0
0	1	1
1	0	1
1	1	1

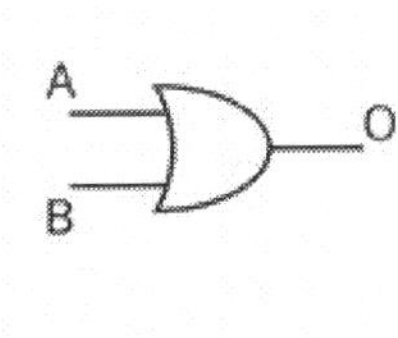

Conversion to Ladder

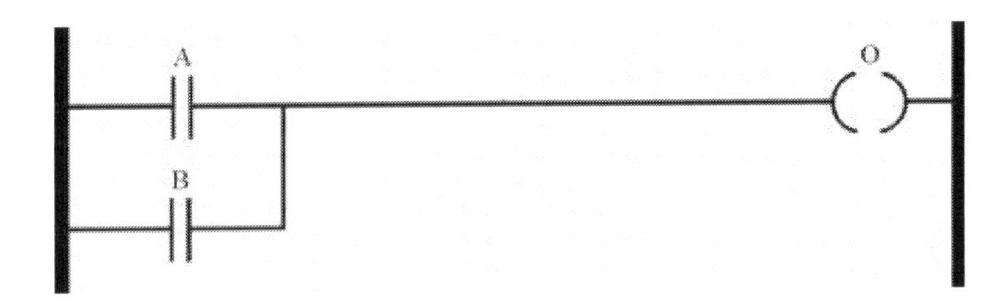

Logic NOT

The truth table of logic NOT is as follows,

A	O
0	1
1	0

Conversion to Ladder Diagram,

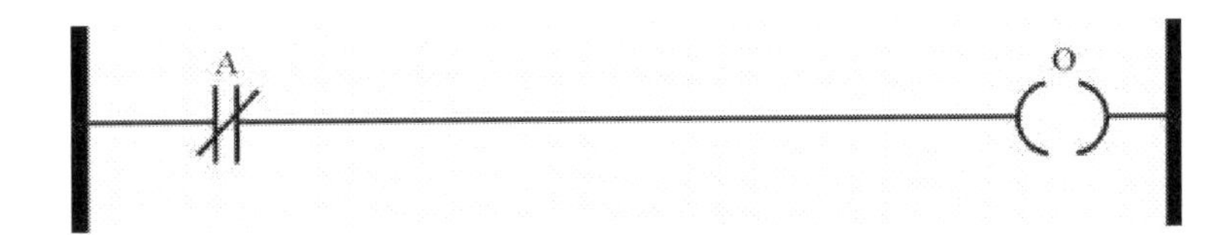

Logic NAND

Logic NAND is a development of logic AND, OR and NOT. The truth table is as follows,

A	B	O
0	0	1
0	1	1
1	0	1
1	1	0

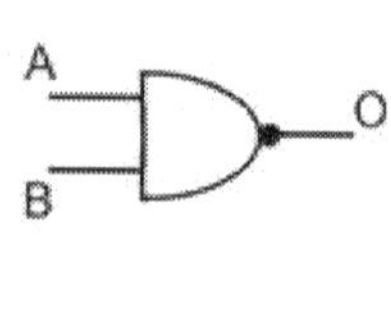

The truth table above have the
Following equation,
O = (A.B) '= A' + B '
The conversion to Ladder Diagram

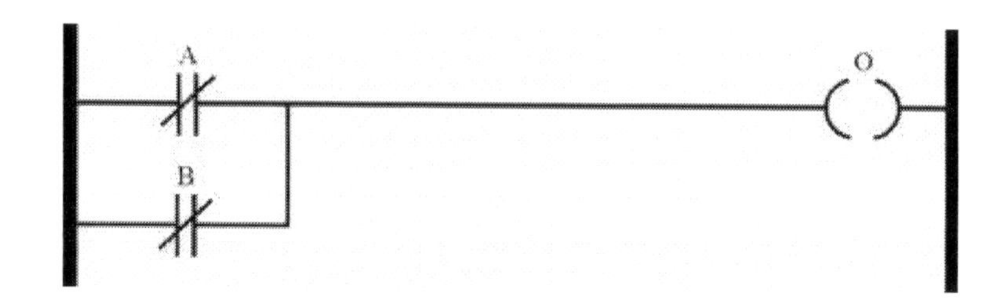

Logic NOR

This logic is also the development of the logic AND, OR and NOT. The truth table is as follows,

A	B	O
0	0	1
0	1	0
1	0	0
1	1	0

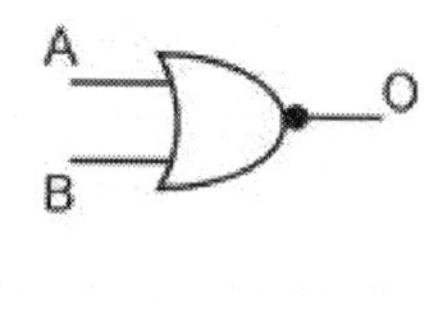

The truth table above have the following equation,
O = (A + B) '= A'. B '
So the conversion to Ladder Diagram,

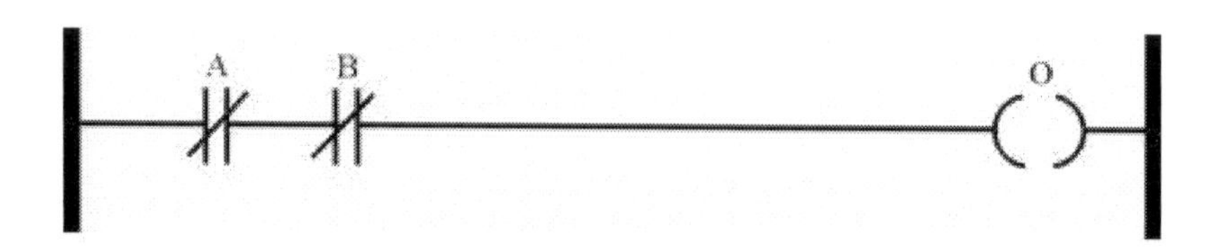

Logic XOR

Similarly the logic of the previous ones this logic is also the development of AND, OR and NOT. This logic is widely used in summing strand (adder). The truth table is as follows,

A	B	O
0	0	0
0	1	1
1	0	1
1	1	0

The truth table above have the following equation,
O = A o B = A '. B + A. B'

So the conversion to Ladder Diagram,

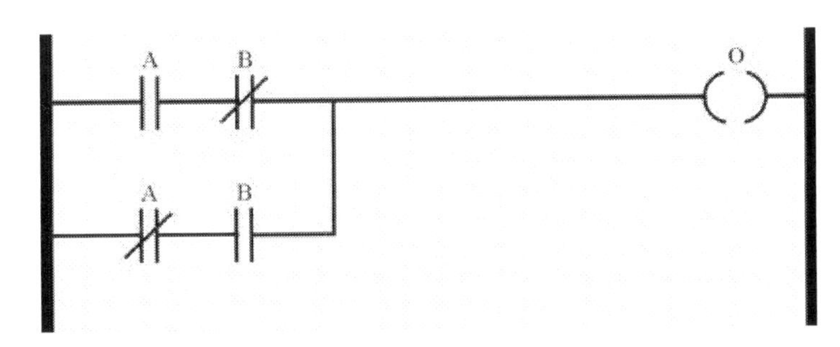

Timer
The timer is used for timing and has coil, contact and register in it. When the coil is On and the estimated time is reached, its contact will be enabled (contact A closed, contact B open). Every timer has its fixed timing period (unit : 1ms/10ms/100ms). Once the coil is Off, the contact be disabled (contact A open, contact B closed) and the present value on the timer will become "0".

Parameters
Accumulator Value (ACC): This is the time elapsed since the timer was last reset. When enabled, the timer updates this continually.

Preset Value (PRE): This specifies the value which the timer must reach before the controller sets the done bit. When the accumulated value becomes equal to or greater than the preset value, the done bit is set. You can use this bit to control an output device.

Time base The time base determines the duration of each time base interval. For Fixed and SLC 5/01 processors, the time base is set at 0.01 second. For SLC 5/02 and higher processors and Micro Logix 1000 controllers, the time base is selectable as 0.01 (10 ms) second or 1.0 second.

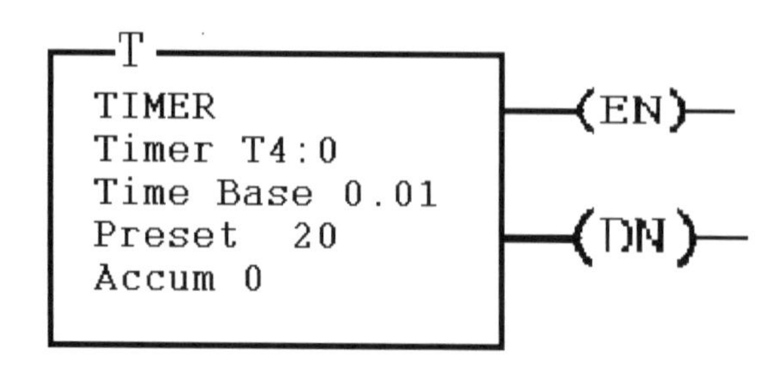

Timer On–Delay (TON)

This Bit	Is Set When	And Remains Set Until One of the Following
Timer Done Bit DN	accumulated value is equal to or greater than the preset value	conditions go false
Timer Timing Bit TT	the accumulated value is less than the preset value	conditions go false or when the done bit is set
Timer Enable Bit EN	conditions are true	conditions go false

Timer off –Delay (TOF)

This Bit	Is Set When	And Remains Set Until One of the Following
Timer Done Bit DN	conditions are true	accumulated value is equal to or greater than the preset value
Timer Timing Bit TT	condition are false and the accumulated value is less than the preset value	conditions are true or the done bit is reset.
Timer Enable Bit EN	conditions are true	conditions go false

The TOF instruction begins to count timebase intervals when the rung makes a true-to-false transition. As long as rung conditions remain false, the timer increments its accumulated value (ACC) each scan until it reaches the preset value (PRE). The accumulated value is reset when rung conditions go true again.

Retentive Timer (RTO)

Use the RTO instruction to turn an output on or off after its timer has been on for a preset time interval. The RTO instruction is a retentive instruction that begins to count timebase intervals when rung conditions become true.

This Bit	Is Set When	And Remains Set Until One of the Following
Timer Done Bit DN instuction	accumulated value is equal to or greater than the preset value	the appropriate RES is enabled.
Timer Timing Bit TT	condition are true and the accumulated value is less than the preset value	conditions go false or when the done bit is set
Timer Enable Bit EN	conditions are true	conditions go false

To reset the retentive timer's accumulated value and status bits after the RTO rung goes false, you must program a reset (RES) instruction with the same address in another rung.

Counter

The counter is used for counting. Before using the counter, you have to give the counter a set value (i.e. the number of pulses for counting). There are coil, contact and registers in the counter. When the coil goes from Off to On, the counter will regard it as an input of 1 pulse and the present value on the counter will plus "1". We offer 16-bit and 32-bit high-speed counters for our users.
The on and off duration of an incoming signal must not be faster than the scan time 2x (assuming a 50% duty cycle).

Entering Parameters

Accumulator Value (ACC)
This is the number of false-to-true transitions that have occurred since the counter was last reset.

Preset Value (PRE)
Specifies the value which the counter must reach before the controller sets the done bit. When the accumulator value becomes equal to or greater than the preset value, the done status bit is set. You can use this bit to control an output device.
Preset and accumulated values for counters range from –32,768 to +32,767, and are stored as signed integers. Negative values are stored in two's complement form.

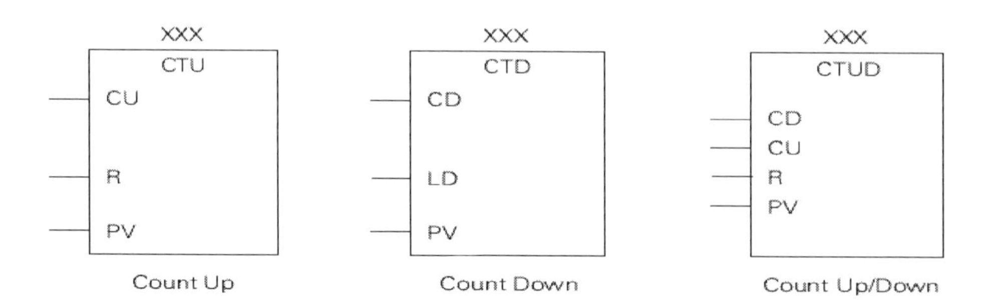

Reset (RES)

Use a RES instruction to reset a timer or counter. When the RES instruction is enabled, it resets the Timer On Delay (TON), Retentive Timer (RTO), Count Up (CTU), or Count Down (CTD) instruction having the same address as the RES instruction.

LIST OF THE COMPARISON INSTRUCTIONS

ADD - Adding

Symbol

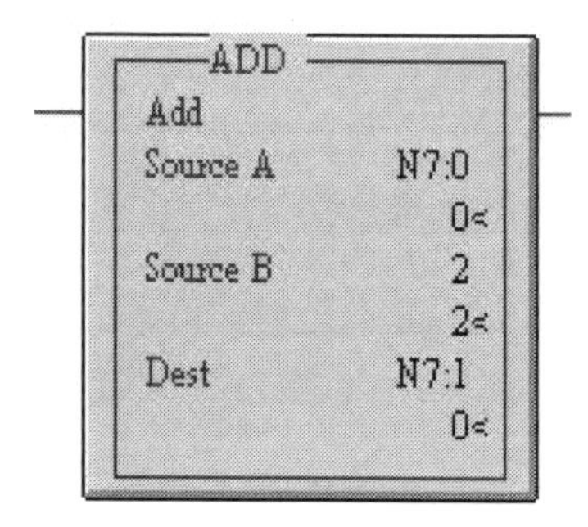

Definition

When rung conditions are true, this output instruction adds Source A to Source B and stores the result at the destination address. Source A and Source B can either be values or addresses that contain values, however Source A and Source B cannot both be constants.

SUB - SUBTRACT

Symbol

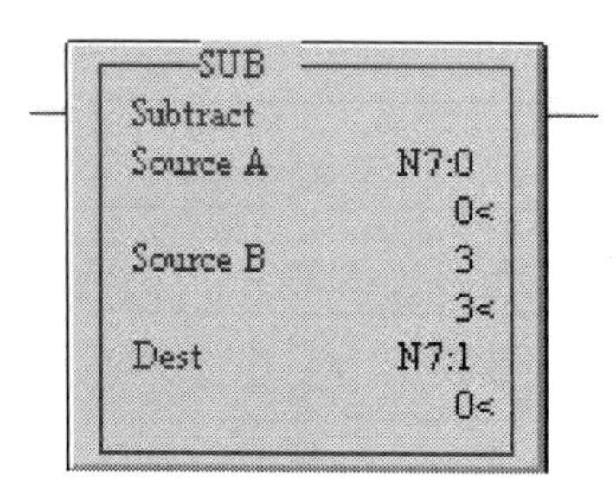

Definition

When rung conditions are true, the SUB output instruction subtracts Source B from Source A and stores the result in the destination. Source A and Source B can either be values or addresses that contain values, however Source A and Source B cannot both be constants.

MUL - MULTIPLY

Symbol

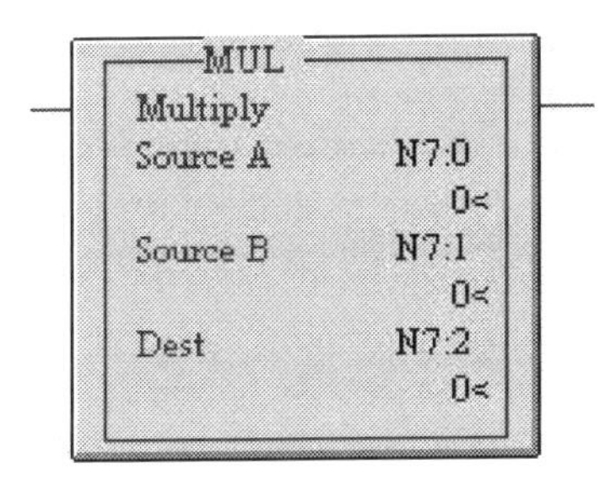

Definition

Use the MUL instruction to multiply one value (source A) by another (source B) and place the result in the destination. Source A and Source B can either be constant values or addresses that contain values however Source A and Source B cannot both be constants.

The math register contains the 32-bit signed integer result of the multiply operation. This result is valid at overflow.

DIV - Divide

Symbol

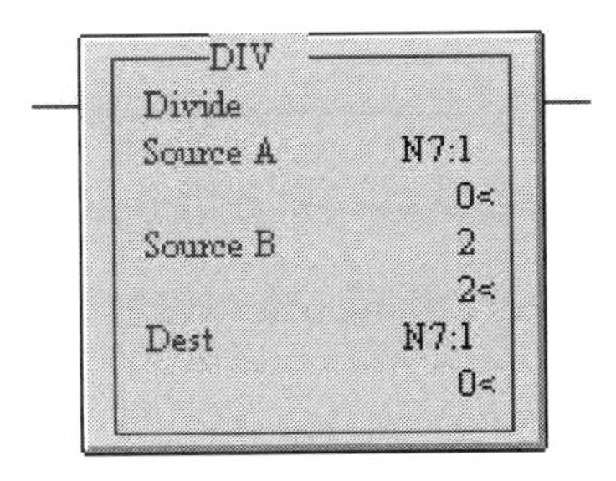

Definition

When rung condition is true, this output instruction divides Source A by Source B and stores the result in the destination and the math register. The value stored in the destination is rounded. The value stored in the math register consists of the unrounded quotient (placed in the most significant word) and the remainder (placed in the least significant word).

Source A and Source B can either be constant values or addresses that contain values, however Source and Source B cannot both be constants.

LIST OF THE COMPARISON INSTRUCTIONS

EQU - Equal

Symbol

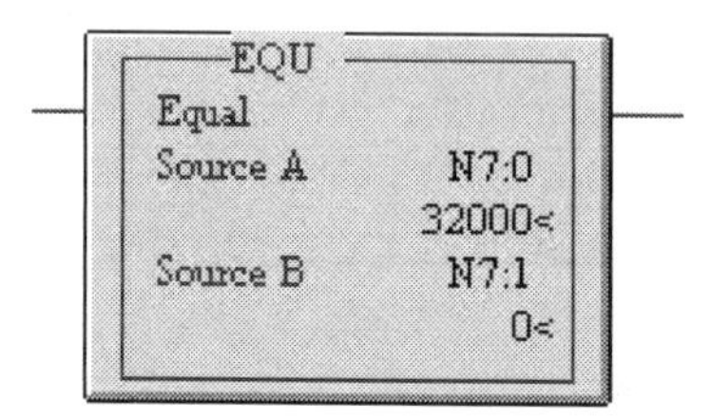

Definition

Test whether two values are equal or not.

If source A and Source B are equal, the instruction is logically true. Source A must be an address. Source B can either be a program constant or an address. Negative integers are stored in two's complement.

NEQ - Not Equal

Symbol

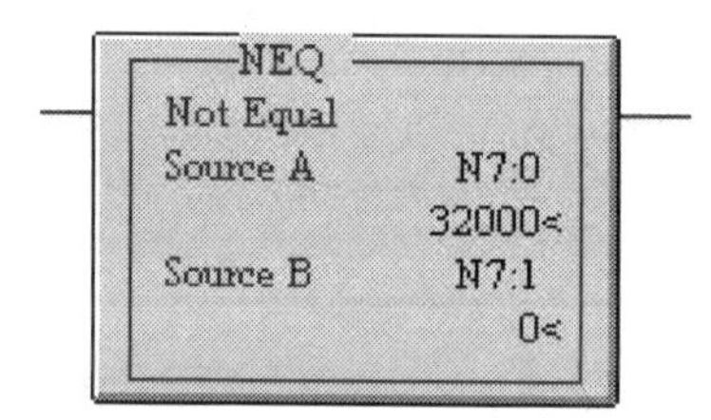

Definition

Test whether one value is not equal to a second value.

If Source A and Source B are not equal, the instruction is logically true. If the two values are equal, the instruction is logically false.

Source A must be an address. Source B can be either a program constant or an address. Negative integers are stored in two's complement.

LES - Less Than

Symbol

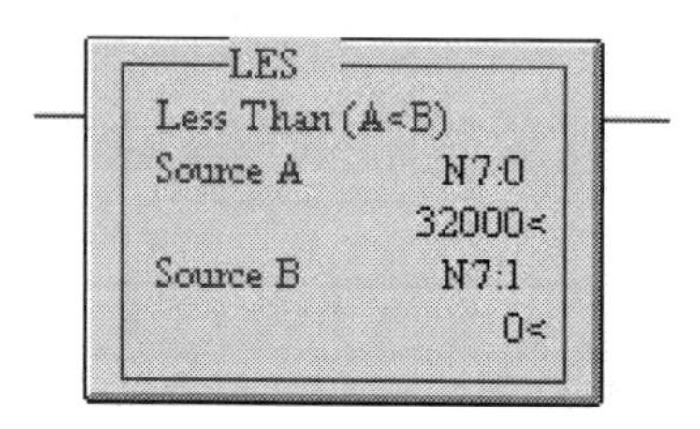

Definition

Test whether one value is less than a second value.

If Source A is less than the value at source B the instruction is logically true. If the value at source A is greater than or equal to the value at source B, the instruction is logically false.

Source A must be an address. Source B can either be a program constant or an address. Negative integers are stored in two's complement.

LEQ - Less Than or Equal

Symbol

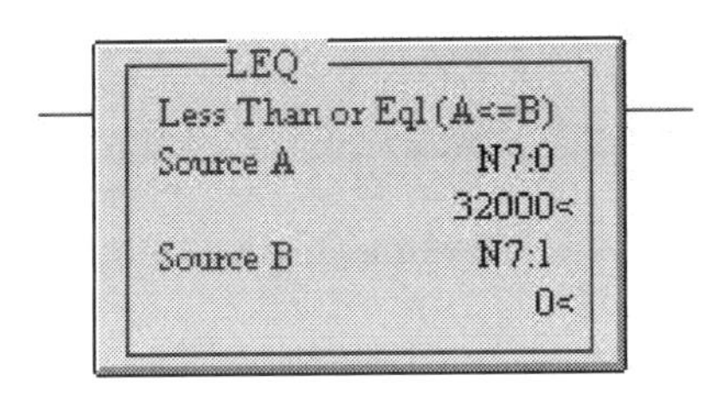

Definition
Test whether one value is less than or equal to a second value.
If value at source A is less than or equal to the value at source B, the instruction is logically true.
If the value at source A is greater than or equal to the value at source B, the instruction is logically false.
Source A must be an address. Source B can either be a program constant or an address.
Negative integers are stored in two's complement.

GRT - Greater Than

Symbol

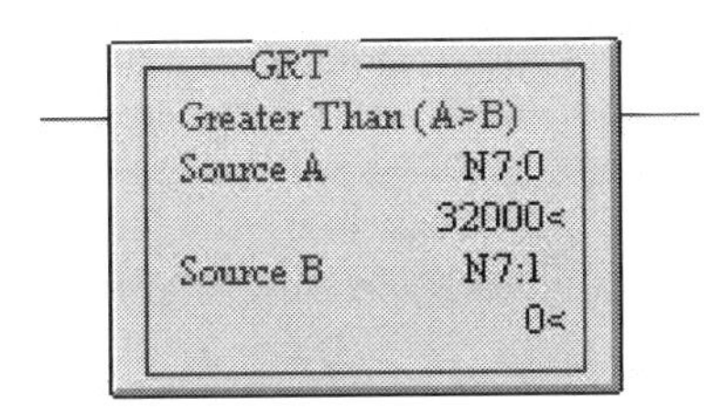

Definition
Test whether one value is greater than the second value.
If the value at source A is greater than the value at source B, the instruction is logically true.
If the value at source A is less than or equal to the value at source B, the instruction is logically false.
Source A must be an address. Source B can either be a program constant or an address.
Negative integers are stored in two's complement.

GEQ - Greater or Equal

Symbol

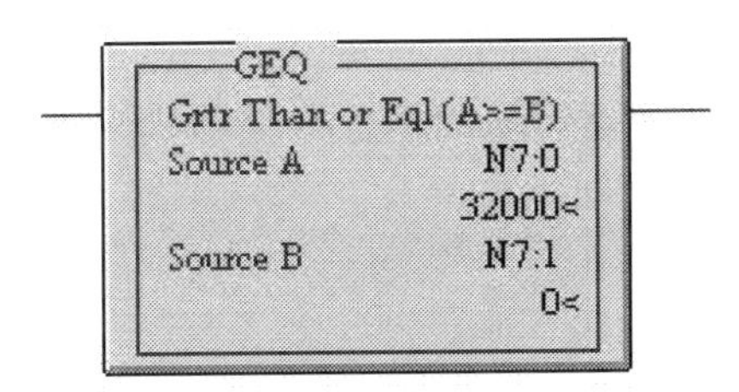

Definition
Test whether one value is greater or equal to a second value.
If the value at source A is greater than or equal the value at source B, the instruction is logically true.

If the value at source A is less than to the value at source B, the instruction is logically false.
Source A must be an address. Source B can either be a program constant or an address.
Negative integers are stored in two's complement.

MEQ - Masked Comparison for Equal

Symbol

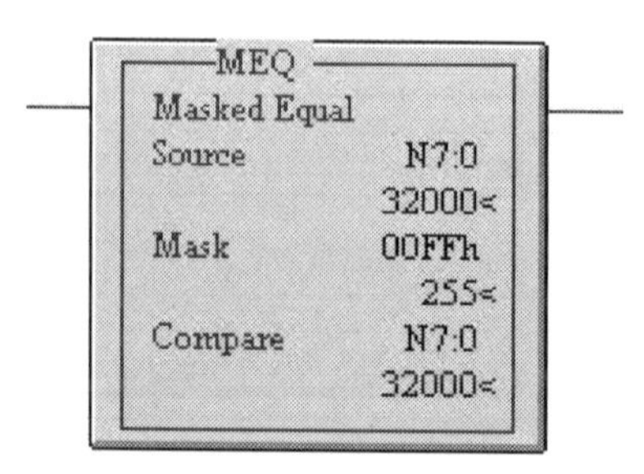

Definition
Test portion of two values to see whether they are equal. Compares 16 bit data of a source address to 16 bit data at a reference address through a mask
Use the MEQ instruction to compare data at a source address with data at a compare address. Use of this instruction allows portions of the data to be masked by a separate word.
Source is the address of the value you want to compare.
Mask is the address of the mask through which the instruction moves data. The mask can be a hexadecimal value.
Compare is an integer value or the address of the reference.
If the 16 bits of data at the source address are equal to the 16 bits of data at the compare address (less masked bits), the instruction is true. The instruction becomes false as soon as it detects a mismatch. Bits in the mask word mask data when reset; they pass data when set.

LIM - Limit Test

Symbol

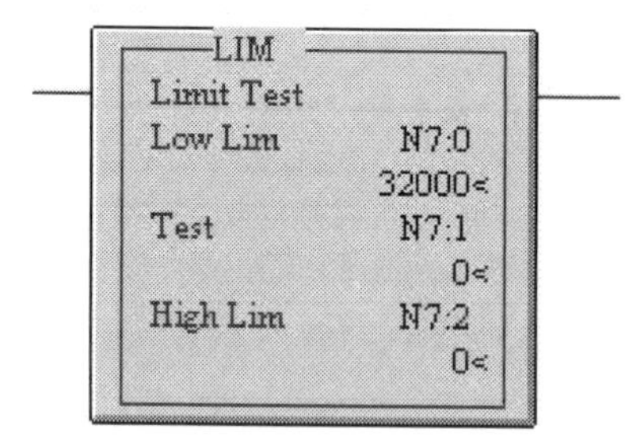

Definition
Test whether one value is within the limit range of two other values
The Low limit, Test, and High Limit values can be word addresses or constants, restricted to the following combination:
If the Test parameter is a program constant, both the Low Limit and High Limit parameters must be word addresses. If the Test parameter is a word address, the Low Limit and High Limit parameters can be either a program constant or a word address.

SCP - Scale with Parameters

Symbol

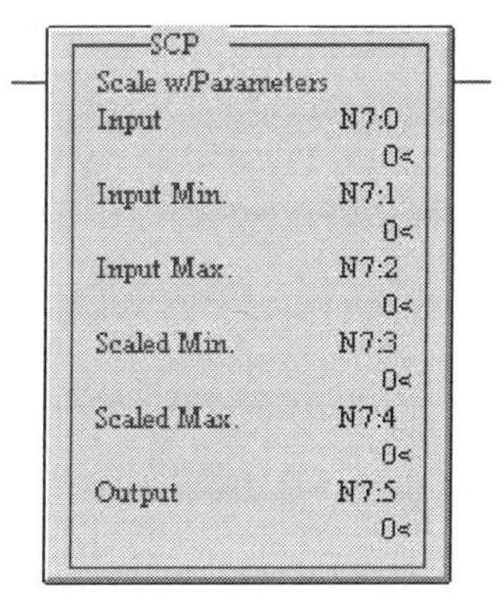

Definition

Produces a scaled output value that has a linear relationship between the input and scaled values. The scaled result is returned to the address indicated by the output parameter.

SCL - Scale Data

Symbol

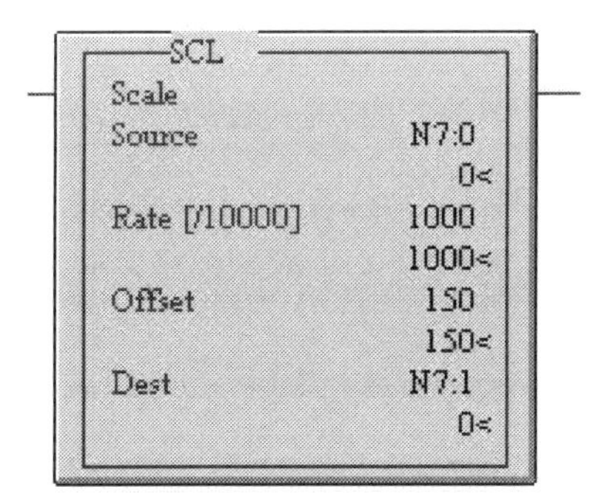

Definition

Use this instruction to scale data from your analog module and bring it into the limits prescribed by the process variable or another analog module. When rung conditions are true, this instruction multiplies the source by a specified rate. The rounded result is added to an offset value and placed in the destination.

PLC Selection criteria consists of:

* System (task) requirements.
* Application requirements.
* What input/output capacity is required?
* What type of inputs/outputs are required?
* What size of memory is required?
* What speed is required of the CPU?
* Electrical requirements.
* Speed of operation.

* Communication requirements.
* Software.
* Operator interface.
* Physical environments.

System requirements

- The starting point in determining any solution must be to understand what is to be achieved.
- The program design starts with breaking down the task into a number of simple understandable elements, each of which can be easily described.

Application requirements

- Input and output device requirements. After determining the operation of the system, the next step is to determine what input and output devices the system requires.
- List the function required and identify a specific type of device.
- The need for special operations in addition to discrete (On/Off) logic.
- List the advanced functions required beside simple discrete logic.

Electrical Requirements

The electrical requirements for inputs, outputs, and system power; When determining the electrical requirements of a system, consider three items:

- Incoming power (power for the control system);
- Input device voltage; and
- Output voltage and current.

Speed of Operation

How fast the control system must operate (speed of operation).
When determining speed of operation, consider these points:
- How fast does the process occur or machine operate?
- Are there "time critical" operations or events that must be detected?
- In what time frame must the fastest action occur (input device detection to output device activation)?
- Does the control system need to count pulses from an encoder or flow-meter and respond quickly?

Communication

If the application requires sharing data outside the process, i.e. communication. Communication involves sharing application data or status with another electronic device, such as a computer or a monitor in an operator's station. Communication can take place locally through a twisted-pair wire, or remotely via telephone or radio modem.

Operator Interface

If the system needs operator control or interaction. In order to convey information about machine or process status, or to allow an operator to input data, many applications require operator interfaces. Traditional operator interfaces include pushbuttons, pilot lights and LED numeric display. Electronic operator interface devices display messages about machine status in descriptive text, display part count and track alarms. Also, they can be used for data input.

Physical Environment

The physical environment in which the control system will be located. Consider the environment where the control system will be located. In harsh environments, house the control system in an appropriate IP-rated enclosure. Remember to consider accessibility for maintenance, troubleshooting or reprogramming.

11 Chapter: SCADA (Supervisory Control and Data Acquisition)

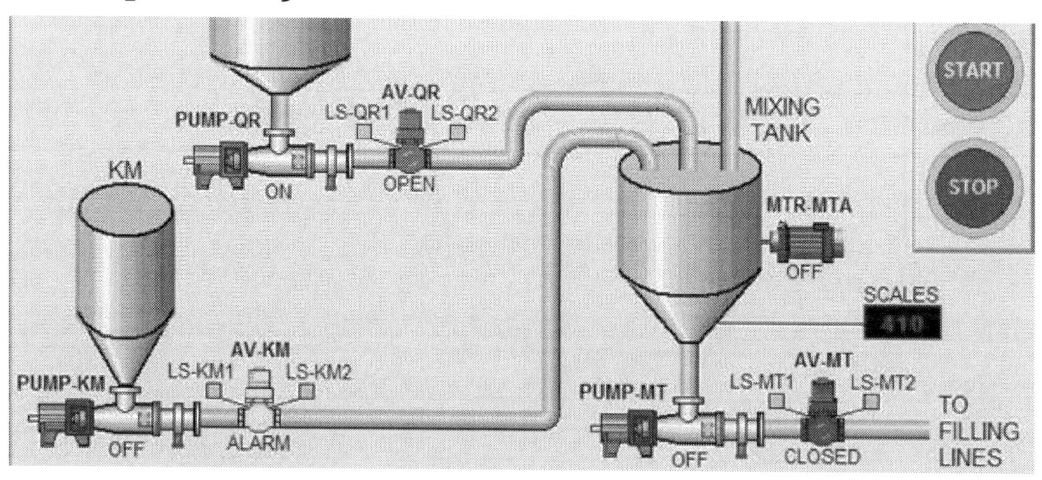

SCADA (supervisory control and data acquisition) is a category of software application program for process control, the gathering of data in real time from remote locations in order to control equipment and conditions. SCADA is used in power plants as well as in oil and gas refining, telecommunications, transportation, and water and waste control. SCADA systems include hardware and software components. The hardware gathers and feeds data into a computer that has SCADA software installed. The computer then processes this data and presents it in a timely manner. SCADA also records and logs all events into a file stored on a hard disk or sends them to a printer. SCADA warns when conditions become hazardous by sounding alarms.

The Complete SCADA solution

- Out of the box, Web Access provides a complete Supervisory Control and Data Acquisition system:
- Alarm Handling Package, including:
- Global Alarm Summary Page showing alarms from all tags on all SCADA nodes in the System,
- Alarm filtering, Sorting and Prioritizing,
- Alarm Logs to printer, local hard drive, Alarm Log Display and Central ODBC database,
- Each Tag has a High-High, High, Low, Low-Low, Deviation and Rate of Change Alarms,
- Text-to-Speech Alarm annunciation reads the Alarm, Value and Description (no pre-recordings!)
- Email, pager and voice notification of alarms are standard features in WebAccess not found in most SCADA software packages.
- Real-time and Historical Trending

- Data Logging and Reports
- Recipe Handler (to change hundreds of setpoints, and control setting with button)
- Scheduler used to schedule setpoint changes and equipment
- Area of Responsibility Security - where users security level changes for different "Areas" of the plant or facility

Key Features:

- Uses real-time graphical interface to develop industrial automation, instrumentation, and embedded systems
- Publishes real-time dynamic and animated graphic screens, trends, alarms, reports, and recipes to standard browsers
- Allows data exchange between wireless and mobile devices
- Provides online configuration, debugging, and remote application management capabilities
- Contains a powerful, flexible tags database with Boolean, Real, String, and Array tags, classes, and indirect pointers
- Provides the tools to configure applications in conformance with the FDA 21 CRF Part 11 regulation
- Advanced math library has more than 100 standard functions
- Programming is via flexible and easy-to-use scripting language
- Provides multi-level security for applications, including use over Intranets and Internet.
- Conforms to industry standards such as Microsoft DNA, OPC, DDE, ODBC, XML, and ActiveX
- Provides automatic language translation at runtime
- Allows internationalization using Unicode

What is tag?

A tag database consists of records called tags. (a tag is similar to a symbol in PLC). In the tag database you define the data you want RSVIEW 32 to monitor. You can organize tags into groups using folders. This speeds up database creation because you can duplicate a folder and its tags in a single operation. For example, if you have several similar machines that require the same tags in the database, you can create a folder called machin1 and define its tags.

Types of tags:

- Digital – stores simple on/off information
- Analog – stores numerical values
- String – stores alphanumeric characters

Different Types of Scripts Used In Intouch SCADA

Scripting language that is used in Wonderware Intouch is called QuickScript and allow to execute commands and logical operations based on specified logic being met. Total number of Scripts in Intouch are 8. And SCADA Application developer can use them at different level.

1. **Application Script** :- These Scripts are linked to the entire SCADA application.
2. **Window Script** :- These Scripts are linked to only specific window of Application or its only applicable to particular SCADA screen.
3. **Key Script** :- These Scripts linked to a specific key or key combination on the keyboard. So you can make any Combination from your keyboard to operate in SCADA.
4. **Touch Action Script** :- These Scripts are associated with an object linked to a Touch Link ,Touch Pushbutton, Action animation link etc.
5. **Data Change Script** :- These Scripts linked to a tag name or tag name field only.
6. **Condition Script** :- These Scripts are linked to a discrete tag name or expression.
7. **ActiveX Event Script** :- These Scripts are those that execute ActiveX control events in Runtime.
8. **Quick Functions Script** :- These Scripts are user made scripts and can be called from other InTouch QuickScripts or animation link expressions. QuickFunctions can be either synchronous or asynchronous, while all other script types are synchronous only.

Types of Trends in SCADA System

1.Real Time Trends

Real-time trends are dynamic i nature. They are updated continuously during run time and you can see the real time data of your Tags which you have configured. Real time trend help us in seeing actual data that is of the process. Like you have make a tag of Water Level in a Sewage Treatment plant and you want to see the the present value of Water level with respect to the time. Then you will use Real Time Trend which can be configured in seconds, Mili-seconds, minutes and hours.

2.Historical Trends

Historical trends as the name indicates are used to view past values of the Tag and you can configure them to see any past date Data. They can be configured in Days, Months, Years etc. You can view your tag value any time.

11.1 Distributed Control Systems (DCS)

What are Distributed Control Systems (DCS)

Various systems are introduced to automate the processes in the manufacturing industry and minimize the human interaction with the machines. These systems not only save the cost but also keep the injuries to minimum. Distributed processes are controlled by decentralized elements in a distributed control system or DCS.

Routine operations are carried out without the need of user intervention. There is an interface known as SCADA (Supervisory Control and Data Acquisition) which lets the user interact with the system. A DCS consists of a remote and a central control panel with a communication medium. Two different names are given to the remote control panels by different suppliers. The names are

1) Remote transmission Unit or RTU
2) Digital Communication Units or DCU

The functions of these remote units are same as they contain I/O modules and communication mediums and processors. These remote control units can be connected to the central control panel or SCADA with the help of a wireless or wired connection. The software used to read the I/O command is of specialized nature.

A detailed analysis of network protocols is required before the selection of DCS is finalized. The systems differ in terms of applications and complexity and the applications depend on the implementation of the system. A DCS with smaller implementation may only consist of a single Programmable Logic Controller or PLC. This controller will be connected to a computer in the remote office.

PLC is also an attribute of the large and complex DCS installations like in electrical grids and in power generation fields. They are also widely used in water treatment plants and in systems for environmental control. Petroleum refineries and petrochemical industry also uses these systems on a mass scale as these are intelligent systems and save all the process data necessary to continue the operations in case of a communication failure.

12 Chapter: HMI (Human Machine Interface)

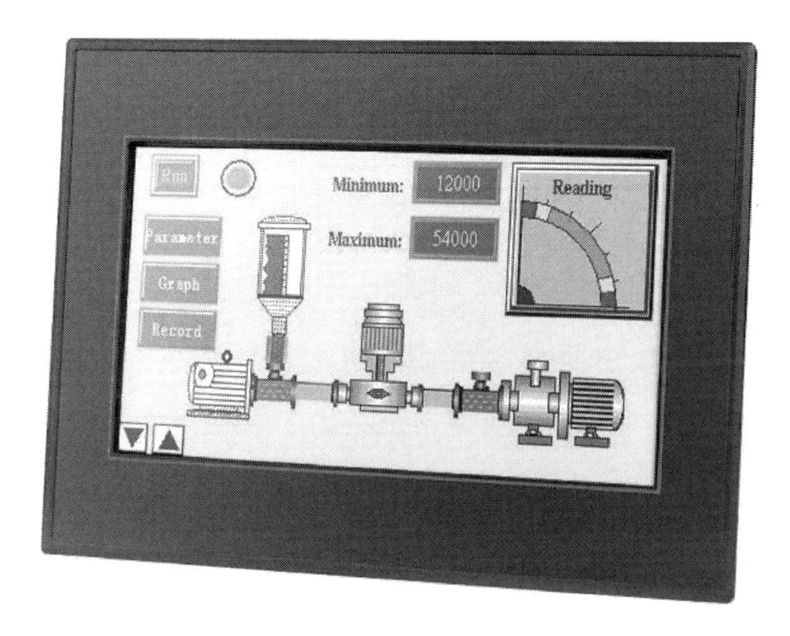

Human Machine Interface aimed at a better Human-machine interaction. Any automation system is said to be blind without HMI. HMI gives the ability to the operator, and the management to view the plant in real time. Add to that the ability to have alarm management that can warn the operator of a problem. It can even log and print all the alarms in real time, which can help the management to improve the production and efficiency.

Today there exists much Human Machine Interface software that could be used to monitor, supervise and control process. What we are presenting here is just an overview of what could be achieved with most of this software.

So what are the main functionalities of an HMI. Well the HMI's main functionality is to monitor, supervise, and control processes. This could be used in a variety of industries such as food processing, sawmills, bottling, semiconductors, oil and gas, automotive, chemical, pharmaceutical, pulp and paper, transportation, utilities, an more. HMI software provides the process knowledge and control needed to perfect the products companies make and the processes they manage. It is said that a control without an HMI is a blind control.

Animated pictures, more importantly it can also display System messages, reports, alarms, trends and manipulate string values and calculate Boolean operations and more complex math operations. This flexibility reduces the task that the PLC Human Machine Interface can display texts, pictures, bar graphs, bitmap and more and more manufacturing designers are recognizing the benefits of using Human Machine Interface to control and to operate their controls.

12.1 Typical Applications

- Machine monitoring and control
- Supervisory Control and Data Acquisition (SCADA)
- Control Center Monitoring, Tracking, and Control
- Building Automation and Security
- Electrical Substation Monitoring
- Pipeline Monitoring and Control
- Transportation Control Systems
- Batch Process Monitoring and Control
- Continuous Process Monitoring and Control
- Heating, Ventilation, and Air Conditioning
- Statistical Process Control (SPC)
- Telecommunications
- Discrete Manufacturing and more...

12.2 Functionality

- Representation of a plant in real time .
- Trending (Real-time / Historical)
- Alarms (Real-time / Historical)
- SPC (Statistical Process Control)
- Recipes
- Reports
- Lop Events
- Historical Data Logging and Browsing
- SQL Server 2000, Oracle, Sybase, ODBC support
- Networking and Redundancy
- Math and Logic
- Password protection and more...

12.3 Control Panel

Panel accessories

Panel accessories like - Voltmeter, Ampere meter, Frequency meter, Red Indicator, Yelow Indicator, Blue Indicator, Hinge, Lock, Knob, Contactor, Relay, Timer, Insulator, Busbar - Support, Selector Switch, Control Fuse, Push Button, Earth Fault Relay, Over Current Relay, CT, Electric Motor Starter, Power Factor Meter can be dragged from the object browser and placed upon the **panel template** as per the design requirement. The library contains various ready components to be used as switchgears & accessories. The designer can easily identify a component as switchgear or accessory.

MCC Panel (Motor Controller Center Panel)

MCC panel is designed to meet a specific requirement of controlling electric motor. All the working parameters of an electric motor are taken into account while designing MCC (Motor controller center). An automated or manual control are fitted onto the MCC panel to stop or start a motor, to regulate speed, to have reverse or forward alternation, to regulate the power and to prevent the electric motor from the hazards of overloads.

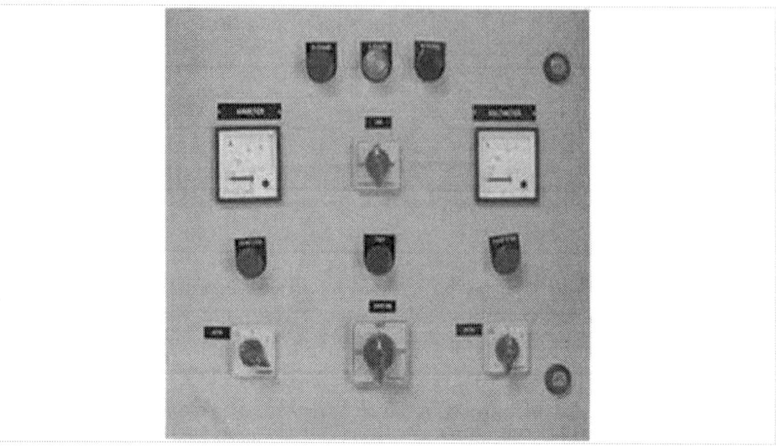

Figure: MCC Panel

We are instrumental in offering a wide range of MCC Panels that are highly demanded in the market for a variety of power distribution processes. Owing to their excellent performance and minimum maintenance, our products are widely demanded in various industries like chemical refineries, pharmaceuticals, textiles, construction, paper and others.

- Manufactured using high grade raw material that makes it sturdy and corrosion resistant
- MCC Panels are fabricated in standard as well as customized sizes meeting all the diversified needs of our clients.

PCC Panel (Power Control Centre Panel)

PCC panel is again a power distribution board to control the power supplied to HT motors, MCC panels and transformers who play vital role in any electrical control system. PCC panels are widely used in refineries, chemical plants, dairies, refrigeration plant, pharmaceuticals and plastic industry. Usually PCC panel is having modular construction with cable inlet is on either top or bottom of the structure

Figure: PCC Panel

The manufacture and distribute PCC panels which are used for attaining maximum and efficient electric power utilization. PCC panel is widely used in power distribution in various process and production industries.

VFD Panel (Variable Frequency Drive Panel)

VFD - Variable frequency drive panel is used to control the rotating speed of AC electric motor by controlling the intensity of electric power provided to the motor. The working of these VFD panels needs to go through speed variations so during their manufacturing they undergo tests on various parameters. The VFD panels are widely used in tube mills, paper mills, extruder plants, rolling Mills, cable industry and CTL Lines. They are even installed in hospitals and big business houses.

Figure: VFD Panel

The manufacturing and supplying a diverse range of VFD panels which is designed using advanced technology. These are available in various dimensions and standard sizes as per the clients' specifications. VFD panel in various specifications and are effectively used in various industrial applications.

PLC Panel (Programmable Logic Control Panel)

PLC panel are designed to control electrical power supply using programmable Control Logic Processor. The processor fitted into the panel structure is connected through wires with reed switches, push buttons and proximity detectors to its input cards. And the same processor is also connected to motor contactors, indication lights and control relays which are the output field devices through wires.

Figure: PLC Panel

PLC panels are commonly called automation panels and are widely used in industrial automation process, commissioning and processing unit. Small to big manufacturing units use these PLC panels to control the production line. PLC panels are also popular in automobile and electronics manufacturing units. PLC Panels with some salient features:

- PLC panels are designed for easy monitoring and controlling which makes process error free
- PLC panel is fitted with high capacity processor which gives reliable performance
- Both the input and output connectors used in the PLC panels are of premium quality

12.3.1 Control Wiring Panel

Control Wiring Panel is cables and wires fabricated to do the wiring of electrical panels and instruments. They are widely used in power distribution process where controlling and observing of circuits, electrical safety and measuring of power allocation is required. Variety of control wires are required to match the varied requirement of market. They are basically used in wiring purpose of small to big manufacturing unit when they opt for automation.

Figure: Control Wiring Panel

Features of Control Wiring Panel are:

Fabricated using quality raw material which makes control wire range superior in quality, performance and is highly durable.

- Control wires are made to stand any climatic condition when used externally.
- Available at reliable cost and in required dimension.

APFC Panel (Automatic Power Factor Correction Panel)

Electrical loads that are usually influential in nature could create heavy loss in case of power cut offs. Such installations include heavy machines, drivers, air cooling plants and motors installed in commercial and industrial units. Such kind of loss in power can create chaos if the production unit is not able to meet deadlines. They are bound to pay penalties in certain cases. To overcome such kind of power loss, APFC – automatic power factor correction panel can be installed to control capacitors used for maintaining high power factor in accordance with variation of load in production unit.

Figure: APFC Panel

Features of APFC panels are:

- Designed to make sturdy for longer life and less breakdowns
- They are economical with ability to maintain constant high power factor along with faster payback period.
- Cables can have inlet either from top or bottom of the panel
- Fitted with spares that are made using advanced technology to meet international industrial standards

AMF Panel (Auto Main Failure Panel)

AMF panel – Auto main failure panel as the name suggest is meant to support the working of a commercial or industrial unit in case of failure in the power supply. AMF panel is fitted with a sensing circuit capable to detect power failure. The AMF panel is connected with a required size generator set. As and when the panel detects a power cut off, the generator gets switched on automatically. The panel is fitted with automatic switch over isolators which change the power supply from mains to gen set. The power is supplied through generator set until the main power supply is resumed.

Figure: AMF Panel

AMF panels are installed for automatic changeover from mains to stand-by generator at the time of power failure. Easy to operate and install, these synchronized AMF panels are used in areas like foundries, apartments, textiles, sugar, chemical industries, hospitals, hotels and other such important businesses. Some of the salient features of AMF panels are:

- Fabricated from premium quality raw material, these automatic AMF panels are durable, corrosion resistant, dust and vermin proof.
- The continuous working of the AMF panel ensures detection of power failure and helps in reducing losses incurred because of such failures
- AMF panels supplied by us are easy to install and operate and is fabricated in desired size and dimensions

13 Chapter: Industrial Data Communications /Networking

13.1 Introduction

Data communication involves the transfer of information from one point to another. Many communication systems handle analog data; examples are telephone systems, radio and television. Modern instrumentation is almost wholly concerned with the transfer of digital data.

Any communications system requires a transmitter to send information, a receiver to accept it, and a link between the two. Types of link include copper wire, optical fiber, radio and microwave.

Digital data is sometimes transferred using a system that is primarily designed for analog communication. A modem, for example, works by using a digital data stream to modulate an analog signal that is sent over a telephone line. Another modem demodulates the signal to reproduce the original digital data at the receiving end. The word 'modem' is derived from modulator and demodulator.

There must be mutual agreement on how data is to be encoded, i.e. the receiver must be able to understand what the transmitter is sending. The structure in which devices communicate is known as a protocol.

13.2 Open Systems Interconnection (OSI) model

The OSI model, developed by the International Organization for Standardization, has gained widespread industry support. The OSI model reduces every design and communication problem into a number of layers as shown in Figure 7.1. A physical interface standard such as RS-232 would fit into the layer 1, while the other layers relate to the protocol software.

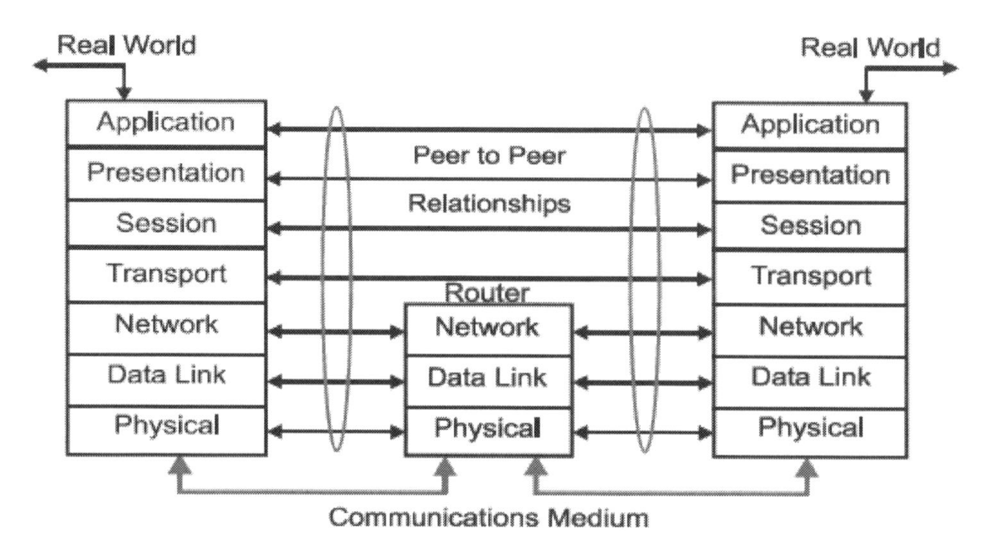

Figure: OSI model representation: two hosts interconnected via a router

The OSI model is useful in providing a universal framework for all communication systems. However, it does not define the actual protocol to be used at each layer. It is anticipated that groups of manufacturers in different areas of industry will collaborate to define software and hardware standards appropriate to their particular industry.

13.2.1 Protocols

As previously mentioned, the OSI model provides a framework within which a specific protocol may be defined. A protocol, in turn, defines a frame format that might be made up of various fields as follows.

Direction of Travel ←

Sync Byte	Destination Address	Source Address	Data to be Transmitted	Error Detection Byte

Figure: Basic structure of an information frame

13.3 RS-232 interface standard

The RS-232 interface standard (officially called TIA-232) is an asynchronous serial communication method. It defines the electrical and mechanical details of the interface between Data Terminal Equipment (DTE) and Data Communications Equipment (DCE), which employ serial binary data interchange.

Figure 7.3 illustrates the signal flows across a simple serial data communications link.

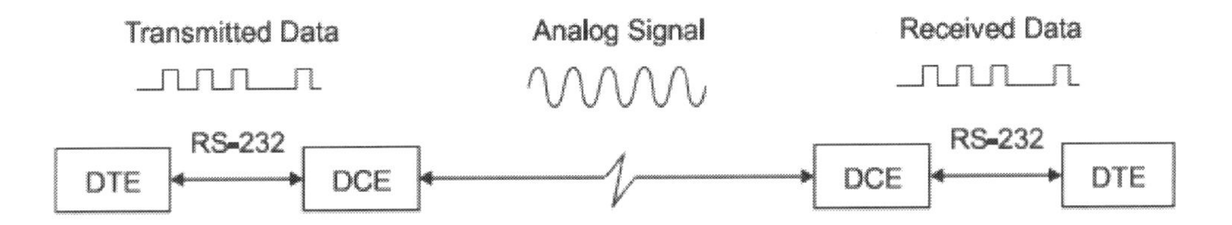

Figure: A typical serial data communications link

The RS-232 has :

- Separate communication line for transmitting and receiving
- Maximum cable length is **50 feet @ 19200 bps baud rate**
- Half duplex as well as full duplex operations

Baud Rate (bps)	Maximum Cable Length (ft)
19200	50
9600	500
4800	1000
2400	3000

13.4 RS 485

The RS485 standard (formally TIA-485) specifies differential data transmission over a terminated twisted pair at up to 10Mbps. The standard specifies electrical characteristics of a driver and receiver, and does not specify any data protocol or connectors. RS485 is popular for inexpensive local networks, multidrop communication links and long haul data transfer over distances of up to 4,000 feet(1200 m). The use of a balanced line means RS485 has excellent noise rejection and is ideal for industrial and commercial applications.

13.5 Data Highway Plus /DH485

Data Highway (DH) and Data Highway Plus (DH+) are industrial bus protocols developed by Rockwell Automation/Allen Bradley.

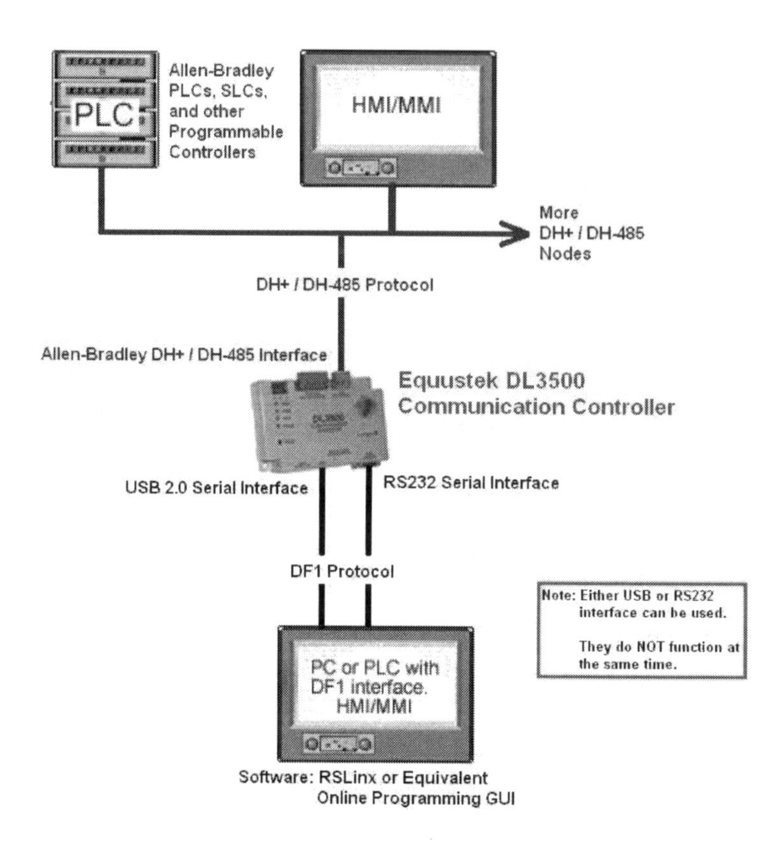

Figure: DH 485

The interconnection between the nodes is accomplished over the **DH or DH+** link. DH and DH+ allow **64 nodes** and the communications protocol used to interconnect the Network link and a PC is called **DF1**.

DH uses a trunk cable that runs up to 10,000 ft., and drop cables to each node that extend to 100 ft. This system uses peer-to-peer communication in which each node bids on being the the Floating Master.

DH+ is appropriate for smaller networks. It uses peer-to-peer communication implementing Token passing. Nominal voltages on the bus are 8 to 12 volts peer-to-peer; and the bus is +/- 200mV sensitive over the two differential lines. Each node on the bus is transformer coupled onto the bus. The bus should be terminated to 150 ohms at each end of the bus. All messages on the bus are either a command or a reply.

Also developed by Rockwell Automation/Allen Bradley is the Data Highway 485 protocol a local area network design for factory-floor applications. DH-485 allows for the connection of up to 32 devices, including controllers, color graphics systems and PCs.

13.6 Modbus

Developed by Modicon, MODBUS is a master/slave serial line communication protocol is used to interconnect intelligent automation devices. This protocol operates at layer 2 of the OSI model.

In a master-slave type system, the "master" node issues explicit commands to one of the "slave" nodes and processes responses. Typically, slave nodes will not transmit data without a request from the master node, and they will not communicate with other slaves.

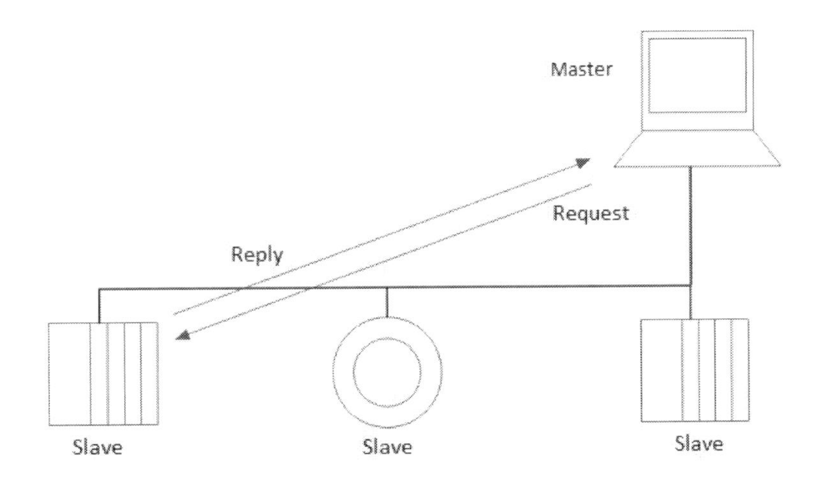

Figure: Basic Modbus Network

At the physical level, MODBUS systems may use different physical interfaces (EIA-485, RS-232). The TIA/EIA-485 (EIA-485) Two-Wire interface is, however, the most common. As an option, the EIA-485 Four-Wire interface may be implemented. And when the system only requires short point-to-point communication, the TIA/EIA-232-E (RS-232) serial interface may be used as an interface.

MODBUS application layer messaging protocol, positioned at layer 7 of the OSI model, provides client/server communication between devices connected on buses or networks. The client role is provided by the Master of the serial bus and the slave nodes act as servers.

Using a RS-232 physical interface, the maximum distance would be 50 ft. at 10kbps. Using a EIA-485 physical interface, the maximum distance would be 4,000 ft. at 100kbps. A maximum of 247 nodes are possible.

13.6.1 Modbus Plus

Modbus Plus was originally developed by Schneider and Modicon and is today managed by the Modbus-IDA user organization. It is a global Fieldbus network with its main application area in Europe and in automation systems from Schneider and Modicon. Modbus-Plus connectivity is available for many different products such as PLC's, Inverters, Drives and I/Os. Modbus Plus is a Master/Slave Fieldbus based on RS-485 transmission technology and a token passing protocol for industrial control applications. Networked devices can exchange messages for the control and monitoring of processes at remote locations in the industrial plant.

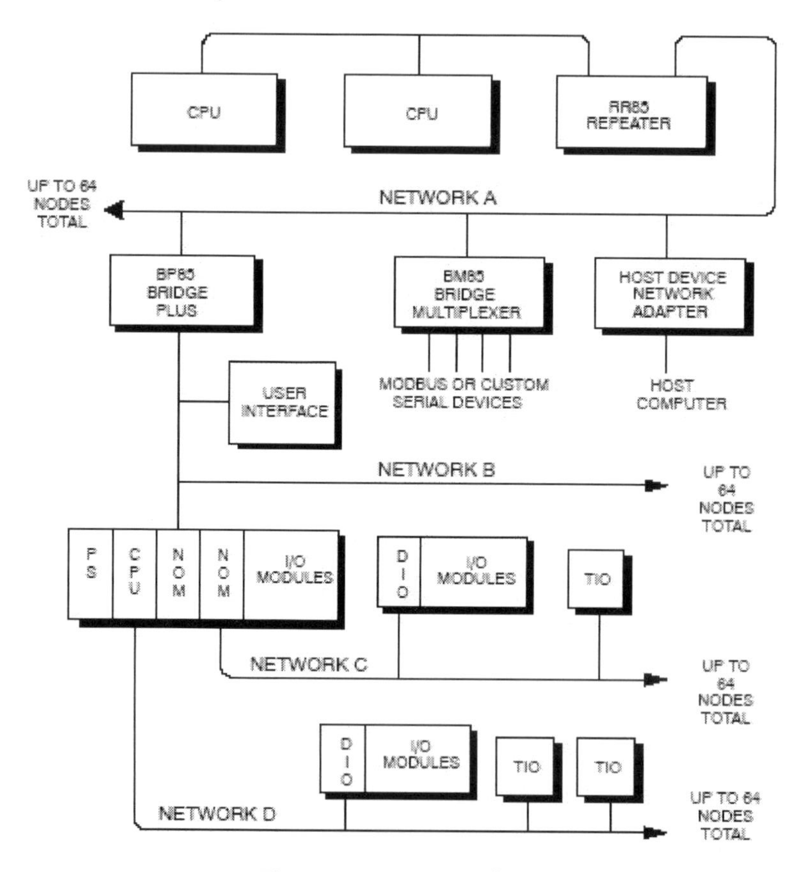

Figure: MODBUS Plus

The network also provides an efficient means for servicing input/output subsystems. Modicon Modbus Plus Distributed I/O (DIO) Drop Adapters and

Terminal Block I/O (TIO) modules can be placed at remote I/O sites to allow the application to control field devices over the network link.

Network Type:	Master/Slave Fieldbus based on RS-485 with Token passing
Topology:	Line topology with segments of up to 32 stations
Installation:	Shielded twisted pair cable with 9-pole D-Sub connector. Cable length per Segment up to 500m extendible with repeaters up to 2.000m.
Speed :	2 Mbit/s
max. Stations :	64
Data:	cyclic I/O and acyclic parameter data
Network Features :	Master/Slave Fieldbus network for Real-Time control applications.

13.7 HART

HART (Highway Addressable Remote Transducer) Protocol is the global standard for sending and receiving digital information across analog wires between smart devices and control or monitoring system.

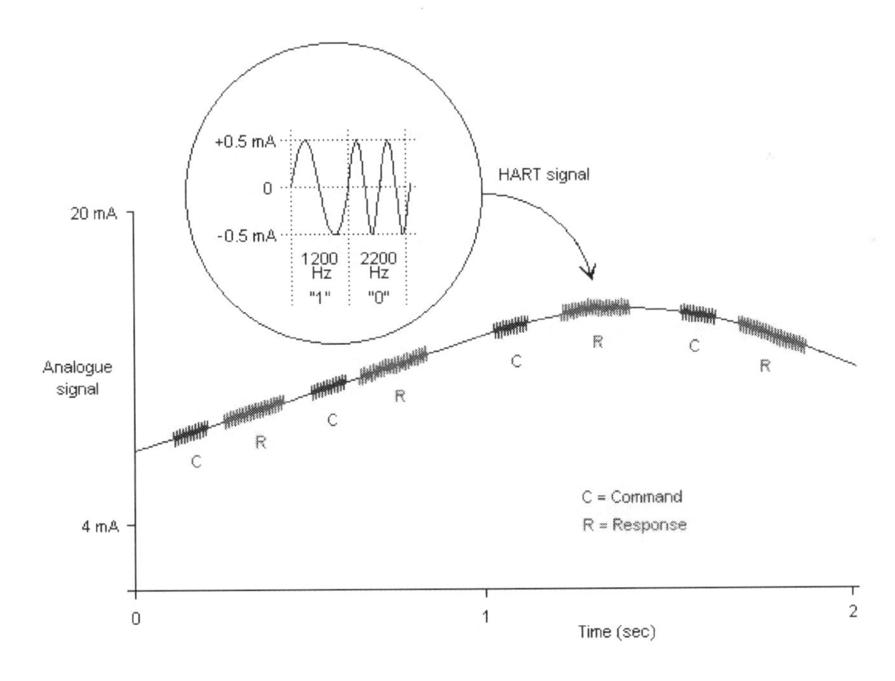

Figure: HART

More specifically, HART is a bi-directional communication protocol that provides data access between intelligent field instruments and host systems. A host can be any software application from technician's hand-held device or laptop to a plant's process control, asset management, safety or other system using any control platform.

The HART system (and its associated protocol) was originally developed by Rosemount and is regarded as an open standard, available to all manufacturers. Because most automation networks in operation today are based on traditional 4-20mA analog wiring, HART technology serves a critical role because the digital information is simultaneously communicated with the 4-20mA signal. HART is a hybrid analog and digital system, as opposed to most field bus systems, that are purely digital. It uses a Frequency Shift Keying (FSK) technique based on the Bell 202 standard. Two individual frequencies of 1200 and 2200 Hz, representing digits '1' and '0' respectively, are used. The average value of the 1200/2400Hz sine wave superimposed on the 4-20mA signal is zero; hence, the 4- 20mA analog information is not affected.

13.8 AS-i

AS-Interface (Actuator Sensor Interface, AS-i) is an industrial networking solution (physical layer, data access method and protocol) used in PLC, DCS and PC-based automation systems. It is designed for connecting simple field I/O devices (e.g. binary ON/OFF devices such as actuators, sensors, rotary encoders, analog inputs and outputs, push buttons, and valve position sensors) in discrete manufacturing and process applications using a single 2-conductor cable.

AS-Interface is an 'open' technology supported by a multitude of automation equipment vendors. According to AS-Interface International there are currently, over 18 Million AS-i field devices are installed globally, growing at about 2 million per year.

AS-Interface is a networking alternative to the hard wiring of field devices. It can be used as a partner network for higher level fieldbus networks such as Profibus, DeviceNet, Interbus and Industrial Ethernet, for whom it offers a low-cost remote I/O solution. It is used in automation applications, including conveyor control, packaging machines, process control valves, bottling plants, electrical distribution systems, airport carousels, elevators, bottling lines and food production lines.

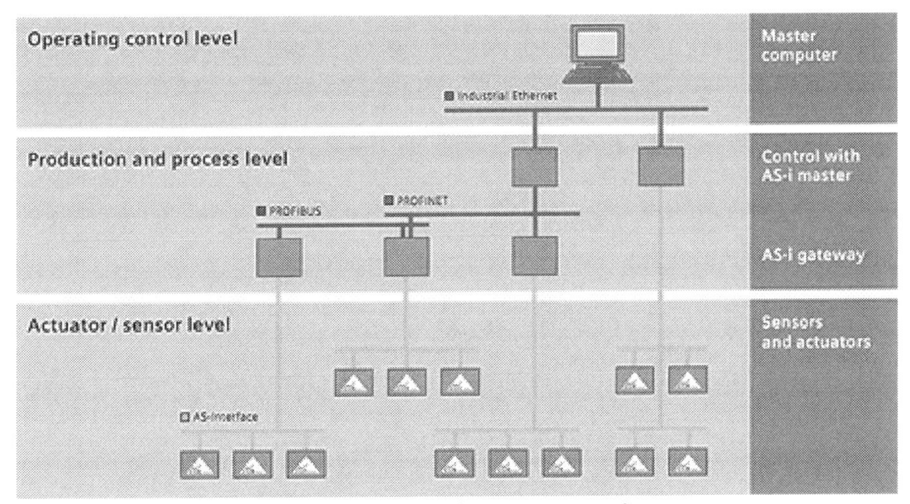

Figure: Typical AS-i network

13.9 DeviceNet

DeviceNet, developed by Allen Bradley, is a low-level device oriented network based on CAN (Controller Area Network) developed by Bosch (GmbH) for the automobile industry. It is designed to interconnect lower level devices (sensors and actuators) with higher level devices (controllers). DeviceNet is classified as a field bus, per specification IEC-62026.

The variable, multi-byte format of the CAN message frame is well suited to this task as more information can be communicated per message than with bit-type systems. The DeviceNet specification is an open specification and available through the ODVA.

DeviceNet can support up to 64 nodes, which can be removed individually under power and without severing the trunk line. A single, four-conductor cable (round or flat) provides both power and data communications. It supports a bus (trunk line drop line) topology, with branching allowed on the drops. Reverse wiring protection is built into all nodes, protecting them against damage in the case of inadvertent wiring errors. The data rates supported are 125, 250 and 500K baud (i.e. bits per second in this case).

Figure7.12 below illustrates the positioning of DeviceNet and CANBUS within the OSI model. CANBUS represents the bottom two layers in the lower middle column, just below DeviceNet Transport. Unlike most other field buses, DeviceNet does implement layers 3 and 4, which makes it a routable system. There are two other products in the same family; Control Net and Ethernet/IP. They share the same upper layer protocols (implemented by CIP, the Control and Information Protocol) and only differ in the lower four layers.

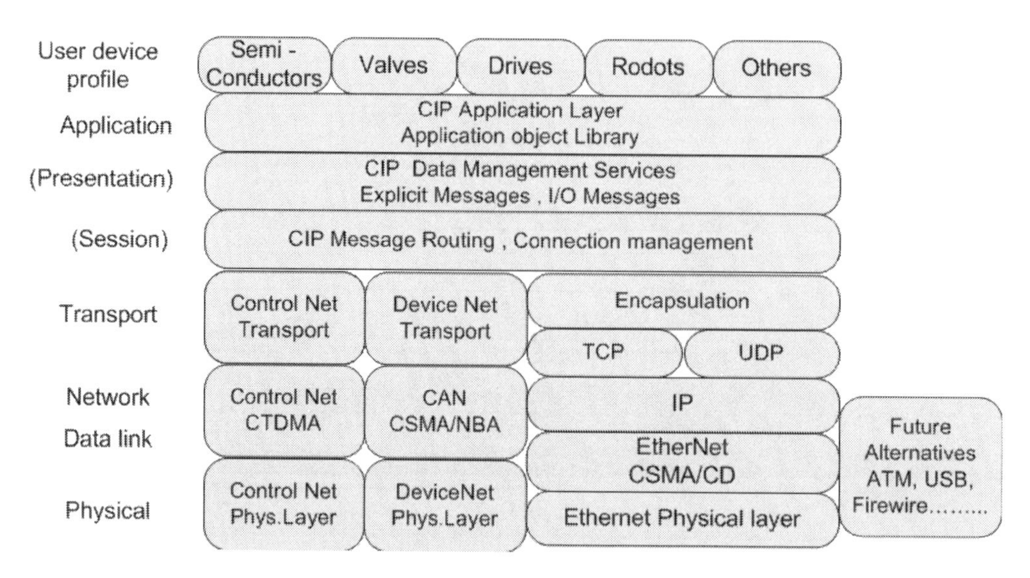

Figure: Devicenet (as well as ControlNet and Ethernet/IP) vs. the OSI model

13.10 Profibus

Profibus is the leading industrial communication system for manufacturing automation in Europe with strong growth in many other markets. Profibus is supported by Siemens and is promoted by the Profibus User Organization. Profibus products are certified by the Profibus User Organization (PNO), guaranteeing worldwide compatibility.

Profibus was originally defined 15 years ago under the German standard DIN 19245 and is today part of the international standard series IEC 61158. Based on the initial functionality, many new features have been added and today, Profibus consists of a family of 3 protocol variations (DP, DPV1, DPV2) that can be used with different physical transmission media.

ProfiBus uses 9-Pin D-type connectors (impedance terminated) or 12mm round (M12-style) quick-disconnect connectors. The number of nodes is limited to 127.

In addition to the protocol definitions, Profibus defines many Application Profiles for specific device types and industries i.e. ProfiDrive for Drives, Profibus-PA for Process Automation and ProfiSafe for Safety Applications. Profibus features real time transmission capabilities, flexible network topologies, advanced data access procedures, good diagnostic concepts and proven configuration mechanisms. ProfiBus supports two main types of devices, namely, masters and slaves.

- Master devices control the bus and when they have the right to access the bus, they may transfer messages without any remote request. These are referred to as active stations
- Slave devices are typically peripheral devices i.e. transmitters/sensors and actuators. They may only acknowledge received messages or, at the request of a master, transmit messages to that master. These are also referred to as passive stations.

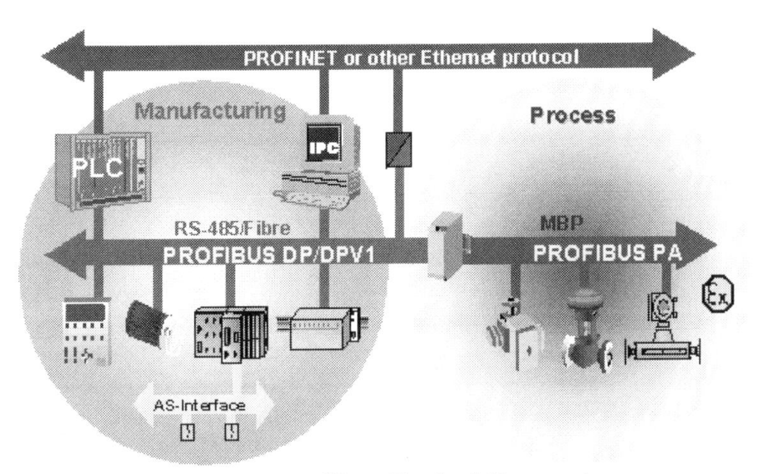

Figure: Profibus Typical Network

13.11 Foundation Fieldbus

FOUNDATION Fieldbus is an all-digital, serial, two-way communications system that serves as the base-level network in a plant or factory automation environment. It is an open architecture, developed and administered by the Fieldbus Foundation.

FOUNDATION fieldbus was originally intended as a replacement for the 4-20 mA standard, and today it coexists alongside other technologies such as Modbus, Profibus, and Industrial Ethernet. FOUNDATION fieldbus today enjoys a growing installed base in many heavy process applications such as refining, petrochemicals, power generation, and even food and beverage, pharmaceuticals, and nuclear applications. FOUNDATION fieldbus was developed over a period of many years by the International Society of Automation, or ISA, as SP50.

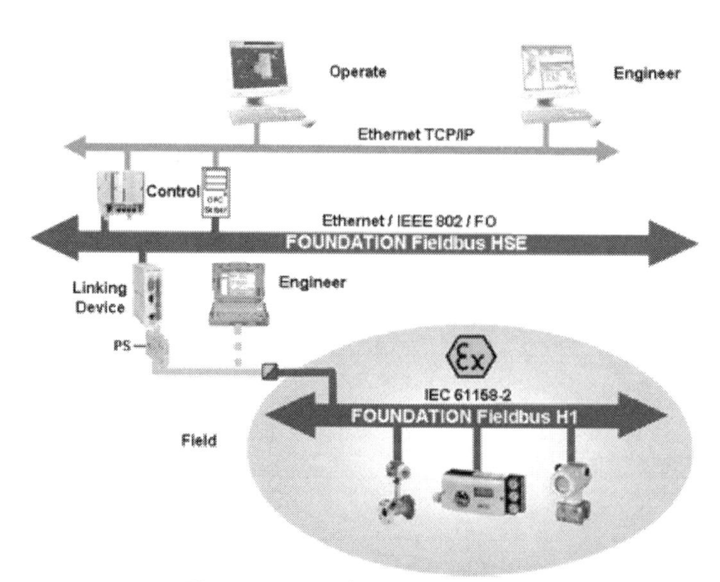

Figure: Foundation Field Bus

Fieldbus is an all-digital, serial, two-way communications protocol that standardizes the interconnection of field devices at a communications rate of 31.25 kB/s. The field devices are made accessible to communication networks (running at 1 or 2.5 Mb/s), through use of linking devices. Because a simple twisted pair of wires can carry the signals between the field devices and the networks, Fieldbus users can reduced the amount of I/O and control equipment (due to distribution of control into the field devices), plus they can use fewer Intrinsically Safe (IS) barriers, cabinets, cables and connectors.

The number of devices possible on a fieldbus link will vary depending on factors such as the power consumption of each device, the use of repeaters, signal attenuation, type of cable used and distance.

The table below provides suggested transmission distances (cable lengths) for each type of fieldbus cable.

Fieldbus Type	Communication Rate	Transmission Distance Ft. / m
A	31.25 Kb/s	6,232 / 1900
B	31.25 Kb/s	3,936 / 1200
High Speed	1.0 Mhz 2.5 Mhz	2,460 / 750 1,640 / 500

13.12 Industrial Ethernet

Industrial Ethernet applies the Ethernet and IP suite of standards developed for data communication to manufacturing control networks. Industrial Ethernet networks that use intelligent switching technology can offer a variety of advantages compared to traditional industrial networks. The technology can be deployed using a switched Ethernet architecture and has proven successful in multiple critical applications in different markets. Because the technology is based on industry standards, Industrial Ethernet enables organizations to save money by moving away from expensive, closed, factory floor- optimized networks.

Early Ethernet was not entirely suitable for control functions as it was primarily developed for office-type environments. Ethernet technology has, however, made rapid advances over the past few years. It has gained such widespread acceptance in Industry that it is becoming the de facto field bus technology for OSI layers 1 and 2. An indication of this trend is the inclusion of Ethernet as the level 1 and 2 infrastructure for Modbus/TCP (Schneider), Ethernet/IP (Rockwell Automation and ODVA), ProfiNet (Profibus) and Foundation Fieldbus HSE.

13.12.1 Ethernet

Ethernet is by far the most widely used LAN technology today, connecting more than 85 percent of the world's LAN connected PCs and workstations. Ethernet refers to the family of computer networking technologies covered by the IEEE 802.3 standard, and can run over both optical fiber and twisted-pair cables. Over the years, Ethernet has steadily evolved to provide additional performance and network intelligence. This continual improvement has made Ethernet an excellent solution for industrial applications. Today, the technology can provide four data rates.

- 10BASE-T Ethernet delivers performance of up to 10 Mbps over twisted-pair copper cable.
- Fast Ethernet delivers a speed increase of 10 times the 10BASE-T Ethernet specification (100 Mbps) while retaining many of Ethernet's technical specifications. These similarities enable organizations to use 10BASE-T applications and network management tools on Fast Ethernet networks.
- Gigabit Ethernet extends the Ethernet protocol even further, increasing speed tenfold over Fast Ethernet to 1000 Mbps, or 1 Gbps. Because it is based upon the current Ethernet standard and compatible with the installed base of Ethernet and Fast Ethernet switches and routers, network managers can support Gigabit Ethernet without needing to retrain or learn a new technology.
- 10 Gigabit Ethernet, ratified as a standard in June 2002, is an even faster version of Ethernet. Because 10 Gigabit Ethernet is a type of Ethernet, it can support intelligent Ethernet-based network services, interoperate with

existing architectures, and minimize users' learning curves. Its high data rate of 10 Gbps makes it a satisfactory solution to deliver high bandwidth in WANs and metropolitan-area networks (MANs).

13.12.2 Ethernet History

Although Xerox's Bob Metcalfe sketched the original Ethernet concept on a napkin in 1973, its inspiration came even earlier. ALOHAnet was a wireless data network created to connect together several widely separated computer systems on Hawaiian college campuses (different islands). The challenge was to enable several independent data radio nodes to communicate on a peer-to-peer basis without interfering with each other. ALOHAnet's solution was a version of the carrier sense, multiple access with collision detection (CSMA/CD) concept. Metcalfe based his Ph.D. work on finding improvements to ALOHAnet. This led to his work on Ethernet.

Ethernet, which later became the basis for the IEEE 802.3 network standard, specifies physical and data link layers of network functionality. The physical layer specifies the types of electrical signals, signaling speeds, media and connector types and network topologies. The data link layer specifies how communications occurs over the media— using the CSMA/CD technique mentioned above—as well as the frame structure of messages transmitted and received.

13.12.3 Industrial Ethernet

Recognizing that Ethernet is the leading networking solution, many industry organizations are porting the traditional fieldbus architectures to Industrial Ethernet. Industrial Ethernet applies the Ethernet standards developed for data communication to manufacturing control networks (Figure below). Using IEEE standards-based equipment, organizations can migrate all or part of their factory operations to an Ethernet environment at the pace they wish.

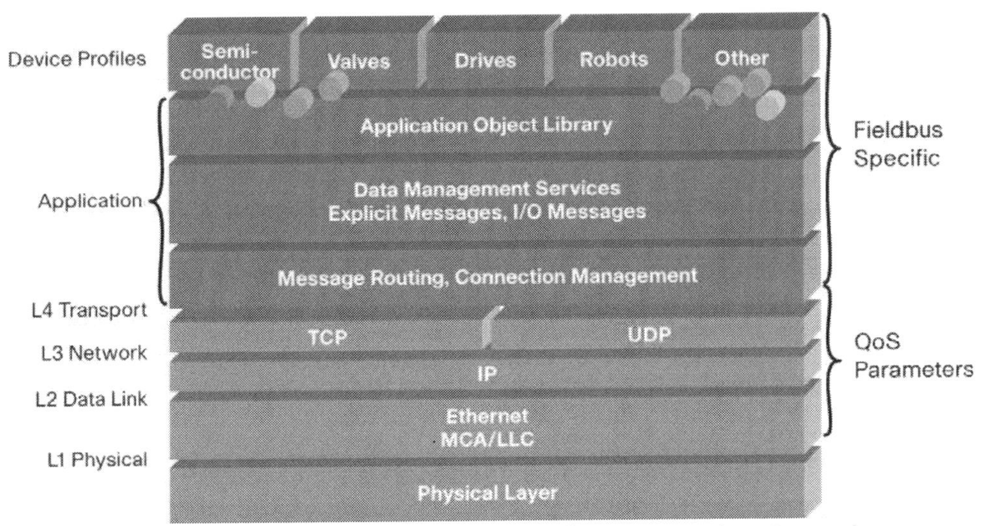

Figure . Using Intelligent Ethernet for Automation Control

For example, Common Industrial Protocol (CIP) has implementations based upon Ethernet and the IP protocol suite (EtherNet/IP), DeviceNet, and ControlNet (among others). Most controllers (with appropriate network connections) can transfer data from one network type to the other, leveraging existing installations, yet taking advantage of Ethernet. The fieldbus data structure is applied to Layers 5, 6, and 7 of the OSI reference model over Ethernet, IP, and TCP/UDP in the transport layer (Layer 4).

The advantage of Industrial Ethernet is that organizations and devices can continue using their traditional tools and applications running over a much more efficient networking infrastructure.

Industrial Ethernet not only gives manufacturing devices a much faster way to communicate, but also gives the users better connectivity and transparency, enabling users to connect to the devices they want without requiring separate gateways.

13.12.4 Ethernet Devices

As mentioned earlier, Ethernet cables are limited in their reach, and these distances (as short as 100 meters) are insufficient to cover medium-sized and large network installations. A repeater in Ethernet networking is a device that allows multiple cables to be joined and greater distances to be spanned. A bridge device can join an Ethernet to another network of a different type, such as a wireless network.

One popular type of repeater device is an **Ethernet hub**. Other devices sometimes confused with hubs are **switches** and **routers**.

Hubs operate at ISO layer 1 - physical layer, Switches operates at ISO layer 2 - data link layer, and Routers operate at ISO layer 3 - network layer.

Hub: A common connection point for devices in a network. Hubs are commonly used to connect segments of a LAN. A hub contains multiple ports. When a packet arrives at one port, it is copied to the other ports so that all segments of the LAN can see all packets.

In a hub, a frame is passed along or "broadcast" to every one of its ports. It doesn't matter that the frame is only destined for one port. The hub has no way of distinguishing which port a frame should be sent to. Passing it along to every port ensures that it will reach its intended destination. This places a lot of traffic on the network and can lead to poor network response times.

Additionally, a 10/100Mbps hub must share its bandwidth with each and every one of its ports. So when only one PC is broadcasting, it will have access to the maximum available bandwidth. If, however, multiple PCs are broadcasting, then that bandwidth will need to be divided between all of those systems, which will degrade performance.

Switch: In networks, a device that filters and forwards packets between LAN segments. Switches operate at the data link layer (layer 2) and sometimes the network layer (layer 3) of the OSI Reference Model and therefore support any packet protocol. LANs that use switches to join segments are called switched LANs or, in the case of Ethernet networks, switched Ethernet LANs.

A switch, on the other hand, keeps a record of the MAC addresses of all the devices connected to it. With this information, a switch can identify which system is sitting on which port. So when a frame is received, it knows exactly which port to send it to, which significantly increasing network response times. And, unlike a Hub, a 10/100Mbps switch will allocate a full 10/100Mbps to each of its ports. So regardless of the number of PCs transmitting, users will always have access to the maximum amount of bandwidth. It's for these reasons why a switch is considered to be a much better choice then a hub.

Router:A device that forwards data packets along networks. A router is connected to at least two networks, commonly two LANs or WANs or a LAN and its ISP.s network. Routers are located at gateways, the places where two or more networks connect. Routers use headers and forwarding tables to determine the best path for forwarding the packets, and they use protocols such as ICMP to communicate with each other and configure the best route between any two hosts.

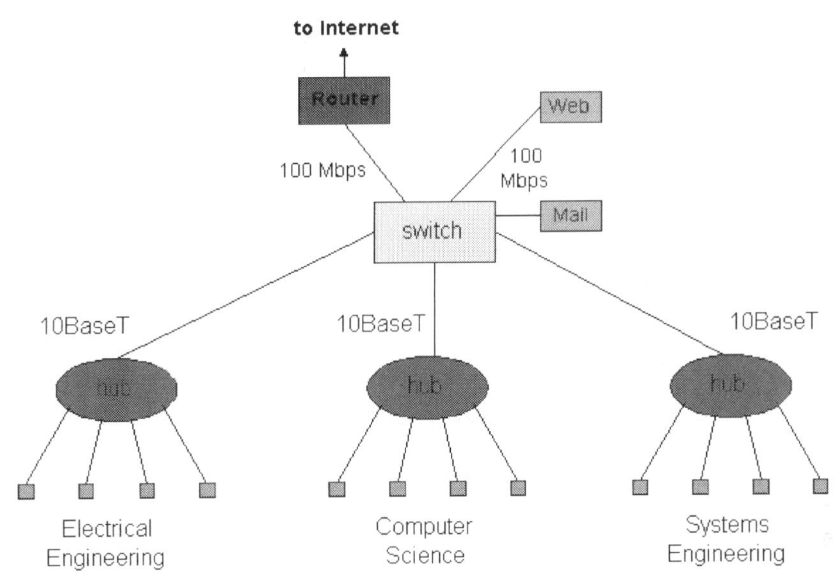

Figure: Ethernet Devices

Above Figure shows how an institution with several departments and several critical servers might deploy a combination of hubs, Ethernet switches and routers. In Figure , each of the three departments has its own 10 Mbps Ethernet segment with its own hub. Because each departmental hub has a connection to the switch, all intra-departmental traffic is confined to the Ethernet segment of the department (assuming the routing tables in the Ethernet switch are complete). The Web and mail servers each have dedicated 100 Mbps access to the switch. Finally, a router, leading to the Internet, has dedicated 100 Mbps access to the switch. Note that this switch has at least three 10 Mbps interfaces and three100 Mbps interfaces.

13.13 TCP/IP

TCP/IP is the de facto global standard for the Internet (network) and host–to–host (transport) layer implementation of internet work applications because of the popularity of the Internet. The Internet (known as ARPANet in its early years), was part of a military project commissioned by the Advanced Research Projects Agency (ARPA), later known as the Defense Advanced Research Agency or DARPA. The communications model used to construct the system is known as the ARPA model.

Whereas the OSI model was developed in Europe by the International Standards Organization (ISO), the ARPA model (also known as the DoD model) was developed in the USA by ARPA. Although they were developed by different bodies and at different points in time, both serve as models for a communications infrastructure and hence provide 'abstractions' of the same reality. The remarkable degree of similarity is therefore not surprising.

Whereas the OSI model has 7 layers, the ARPA model has 4 layers. The OSI layers map onto the ARPA model as follows.

- The OSI session, presentation and applications layers are contained in the ARPA process and application layer.
- The OSI transport layer maps onto the ARPA host–to–host layer (sometimes referred to as the service layer).
- The OSI network layer maps onto the ARPA Internet layer.
- The OSI physical and data link layers map onto the ARPA network interface layer.

The relationship between the two models is depicted in Figure 10.13.

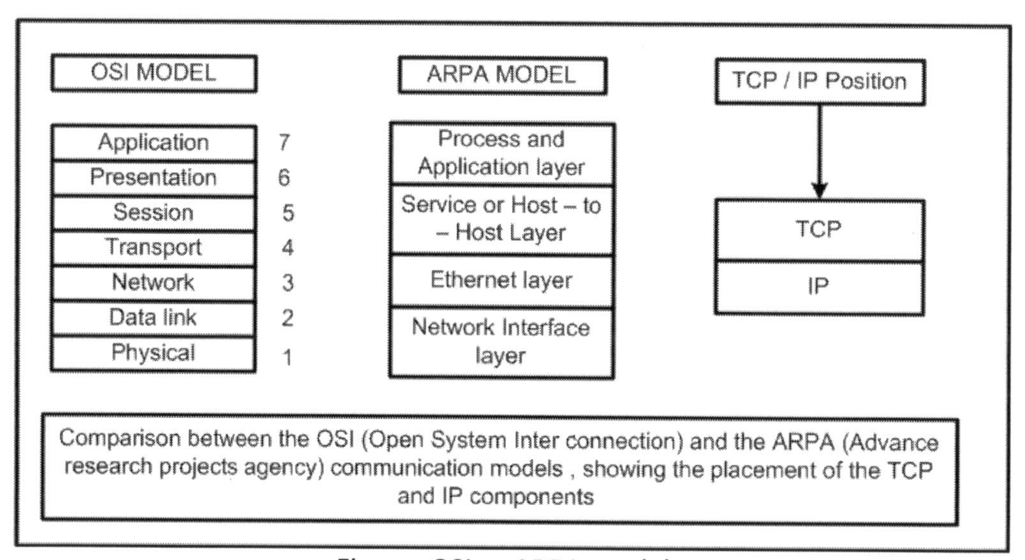

Figure: OSI vs ARPA models

TCP/IP, or rather the TCP/IP protocol suite is not limited to the TCP and IP protocols, but consists of a multitude of interrelated protocols that occupy the upper three layers of the ARPA model. TCP/IP does NOT include the bottom network interface layer, but depends on it for access to the medium.

As depicted in Figure 10.14, an Internet transmission frame originating on a specific host (computer) would contain the local network (for example, Ethernet) header and trailer applicable to that host. As the message proceeds along the Internet, this header and trailer could be replaced depending on the type of network on which the packet finds itself - be that X.25, frame relay or ATM. The IP datagram itself would remain untouched, unless it has to be fragmented and reassembled along the way.

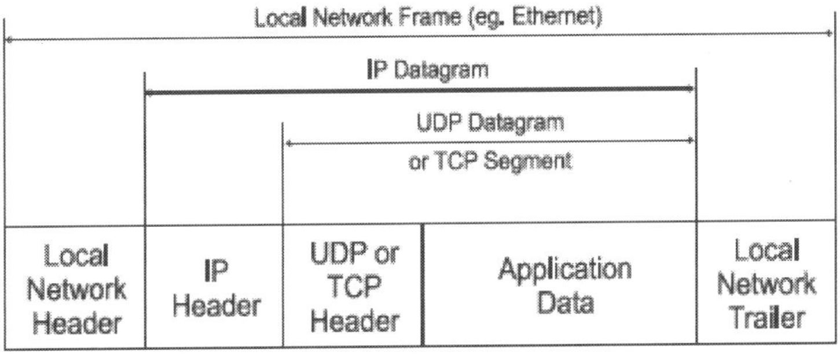

Figure: Internet frame

The Internet layer: This layer is primarily responsible for the routing of packets from one host to another.

The host–to–host layer: This layer is primarily responsible for data integrity between the sender host and receiver host regardless of the path or distance used to convey the message.

The process/application layer: This layer provides the user or application programs with interfaces to the TCP/IP stack.

Internet layer protocols (packet transport): Protocols like internet protocol (IP), the internet control message protocol (ICMP) and the address resolution protocol (ARP) are responsible for the delivery of packets (datagrams) between hosts.

Routing: Unlike the host–to–host layer protocols (for example, TCP), which control end–to–end communications, IP is rather 'shortsighted.' Any given IP node (host or router) is only concerned with routing (switching) the datagram to the next node, where the process is repeated.

13.13.1 Internet Protocol Version 4 (IPv4)

Internet Protocol is one of the major protocols in the TCP/IP protocols suite. This protocol works at the network layer of the OSI model and at the Internet layer of the TCP/IP model. Thus this protocol has the responsibility of identifying hosts based upon their logical addresses and to route data among them over the underlying network.

IP provides a mechanism to uniquely identify hosts by an IP addressing scheme. IP uses best effort delivery, i.e. it does not guarantee that packets would be delivered to the destined host, but it will do its best to reach the destination. Internet Protocol version 4 uses 32-bit logical address. IPv4 provides an addressing capability of approximately 4.3 billion addresses.

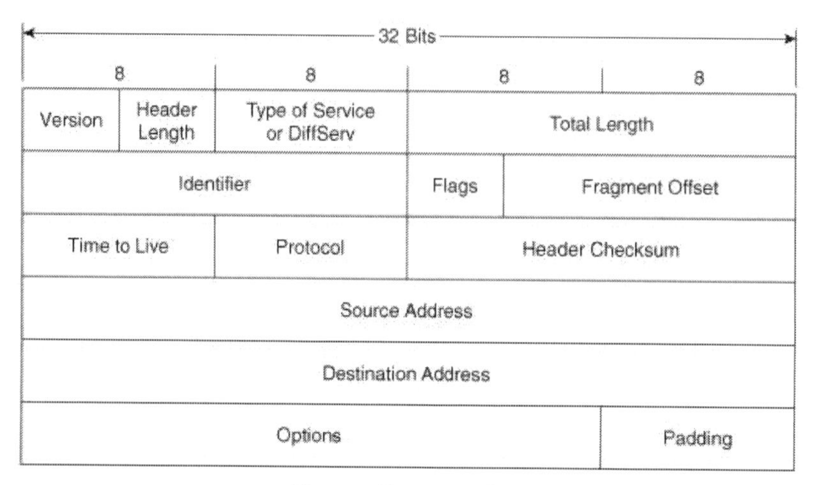

Figure: IPv4 Header

IP header includes many relevant information including Version Number, which, in this context, is 4. Other details are as follows:

- **Version:** Version no. of Internet Protocol used (e.g. IPv4).
- **IHL:** Internet Header Length; Length of entire IP header.
- **DSCP:** Differentiated Services Code Point; this is Type of Service.
- **ECN:** Explicit Congestion Notification; It carries information about the congestion seen in the route.
- **Total Length:** Length of entire IP Packet (including IP header and IP Payload).
- **Identification:** If IP packet is fragmented during the transmission, all the fragments contain same identification number. to identify original IP packet they belong to.
- **Flags:** As required by the network resources, if IP Packet is too large to handle, these 'flags' tells if they can be fragmented or not. In this 3-bit flag, the MSB is always set to '0'.
- **Fragment Offset:** This offset tells the exact position of the fragment in the original IP Packet.
- **Time to Live:** To avoid looping in the network, every packet is sent with some TTL value set, which tells the network how many routers (hops) this packet can cross. At each hop, its value is decremented by one and when the value reaches zero, the packet is discarded.

- **Protocol:** Tells the Network layer at the destination host, to which Protocol this packet belongs to, i.e. the next level Protocol. For example protocol number of ICMP is 1, TCP is 6 and UDP is 17.
- **Header Checksum:** This field is used to keep checksum value of entire header which is then used to check if the packet is received error-free.
- **Source Address:** 32-bit address of the Sender (or source) of the packet.
- **Destination Address:** 32-bit address of the Receiver (or destination) of the packet.
- **Options:** This is optional field, which is used if the value of IHL is greater than 5. These options may contain values for options such as Security, Record Route, Time Stamp, etc.

13.13.2 Internet Protocol Version 6 (IPv6)

IPv6 is the sixth revision to the Internet Protocol and the successor to IPv4. It functions similarly to IPv4 in that it provides the unique, numerical IP addresses necessary for Internet-enabled devices to communicate. However, it does sport one major difference: it utilizes 128-bit addresses.

The Internet Protocol version 6 (IPv6) is more advanced and has better features compared to IPv4. It has the capability to provide an infinite number of addresses. It is replacing IPv4 to accommodate the growing number of networks worldwide and help solve the IP address exhaustion problem.

One of the differences between IPv4 and IPv6 is the appearance of the IP addresses. IPv4 uses four 1 byte decimal numbers, separated by a dot (i.e. 192.168.1.1), while IPv6 uses hexadecimal numbers that are separated by colons (i.e. fe80::d4a8:6435:d2d8:d9f3b11).

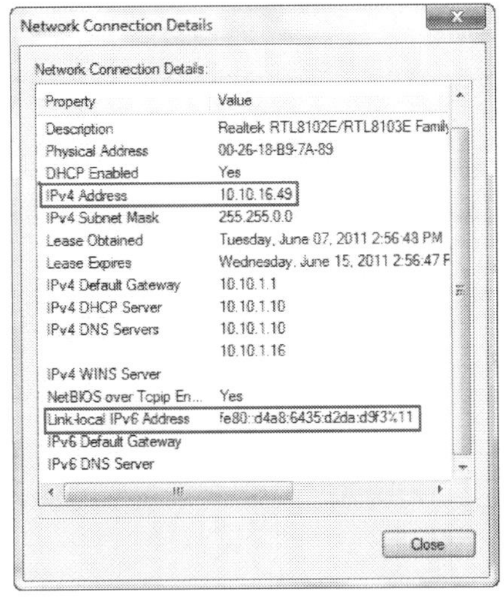

Figure: IPv4 and IPv6 in Windows OS Network Connection details

Below is the summary of the differences between the IPv4 and IPv6:

	IPv4	IPv6
No. of bits on IP Address	32	128
Format	decimal	hexadecimal

14 Chapter: Project Management

Project management is the transformation of an identifiable business need into a beneficial solution, typically a physical asset. When this transformation features effective and efficient use of modern processes, tools, techniques, skills and experience, it can be called "up-to-date" project management.

Topics addressed in this chapter include a summary of up-to-date project management methodology, expectations of today's customers and employees, and handicaps existing within organizations and their infrastructure. Issues surrounding planning, estimating, control, investment in a project's "social capital", and required project management skills are briefly reviewed. Ways to mitigate the results of project over-planning and over-control are also discussed. For a more complete picture of project management, refer to materials available from the Project Management Institute (PMI) and the Construction Industry Institute (CII).

Project planning and control versus. people is identified as one of the most critical factors in successful project execution. It has been routinely said that people, not systems, are the key to a successful project. This paper addresses some practical "how to's" of this statement as applied to Instrumentation projects.

Project management requirements start with the identification of business opportunities and drivers, and evolve into project planning and execution. Up-to-date project methodology should reflect specific, company-tailored, industry-recognized standards, and should take the form of a project management process.

Figure 1 shows a typical five-phase project management process with objectives identified for each phase .

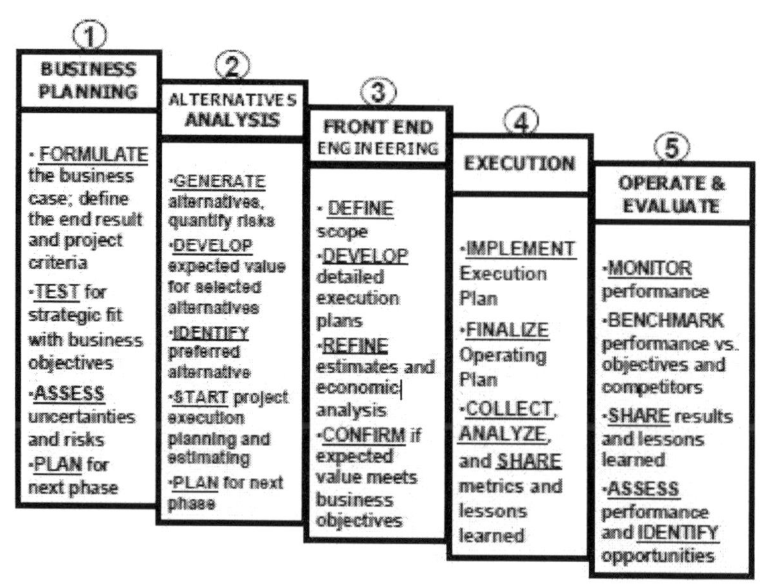

Figure: Project Management Process Phases & Their Objectives

In the first phase, Business Planning, the business need is identified. This is the most important aspect of any project. This well-defined and documented set of business goals or targets is called the business case, business statement or project charter. Armed with the business case (the project's bible), the project manager selects the project team and sets the roles and responsibilities of all participants.

In the second phase, called Alternative Analysis, the project manager (supported by the project team) evaluates various capital and non-capital alternatives, and selects the one that best meets the business needs. Project execution, planning and estimating should begin as early as possible in this phase, especially if the key business driver is speed or time to market. The preliminary project execution plan should include a degree of project control that matches the project's planning.

In this phase, Instrumentation project requirements must be established and should include system requirements for all users (stakeholders) as well as Instrumentation project constraints (ref.2). These requirements, organized and approved by the users, are sometimes called User Requirement Specifications and will be updated in all further phases, starting with the front-end engineering.

In the third or Front-End Engineering phase, the project scope is fully defined, and a detailed Execution Plan is developed. This plan becomes a baseline for project execution and costs against which all project decisions will be evaluated. The project execution plan must address all aspects of the project including "who", "when", "where" and "how much". The plan has to be communicated to the project's personnel via dated paper or electronic copies through the project website. This plan forms part of the project authorization request that goes to management for approval.

Instrumentation front-end engineering should be based on the approved I&C project requirements, where stakeholders decide what they need and expect. It culminates with full functional specifications and revised, re-approved I&C project requirements. Management must approve the project's costs and schedules and commit their support to the execution strategy. Management approval places the project team in a new mode, namely execution-and-control with no deviation from the plan. At this point, only critical safety and operability changes should be accepted.

In the fourth phase, Execution, the Execution Plan is implemented, and the Operating Plan is finalized. Project control supports the project management effort by evaluating collected actual data and relating it to the plan.

These findings, along with trends and forecasts, are reported to the project manager who takes corrective action, if required. In this phase, metrics are collected, analyzed and shared. Project control is described in more detail later.

Instrumentation implementation of the execution plan starts with the design specification and production of the detailed documentation followed by installation, start up and commissioning.

In the fifth and last phase, Operate and Evaluate, the project team is placed in an operation and performance-monitoring mode.

Performance is assessed and benchmarked against objectives and competitors. Results and lessons learned are shared and opportunities are identified.

14.1 The Project Workflow and Control Process

Figure shows a typical workflow for a project with a simple reiteration procedure in the project execution. Fast-track revamp work definitions may constantly change, necessitating continuous feedback on the possibilities of each technical solution and its dissemination throughout the project work team. Continuous feedback ensures the involvement of all disciplines throughout the revamp. This requires that the team members remain highly flexible and trust each other. Trust is essential since it enables team members to freely carry out discussions and to offer constructive analysis and exchange of ideas with other team members.

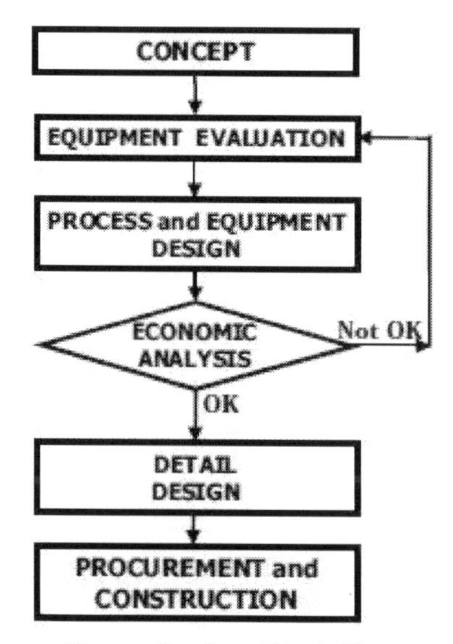

Figure: Project Work Flow

Figure below depicts the project control process. Defining control requirements in connection with the selection of parameters and tolerances (as dictated by project planning) is extremely important. Special attention must be paid to minimizing control, as explained in detail later. The collection of data follows; this is another critical item in project control. If data is not collected properly and on time, the data processing and reporting that follows are flawed. The next stage is performance analysis; data is compared with tolerances and evaluated. These findings, along with trends and forecasts, are reported to the project manager. The project manager takes corrective action, if required, and provides feedback to the project control group. The last part of process control is follow-up until corrective actions are evaluated as having been effective.

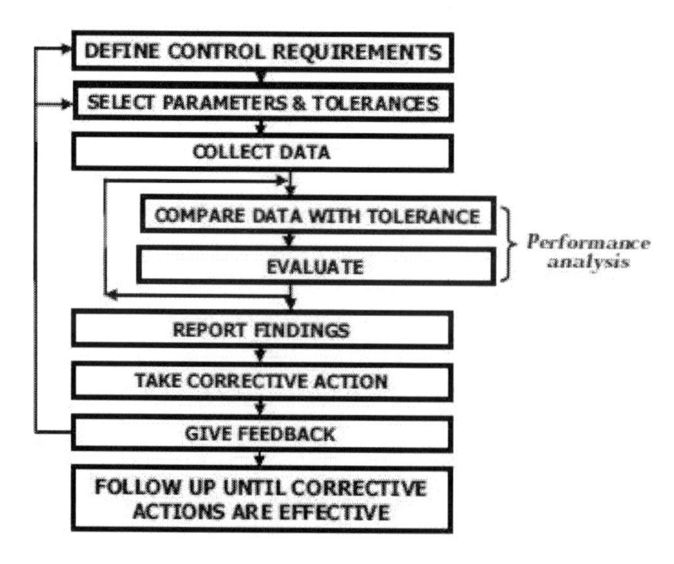

Figure: Project Control Process

On poorly executed projects, project control is limited mainly to cost and schedule tracking and reporting. Project control should allow tracking of the project against the established Execution Plan. It should allow the project team to avoid delays and to develop alternate plans to recover from potential delays.

Notice also that on larger projects, a project audit (called also a Technical Audit or an IPT - Independent Project Review) should be used to verify progress, typically at 30% and 80% of engineering completion, and later at 50% of construction completion. The Independent Project Review is a spot check by an independent third party. Its purpose is to ensure that the project is on schedule and within budget with the goal of reviewing its status and recommending corrective actions, if required.

14.2 Project S-Curve

S-curves are an important project management tool. They allow the progress of a project to be tracked visually over time, and form a historical record of what has happened to date. Analyses of S-curves allow project managers to quickly identify project growth, slippage, and potential problems that could adversely impact the project if no remedial action is taken.

14.2.1 Determining Growth

Comparison of the Baseline and Target S-curves quickly reveals if the project has grown (Target S-curve finishes above Baseline S-curve) or contracted (Target S-curve finishes below Baseline S-curve) in scope. A change in the project's scopes implies a re-allocation of resources (increase or decrease), and the very possible requirement to raise contract variations. If the resources are fixed, then the duration of the project will increase (finish later) or decrease (finish earlier), possibly leading to the need to submit an extension of time claim.

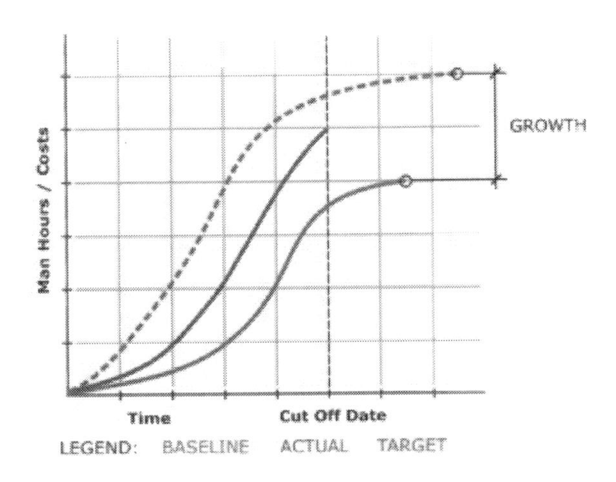

Figure: Calculating Project Growth using S-curves

14.2.2 Determining Slippage

Slippage is defined as:

"The amount of time a task has been delayed from its original baseline schedule. The slippage is the difference between the scheduled start or finish date for a task and the baseline start or finish date. Slippage can occur when a baseline plan is set and the actual dates subsequently entered for tasks are later than the baseline dates or the actual durations are longer than the baseline schedule durations".

Comparison of the Baseline S-curve and Target S-curve quickly reveals any project slippage (i.e. the Target S-curve finishes to the right of the Baseline S-curve). Additional resources will need to be allocated or additional hours worked in order to eliminate (or at least reduce) the slippage. An extension of time claim may need to be submitted if the slippage cannot be eliminated or reduced to an acceptable level.

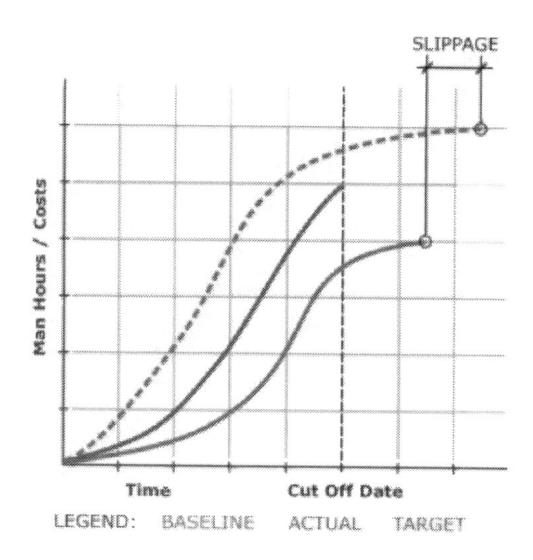

Figure: Calculating Project Slippage using S-curves

S-curve can be applied using roughly summarized data to gain an overall picture of the project's progress. It is supplemented by a simple table showing work-hours per reporting period (typically a month), total budgeted work-hours (to note budget changes), and total work-hours spent to date.

14.3 Project Risk Management

Risk is inevitable in a business organization when undertaking projects. However, the project manager needs to ensure that risks are kept to a minimal. Risks can be mainly between two types; negative impact risk and positive impact risk.

Not all the time would project managers be facing negative impact risks as there are positive impact risks too. Once the risk has been identified, project managers need to come up with a mitigation plan or any other solution to counter attack the risk. Managers can plan their strategy based on four steps of risk management which prevails in an organization. Following are the steps to manage risks effectively in an organization.

- Risk Identification
- Risk Quantification
- Risk Response
- Risk Monitoring and Control

Let's go through each of the step-in project risk management:

14.3.1 Risk Identification

Managers face many difficulties when it comes to identifying and naming the risks that occur when undertaking projects. These risks could be resolved through structured or unstructured brainstorming or strategies. It's important to understand that risks pertaining to the project can only be handled by the project manager and other stakeholders of the project.

Risks, such as operational or business risks will be handled by the relevant teams. The risks that often impact a project are supplier risk, resource risk, and budget risk. Supplier risk would refer to risks that can occur in case the supplier is not meeting the timeline to supply the resources required.

Resource risk occurs when the human resource used in the project is not enough or not skilled enough. Budget risk would refer to risks that can occur if the costs are more than what was budgeted.

14.3.2 Risk Quantification

Risks can be evaluated based on quantity. Project managers need to analyze the likely chances of a risk occurring with the help of a matrix.

The Risk Matrix is also popularly known as the Probability and Impact Matrix. The Risk Matrix is used during Risk Assessment and is born during Qualitative Risk Analysis in the Risk Management process. It is a very effective tool that could be used successfully with Senior Management to raise awareness and increase visibility of risks so that sound decisions on certain risks can be made in context.

Impact →	1	2	3	4	5
Probability ↓	Negligible	Minor	Moderate	Significant	Severe
(81-100)%	Low Risk	Moderate Risk	High Risk	Extreme Risk	Extreme Risk
(61-80)%	Minimum Risk	Low Risk	Moderate Risk	High Risk	Extreme Risk
(41-60)%	Minimum Risk	Low Risk	Moderate Risk	High Risk	High Risk
(21-40)%	Minimum Risk	Low Risk	Low Risk	Moderate Risk	High Risk
(1-20)%	Minimum Risk	Minimum Risk	Low Risk	Moderate Risk	High Risk

Figure: Risk Quantification

A risk is "rated" for its Probability and Impact on a scale to understand where on the Risk Matrix it lies. Which Risks in the process move forward into the Risk Management process will depend on the industry, company, project and people. Some are by nature more risk tolerant than others. For example, we can have a project where the team agrees that any risk that is in the yellow, orange or red cell can move forward in the Risk Management process. The rest remain in the watch list and are "accepted". Another project could have different criteria.

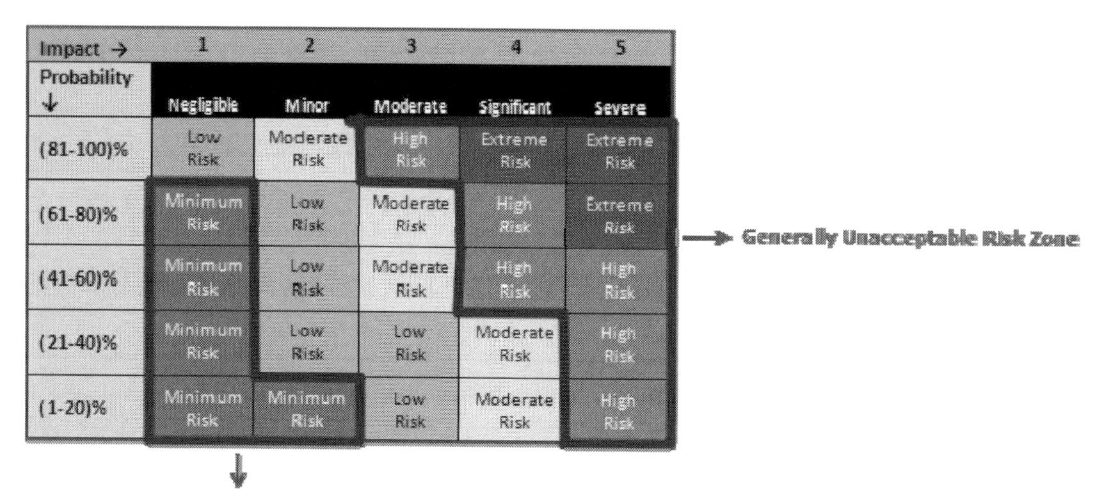

Impact →	1	2	3	4	5
Probability ↓	Negligible	Minor	Moderate	Significant	Severe
(81-100)%	Low Risk	Moderate Risk	High Risk	Extreme Risk	Extreme Risk
(61-80)%	Minimum Risk	Low Risk	Moderate Risk	High Risk	Extreme Risk
(41-60)%	Minimum Risk	Low Risk	Moderate Risk	High Risk	High Risk
(21-40)%	Minimum Risk	Low Risk	Low Risk	Moderate Risk	High Risk
(1-20)%	Minimum Risk	Minimum Risk	Low Risk	Moderate Risk	High Risk

- **The Probability and Impact "Scales":**

It is recommended that companies have standard scales on their projects for how risks can be rated. This will help everyone be on the same page. A sample scale is provided below.

- **Probability:** A scale of 1%-100% will be used for Probability.

(1-20)% means very low
(21-40)% means low
(41-60)% means medium

(61-80)% means high
(81-100)% means it is a fact

- **Impact:** A scale of 1-5 is normally used for impact ratings where;

1 means negligible
2 means minor
3 means moderate
4 means significant
5 means severe

14.3.3 Risk Response

When it comes to risk management, it depends on the project manager to choose strategies that will reduce the risk to minimal. Project managers can choose between the four risk response strategies which are outlined below.

- Risks can be avoided
- Pass on the risk
- Take corrective measures to reduce the impact of risks
- Acknowledge the risk

14.3.4 Risk Monitoring and Control

Risks can be monitored on a continuous basis to check if any change is made. New risks can be identified through the constant monitoring and assessing mechanisms.

14.3.5 Risk Management Process

Following are the considerations when it comes to risk management process.

- Each person involved in the process of planning needs to identify and understand the risks pertaining to the project.
- Once the team members have given their list of risks, the risks should be consolidated to a single list in order to remove the duplications.
- Assessing the probability and impact of the risks involved with a help of a matrix.
- Split the team into subgroups where each group will identify the triggers that lead to project risks.
- The teams need to come up with a contingency plan whereby to strategically eliminate the risks involved or identified.
- Plan the risk management process. Each person involved in the project is assigned a risk in which he/she looks out for any triggers and then finds a suitable solution for it.

14.3.6 Risk Register

Often project managers will compile a document which outlines the risk involved and the strategies in place. This document is vital as it provides a huge deal of information. Risk register will often consist of diagrams to aid the reader as to the types of risks that are dealt by the organization and the course of action taken. The risk register should be freely accessible for all the members of the project team.

14.3.7 Project Risk; an Opportunity or a Threat?

As mentioned above risks contain two sides. It can be either viewed as a negative element or a positive element. Negative risks can be detrimental factors that can haphazard situations for a project.

Therefore, these should be curbed once identified. On the other hand, positive risks can bring about acknowledgements from both the customer and the management. All the risks need to be addressed by the project manager.

An organization will not be able to fully eliminate or eradicate risks. Every project engagement will have its own set of risks to be dealt with. A certain degree of risk will be involved when undertaking a project.

The risk management process should not be compromised at any point, if ignored can lead to detrimental effects. The entire management team of the organization should be aware of the project risk management methodologies and techniques.

Enhanced education and frequent risk assessments are the best way to minimize the damage from risks.

15 Chapter : IoT & IIoT

The IoT (Internet of Things) is a network of connected devices which communicate over the Internet, and they do so autonomously, machine to machine, without the need for human intervention.

The first reference to the IoT was in 1982, when researchers at Carnegie Mellon University developed the world's first IoT-enabled Coke Machine. Mark Weiser developed the concept further in the early 90s; and Kevin Ashton coined the term 'Internet of Things' around 1999.

It would be easy to think that a smart phone is itself an Internet of Things device. It is able to connect to the Internet, and it is able to sense its environment using a myriad of sensors inside itself: location sensors, motion sensors, touch sensors, and many more. However, what is important is that currently these sensors do not communicate with each other. So we can call these smart phone as "Intranet of Things", which will become Internet of Thing device sooner or later.

The smartness has spread to a variety of devices including humans, animals, buildings, energy stations, smartphones, tablets, bicycle, sensors, cameras, vehicles, and so on. Smartness refers to the ability of the object to provide some form of Sensing/Actuation together with processing, storage and communications.

Cisco predicts 50 billion interconnected devices by 2020. Intel, more optimistically, predicts 200 billion by that same year.

15.1 IIoT

The IoT includes everything from smart homes, mobile fitness devices, and connected toys to the Industrial Internet of Things (IIoT) with smart agriculture, smart cities, smart factories, and the smart grid. IIoT connects devices such as industrial equipment on to the network, enabling information gathering and management of these devices via software to increase efficiency, enable new services (factory automation, robotics, supply chain efficiency.

Figure: Example of IIoT , Interconnected Manufacturing

Industrial systems that interface the digital world to the physical world through sensors and actuators that solve complex control problems are commonly known as cyber-physical systems. There really are only three main distinctions: ubiquitous sensing, advanced analytics, and IT methodologies. Each of these is described briefly below.

15.1.1 Ubiquitous Sensing

Analogous to the broader IoT space, which envisions ubiquitous connectivity of intelligent devices, Industrial IoT is characterized by ubiquity of connected sensors and actuators. Where traditional automation employed sensors and actuators primarily for the most critical elements of control, IIoT includes sensors and actuators for facility operations, machine health, ambient conditions, quality, and a variety of other functions. Virtually everything that can be measured and controlled within the industrial context is fair game for IIoT. The ubiquity of sensing and control is key to enabling the next cornerstone of IIoT – advanced analytics.

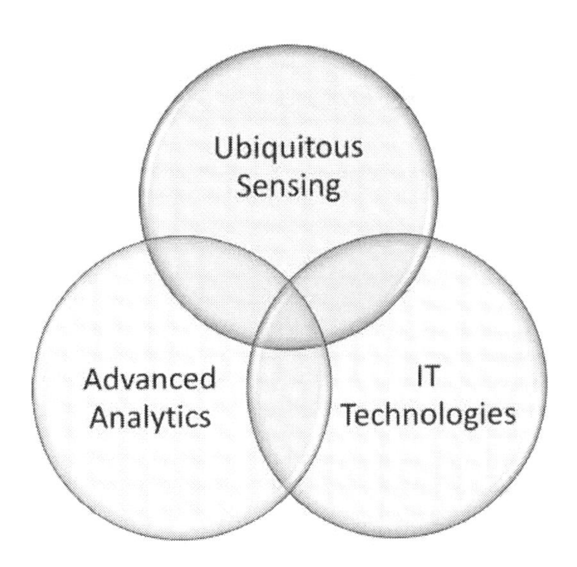

Figure : Key distinctions of IIoT as compared to traditional industrial automation

15.1.2 Advanced Analytics

Advanced analytics enables the IIoT system to realize higher levels of operational efficiency by extracting meaning from a vast array of deployed sensors. Similar to cloud datacenters, where sensors data is used to optimize virtually every aspect of operational efficiency, smart factories and other IIoT applications utilize analytics to improve uptime, optimize asset utilization, and reduce overhead costs. Improved operational efficiency provided by advanced analytics is the primary motivator for IIoT adoption today.

15.1.3 IT methodologies

The third defining characteristic of Industrial Internet of Things is the transformation of traditional automation techniques to utilize technologies that have been historically associated with information technology. This transformation has three key benefits. First, migration to IT technology enables the IIoT operator to utilize the large IT talent pool to deploy, monitor, and optimize their IIoT application. Second, standardization around IT practices helps to eliminate islands of proprietary equipment within the installation and provide tighter integration between the control domain and the operations domain. Lastly, adoption of IT methodologies enables IIoT companies to leverage the large existing base of IT hardware and software solutions when appropriate. Each of these benefits offers significant potential for capital and operational savings.

15.2 IIoT Deployment Barriers to Adoption

A recent study by Morgan Stanley (Morgan Stanley, 2017) indicates the top three challenges to IIoT deployment in order are: cybersecurity, lack of standardization, and legacy installed base. Each of these is summarized briefly below.

15.2.1 Cybersecurity

Cybersecurity in IIoT takes on new dimensions because the connected devices interact and control real world activities. Connected factories, power plants, aircraft and other vehicles pose significant threats to public safety if hacked. Corporate and national economic impacts also cannot be overlooked. The collapse of a power grid or national transportation system has much farther reaching impacts than even the largest consumer data breaches. For these reasons, robust cybersecurity is an absolute essential in IIoT. It is expected that most IIoT applications will run on private, dedicated networks with strict physical access control protocols.

15.2.2 Lack of Standardization

Historically, industrial automation has been accomplished using a variety of proprietary, vertically integrated automation solutions, or by open standards -based industrial computing solutions. While the first of these solutions offers convenience of an integrated approach, each vendor's equipment may not work well with others. This causes islands of isolated equipment within the industrial deployment that is difficult to integrate and manage as a whole. The second solution, while offering many benefits such as scalability, flexibility, and less risk, often puts the burden of software creation on the operator. This can be cumbersome when attempting to assimilate the large number of dissimilar sensor types associated with IoT deployment.

Standardization of the upstream interfaces for controller devices and met a-data models for sensors would go a long way toward eliminating both of these problems. Standardized interfaces would allow dissimilar pieces of hardware to communicate with the IIoT command center in a uniform fashion and eliminate isolated islands within the installment. Likewise, an extensible standardized meta-data model for sensors would allow for systematic detection and control of sensors and control points without extensive code re-writes. From a hardware standpoint, IIoT would also benefit from greater standardization around communications interfaces, power, and environmental requirements.

Large industrial automation suppliers are not incentivized to embark on open standardization because it loosens the customer's dependence upon their proprietary solutions. Smaller automation suppliers lack the industry clout or size to take on such an ambitious undertaking. This is a task best suited for an industry standards organization, and one which PICMG is well equipped to handle.

15.2.3 Legacy Installed Base

Very few technology transformations occur overnight. As a result, legacy equipment must be able to coexist with the new. Any successful IIoT strategy must incorporate this reality. Standardization can help bridge the gap in the short term and standardization agency can prepare documents of backward compatibility and interoperability toward alleviating the worst of these issues.

15.3 IIoT Architectural Overview

Because factory automation is projected to be the largest and fastest growing segment of the Industrial Internet of Things market, this section of the document focuses on an architecture for the smart factory. This selection was chosen merely as an example of a relevant application to which the IIoT architecture may be applied.

15.3.1 Smart Factory

For years, traditional factories have been operating at a disadvantage, impeded by production environments that are "disconnected" or, at the very least, "strictly gated" to corporate business systems, supply chains, and customers and partners. Managers of these traditional factories are essentially "flying blind" and lack visibility into their operations. These operations are composed of plant floors, front offices, and suppliers operating in independent silos. Consequently, rectifying downtime issues, quality problems, and the root causes of various manufacturing inefficiencies is often difficult.

The main challenges facing manufacturing in a factory environment today include the following:

- Accelerating new product and service introductions to meet customer and market opportunities
- Increasing plant production, quality, and uptime while decreasing cost
- Mitigating unplanned downtime (which wastes, on average, at least 5% of production)
- Securing factories from cyber threats
- Decreasing high cabling and re-cabling costs (up to 60% of deployment costs)
- Improving worker productivity and safety

While we tend to look at IoT as an evolution of the Internet, it is also sparking an evolution of industry. In 2016 the World Economic Forum referred to the evolution of the Internet and the impact of IoT as the "fourth Industrial Revolution."5 The first Industrial Revolution occurred in Europe in the late eighteenth century, with the application of steam and water to mechanical production. The second Industrial Revolution, which took place between the early 1870s and the early twentieth century, saw the introduction of the electrical grid and mass production. The third revolution came in the late 1960s/early 1970s, as computers and electronics began to make their mark on manufacturing and other industrial systems. The fourth Industrial Revolution is happening now, and the Internet of Things is driving it. Figure below summarizes these four Industrial Revolutions as Industry 1.0 through Industry 4.0.

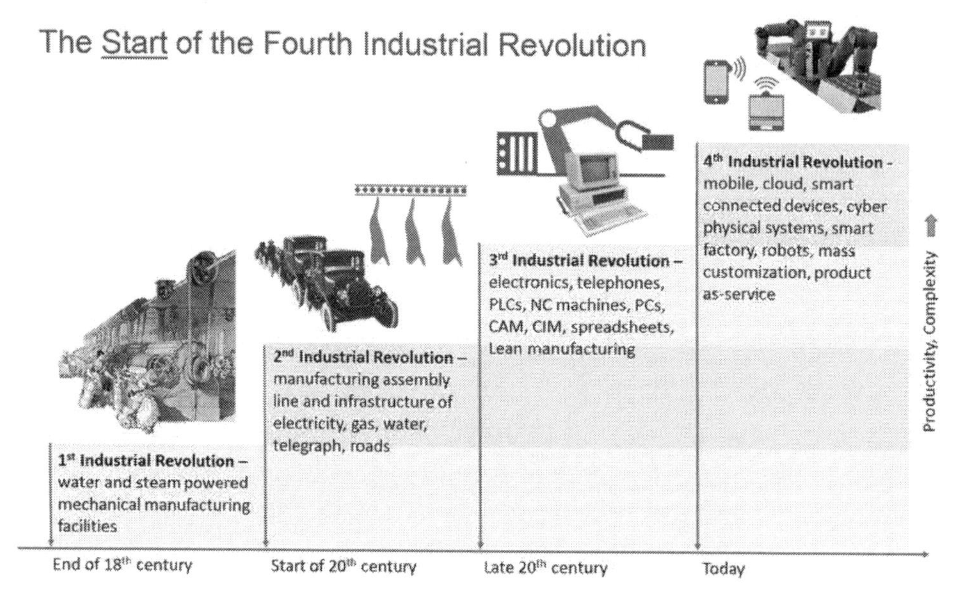

Figure: The Four Industrial Revolutions

The factory floor is the heart of the smart IIoT application. It contains multiple robotic assembly machines, automated test equipment and various other process -related pieces of equipment. Each of these is fully automated and integrated utilizing the same network interfaces and common data model and protocols. In addition to control and monitoring of the actual manufacturing process , the machines are also instrumented with other sensors to help assess the health of the equipment and correlate operational dynamics with factory output quality.

In order to feed the automated factory, the warehouse and stockroom is also fully instrumented. Because the factory control and the inventory control systems both leverage IT methodologies, integration and analysis between the two domains is easily achievable, allowing actual factory production rates to factor into intelligent purchasing and inventory management algorithms.

Environmental conditions are monitored in real time providing useful information regarding energy usage from air conditioning, lighting and other resources. This function also monitors and controls other resources such as on-site power generation and backup generator status. This information, combined with deep analytics , may be used to prioritize workloads in order to optimize resource utilization and minimize operational costs.

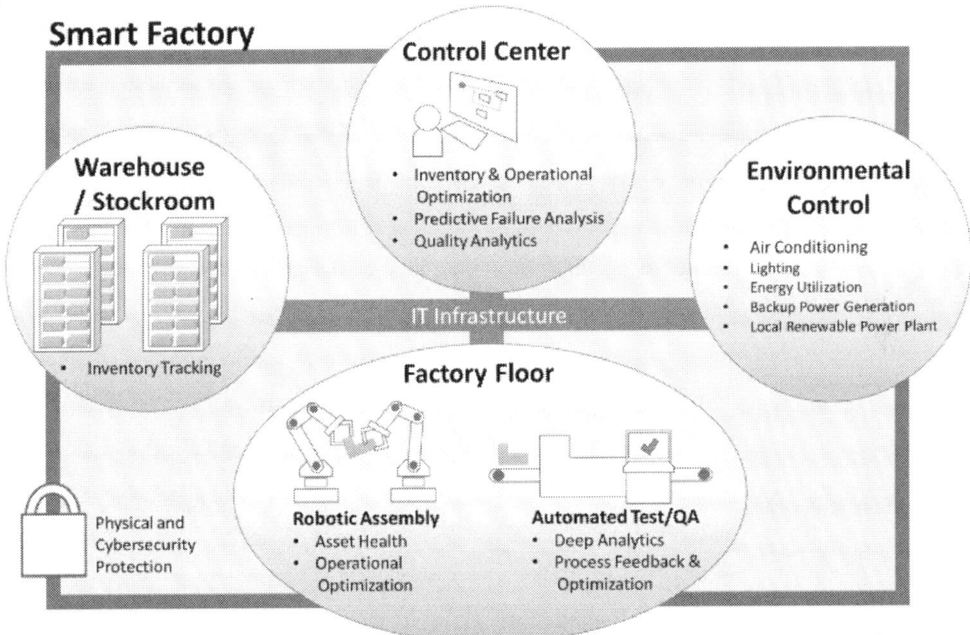

Figure: Smart Factory Layout

All these functions are interconnected with the factory control center via Ethernet (or industrial Ethernet when required). The control center provides visualization and control of the entire factory operations utilizing standard IT technologies.

15.3.2 Architectural Decomposition for IIoT

Figure below shows an architectural decomposition for Industrial Internet of Things. All components are connected via Ethernet (or industrial Ethernet) unless otherwise shown. Legacy equipment co-exists with newer equipment, though potentially at a lower level of functionality, and a common metadata model enables discovery and control of IIoT devices in a flexible and extensible fashion.

At the lowest level of the architecture, sensors and control points provide connectivity to physical phenomena within the factory. IIoT sensors present themselves as intelligent, managed devices over the factory network using the common meta -data model. Using RESTful application programming interfaces, sensors may be monitored and controlled using standard IT methodologies. Because these sensors operate in a live factory environment, ruggedization is an expected requirement.

For sensors and actuators that must respond in a hard, real-time fashion, it may be necessary to place a controller close to devices in order to monitor the devices locally. This reduces the latency and improves determinism over having the devices remotely controlled through the factory control center. These local IIoT controllers would present the connected sensor data models to the upstream control center. They would also introduce programmable "listener" functions that implement local policies when sensor events occur. Listener functions may also be directly implemented in sensors and actuators.

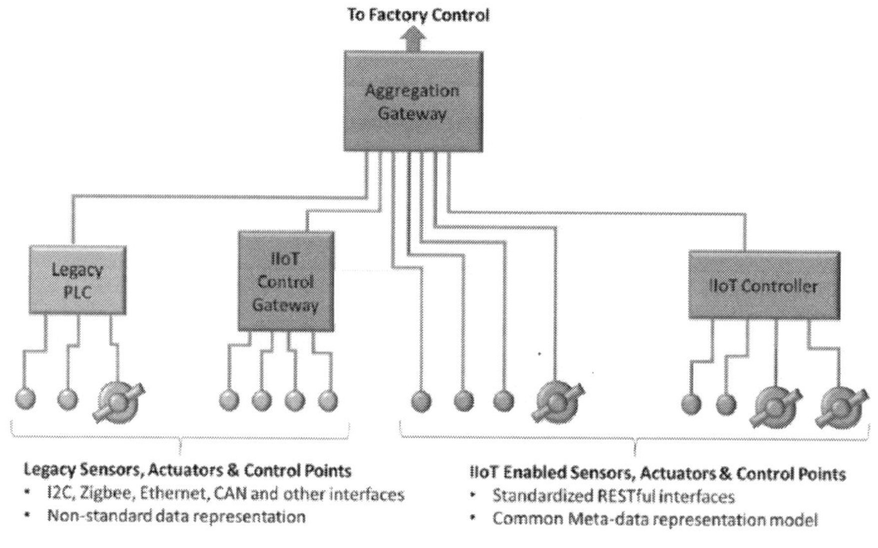

Figure: IIoT Components

Legacy sensors and controllers may be connected to the IIoT control center. Initially, PLCs can be connected over their existing interfaces and be managed through legacy software. As an intermediate step to full IIoT functionality, the PLC can later be replaced by an IIoT control gateway. This device "translates" the sensor's native protocols to a RESTful data interface using the common meta-data model. This allows the same sensors to be used while the control architecture is migrated to IIoT technologies. As a final step, sensors can later be replaced with fully IIoT-enabled sensors.

The final piece of the IIoT architecture is an aggregating network gateway. This device serves to aggregate and isolate traffic between zones on the factory floor and the rest of the network. In many cases, the bandwidth of traffic from the factory devices will be low so a ruggedized, 10/100/1000 switch will typically be more than sufficient.

15.3.3 RESTful APIs

REST, which stands for REpresentational State Transfer, is a communication architecture style based on the IETF RFC 2616 protocol (Fielding, 1999). REST refers to a stateless software architecture that provides many underlying characteristics and protocols that govern the behavior of clients and servers.

15.3.3.1 What Is REST?

REST stands for Representational State Transfer. It is a stateless software architecture that provides many underlying characteristics and protocols that govern the behavior of clients and servers.

15.3.3.2 What Is Meant by RESTful?

An API that has following features is known as RESTful API:

- Client-server architecture: The client is the front-end and the server is the back-end of the service. It is important to note that both of these entities are independent of each other.
- Stateless: No data should be stored on the server during the processing of the request transfer. The state of the session should be saved at the client's end.
- Cacheable: The client should have the ability to store responses in a cache. This greatly improves the performance of the API.

15.3.3.3 What Is a RESTful API?

A RESTful API (also known as a RESTful web service) is a web service implemented using HTTP protocol and the principles of REST. It is a collection of resources that employ HTTP methods (GET, PUT, POST, DELETE).

The collection of the resources is then represented in a standardized form (usually XML) that can be any valid Internet media type, if it is a valid hypertext standard.

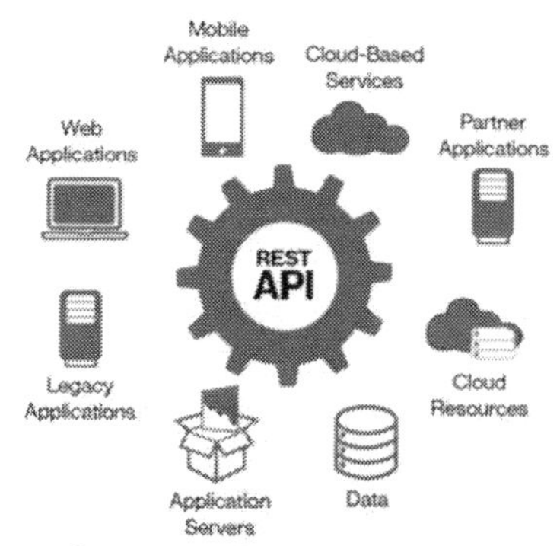

Figure: Collection of resource for RESTful API

Popular RESTful API request formats:

- REST
- XML-RPC
- SOAP

Popular RESTful API response formats:

- REST
- XML-RPC
- SOAP
- JSON
- PHP

15.3.3.4 Why Use a RESTful API?

A RESTful API is used to make applications distributed and independent over the internet with the aim of enhancing the performance, scalability, simplicity, modifiability, visibility, portability, and reliability of the application.

15.3.4 REST APIs for IIoT

Traditional industrial system architecture is built around a bus topology. Assets are connected to the bus and speak the same protocol. The problem with leveraging this architecture in IIoT applications is that systems that are not a part of the bus and do not understand the bus protocol cannot leverage the data and resources available on the bus.

But for the IIoT to be viable, IIoT hardware and software assets must connect and start talking to each other. REST APIs offer a standard form of sharing data and resources between IIoT devices and IIoT software. Ethernet and TCP/IP were the first step toward the IIoT. REST APIs are the next step in moving up the OT (Operational Technology)/ IT (Information technology) convergence stack.

REST APIs are the tools that allow engineers to connect real-world physical assets like sensors, motors, pumps, relays and control systems to the digital world and communicate directly to the cloud—no middleware, protocol conversion, or edge gateways required. REST APIs are used across the Internet today. They're the technology that stitches the IIoT together.

Currently no RESTful APIs are defined to specifically meet the needs of IIoT.

15.3.5 Industrial Automation using IOT

IoT is the collection of the sensors data through embedded system and this embedded system upload the data on internet. There are many challenges to IoT and Industrial Automation for example Data and service security, Trust, data integrity, information privacy, scalability and interoperability Automation Domain Constrains.

This section covers the concept of Industrial workstation, a large number of boards, including micro-controllers, field programmable gate arrays (FPGAs) and single-board computers. Among these, Arduino and , Raspberry Pi etc are open-source devices, with components available from a variety of suppliers, and both require a high level of programming skills and some imagination before they can be used for real-time industrial control applications and Industrial Automation using IoT.

15.3.5.1 Arduino

Figure: Arduino board

The Arduino platform was created back in 2005 by the Arduino company and allows for open source prototyping and flexible software development and back-end deployment while providing significant ease of use to developers, even those with very little experience building IoT solutions.

Arduino is sensible to literally every environment by receiving source data from different external sensors and is capable to interact with other control elements over various devices, engines and drives. Arduino has a built-in micro controller that operates on the Arduino software.

Projects based on this platform can be both standalone and collaborative, i.e. realized with use of external tools and plugins. The integrated development environment (IDE) is composed of the open source code and works equally good with Mac, Linux and Windows OS. Based on a processing programming language, the Arduino platform seems to be created for new users and for experiments. The processing language is dedicated to visualizing and building interactive apps using animation and Java Virtual Machine (JVM) platform.

Let's note that this programming language was developed for the purpose of learning basic computer programming in a visual context. It is an absolutely free project available to every interested person. Normally, all the apps are programmed in C/C++, and are wrapped with avr-gcc (WinAVR in OS Windows).

Arduino offers analogue-to-digital input with a possibility of connecting light, temperature or sound sensor modules. Such sensors as SPI or I2C may also be used to cover up to 99% of these apps' market.

Arduino is a microcontroller (generally it is the 8-bit ATmega microcontroller), but not a mini-computer, which makes Arduino somehow limited in its features for advanced users. Arduino provides an excellent interactivity with external devices and offers a wide range of user manuals, project samples as well as a large community of users to learn from / share knowledge with.

15.3.5.2 Raspberry Pi

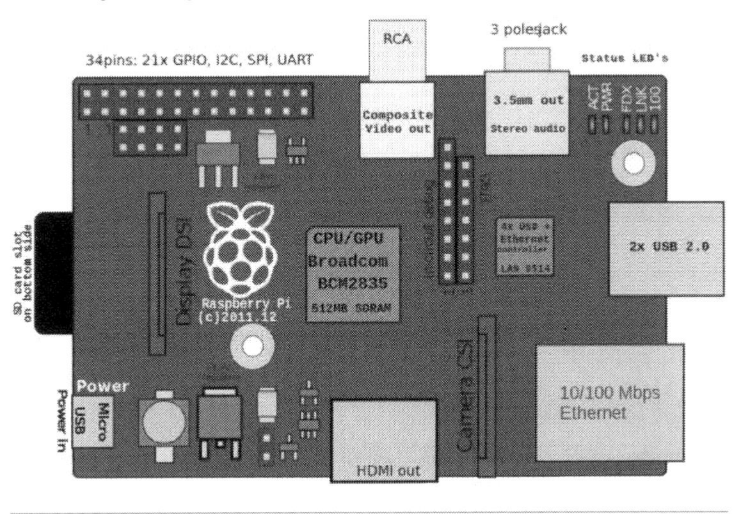

Figure: Raspberry Pi

Raspberry Pi is a mono-board computing platform that's as tiny as a credit card. Initially it was developed for computer science education with later on progress to wider functions.

Since the inception of Raspberry, the company sold out more than 8 million items. Raspberry Pi 3 is the latest version and it is the first 64-bit computing board that also comes with built-in Wi-Fi and Bluetooth functions. According to Raspberry Pi Foundation CEO Eben Upton, "it's been a year in the making". The Pi3 version is replaced with a quad-core 64-bit 1.2 GHz ARM Cortex A53 chip, 1GB of RAM, VideoCore IV graphics, Bluetooth 4.1 and 802.11n Wi-Fi. The developers claim the new architecture delivers an average 50% performance improvement over the Pi 2.

Another peculiarity of Raspberry Pi is the GPIO (General Purpose Input-Output), which is a low-level interface of self-operated control by input-output ports. Raspberry has it as a 40-pin connector.

Chapter 1 Raspberry Pi uses Linux as its default operating system (OS). It's also fully Android compatible. Using the system on Windows OS is enabled through any virtualization system like XenDesktop. If you want to develop an application for Raspberry Pi on your computer, it is necessary to download a specific toolset comprised of ARM-compiler and some libraries complied down to ARM-target platform like glibc.

15.3.5.3 Intel Galileo

Figure: Intel Galileo

Frequently referred to as a "reliable ally" of Arduino, Intel Galileo is a highly integrated board that's just a little larger than a credit card. The microcomputer is equipped with Intel® Quark™ SoC X1000, operating at speeds of up to 400 MHz, a motherboard with up to 8 Mb of flash memory and 256 RAM. The device also has a VLAN port available with the 100Mb capacity, a microSD card and mini PCI express slots, RS 232, USB 2.0 ports with a possibility to connect up to 128 devices. This platform works with a very light distribution of Linux and a standard environment of Arduino. Intel Galileo has such features as its own USB controller and data exchange without SPI components. Another cool feature is that there is an expansion slot for PCI Express for Wi-Fi, Bluetooth and 3G installation. Intel Galileo supports the Arduino IDE.

It's worth mentioning that such a microprocessor as Galileo can be used for a wide variety of functions; among them are robotic engineering and IoT technologies. Intel has released two versions of Galileo - Intel Galileo and Galileo Gen 2 - in order to expand its own solutions.

15.3.5.4 UDOO NEO

Figure: UNDOO NEO

UDOO NEO is an all-in-one open hardware low-cost computer equipped with a NXP™ i.MX 6SoloX applications processor for Android and Linux.

UDOO NEO embeds two cores on the same processor: a powerful 1GHz ARM® Cortex-A9, and an ARM Cortex-M4 I/O real-time co-processor that can run up to 200Mhz.

While the Cortex-A9 can run both Android Lollipop and UDOObuntu 2 - a dedicated Ubuntu-based Linux distro - the Cortex-M4 allows easy access to a Arduino™ environment. The snap-in connector ensures a plug-and-play interaction with most sensors and actuators.

Thanks to its embedded 9-axis motion sensors and a Wi-Fi + Bluetooth 4.0 module, the board is ideal to create robots, drones and rovers as well as any Mobile IoT project you can imagine.

15.3.6 Cloud-based IoT Platforms

In addition to standard IoT platforms, Cloud-based platforms are taking their way, which provides a convenient "on demand" network access to a set of configurable computing resources available to a specific group of users for data storage, servers, applications and/or services, and networks.

There are various platform models for work with clients through Cloud; the most commonly used ones are Cloud Software as a Service (SaaS), Cloud Platform as a Service (PaaS), Cloud Infrastructure as a Service (IaaS) and others. A user can choose any of these models with respect to demands of his business.

The most popular cloud services available for IoT development include, but aren't limited to: Amazon S3, Microsoft Azure, Google App Engine, Salesforce1 Platform, Heroku, etc.

16 Chapter: IoT Protocols and Standards

This section covers the different standards offered by IEEE, IETF and ITU to enable technologies matching the rapid growth in IoT. These standards include communication, routing, network and session layers of the networking stack that are being developed just to meet requirements of IoT. The section also includes management and security protocols.

IoT covers a huge range of industries and use cases that scale from a single constrained device up to massive cross-platform deployments of embedded technologies and cloud systems connecting in real-time.

Tying it all together are numerous legacy and emerging communication protocols that allow devices and servers to talk to each other in new, more interconnected ways.

At the same time, dozens of alliances and coalitions are forming in hopes of unifying the fractured and organic IoT landscape.

Internet of Things (IoT) and its protocols are among the most highly funded topics in both industry and academia. The rapid evolution of the mobile internet, mini-hardware manufacturing, micro-computing, and machine to machine (M2M) communication has enabled the IoT technologies. According to Gartner, IoT is currently on the top of their hype-cycle, which implies that a large amount of money is being invested on it by the industry. Billions of dollars are being spent on IoT enabling technologies and research while much more is expected to come in the upcoming years

16.1 IoT Ecosystem

Figure below shows a **7-layer model of IoT ecosystem**. At the bottom layer is the market or application domain, which may be smart grid, connected home, or smart health, smart factory etc. The second layer consists of sensors that enable the application. Examples of such sensors are temperature sensors, humidity sensors, electric utility meters, or cameras. The third layer consists of interconnection layer that allows the data generated by sensors to be communicated (thru wireline or wireless connectivity), usually to a computing facility, data center, or a cloud. There the data is aggregated with other known data sets such as geographical data, population data, or economic data. The combined data is then analyzed using machine learning and data mining techniques. To enable such large distributed applications, we also need the latest application level collaboration and communication software, such as, software defined networking (SDN), services oriented architecture (SOA), etc. Finally, the top layer consists of services that enable the market and may include energy management, health management, education, transportation etc.

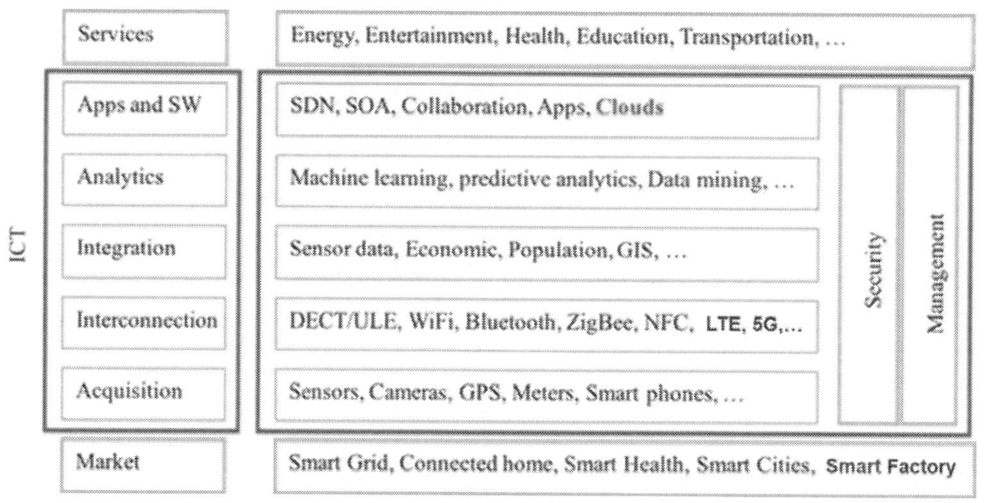

Figure: IoT Ecosystem

In addition to these 7 layers that are built on the top of each other, there are security and management applications that are required for each of the layers and are, therefore, shown on the side.

In the next section we concentrate on the interconnection layer. This layer itself can be shown in a multi-layer stack as shown in Figure below.

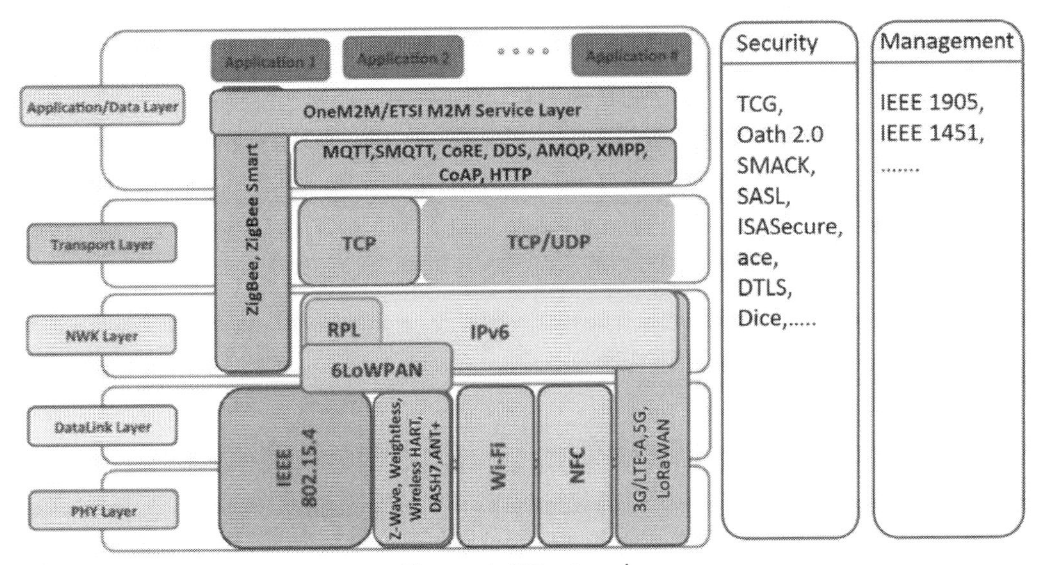

Figure: IoT Protocols

We are discussing only the datalink, network, and Application/Data Layers. The datalink layer connects two IoT elements which generally could be two sensors or the sensor and the gateway device that connects a set of sensors to the Internet. Often there is a need for multiple sensors to communicate and aggregate information before getting to the Internet. Specialized protocols have been designed for routing among sensors and are part of the network layer. The Application layer protocols enable messaging among various elements of the IoT communication subsystem. A number of security and management protocols have also been developed for IoT as shown in the figure.

16.2 Protocols for IoT

16.2.1 IoT Data Link Layer Protocol

In this section, we discuss the datalink layer protocol standards. The discussion includes physical (PHY) and MAC layer protocols which are combined by most standards.

16.2.1.1 IEEE 802.15.4

The original IEEE 802.15.4 standard was released in 2003. The original version supported two physical layers, one of them working in the 868 and 915 MHz frequency bands and the other working in the 2.4GHz band. Later on, there was another revision released in 2006, which improved the transfer speeds. Additional bands were added in the subsequent revisions.

The IEEE 802.15.4 supports two classes of devices: Fully functional devices (FFD), which have full network functionalities and the Reduced functional devices (RFD), which possess limited functionalities. All personal area networks (PAN) consists of at least one FFD which acts as the PAN coordinator which is responsible for maintaining the PAN. RFDs are responsible for directly obtaining data from the environment and sending them to a PAN coordinator.

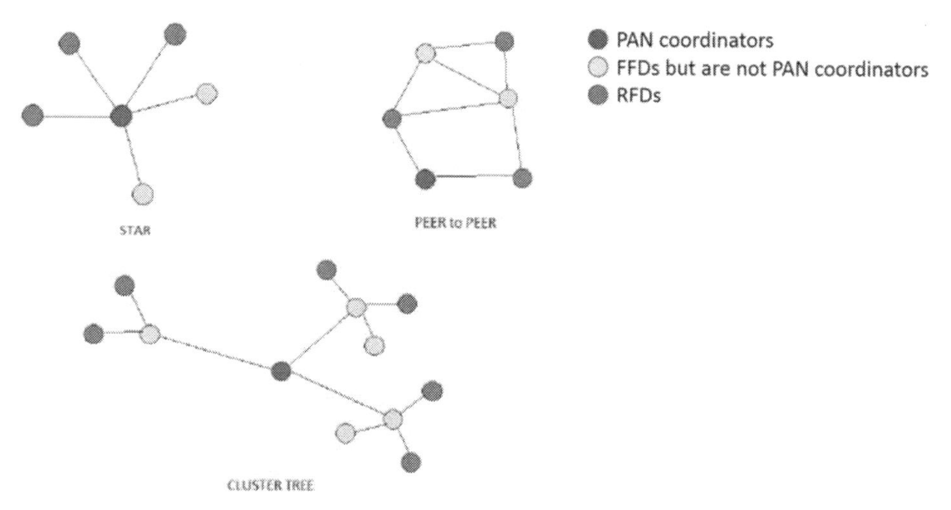

Figure: IEEE 802.15.4 Network Topologies

As seen from the figure, in a star topology, all devices directly interact with only the PAN coordinator. In a peer-to-peer topology, the FFDs can communicate with each other. In a cluster tree topology, the RFDs communicate with an FFD which in turn communicate with the PAN coordinator.

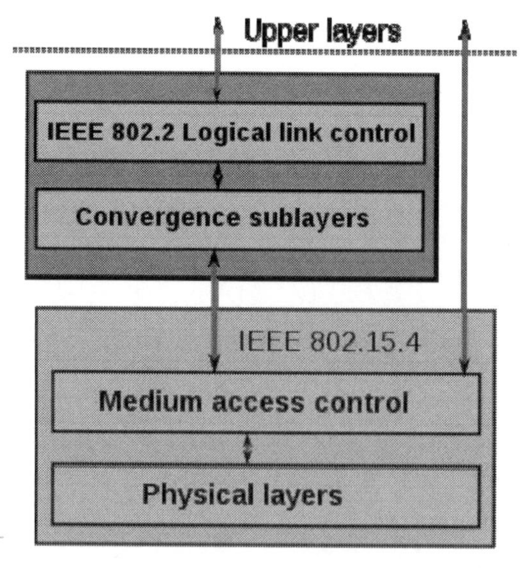

Figure: IEEE 802.15.4 protocol stack

Devices are conceived to interact with each other over a conceptually simple wireless network. The definition of the network layers is based on the OSI model; although only the lower layers are defined in the standard, interaction with upper layers is intended, possibly using an IEEE 802.2 logical link control sublayer accessing the MAC through a convergence sublayer. Implementations may rely on external devices or be purely embedded, self-functioning devices.

16.2.1.2 ZigBee

ZigBee is a IEEE 802.15.4- based specification for a suite of high-level communication protocols used to create personal area networks with small , low-power digital radios

It uses the IEEE 802.15.4 standard to define its physical and MAC layer.

The network layer (NWK) is responsible for functions such as starting a network, joining and leaving a network, routing, configuration of new devices, addressing and security

The application layer is divided into the application support sub-layer (APS), and the manufacturer defined application objects. Each application object contains software that is written by the application developer.

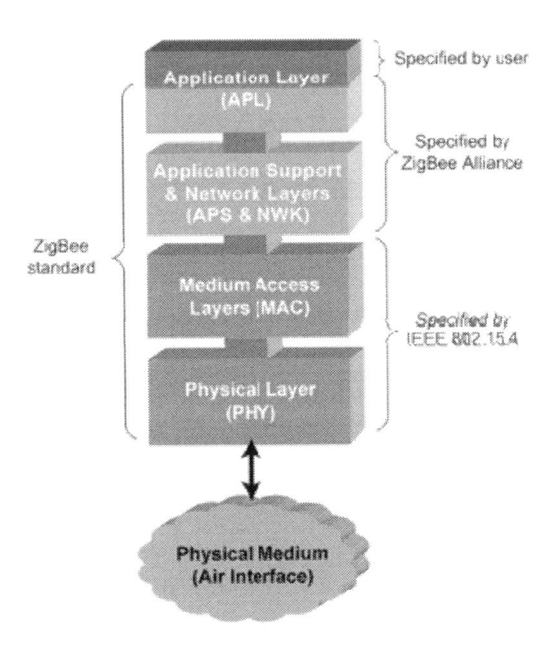

Figure: ZigBee protocol stack

16.2.1.3 ZigBee Smart

ZigBee smart energy is designed for a large range of IoT applications including smart homes, remote controls and healthcare systems. It supports a wide range of network topologies including star, peer-to-peer, or cluster-tree. A coordinator controls the network and is the central node in a star topology, the root in a tree or cluster topology and may be located anywhere in peer-to-peer.

ZigBee standard defines three stack profiles: ZigBee, ZigBee Pro and ZigBee IP. These stack profiles support full mesh networking and work with different applications allowing implementations with low memory and processing power.

As previously mentioned, the ZigBee MAC and PHY layers have been defined as part of the IEEE 802.15.4 work. The three network layers in ZigBee are RF4CE, PRO and Green Power, where Green Power essentially is a feature of PRO. Also a fourth network layer has just been released (April 2013) for smart grid/utility applications: ZigBee IP (ZIP). These network layers are quite complementary.

RF4CE is intended for devices that require a lot of human interfaces (like keyboards, or remote controls), and low latency and low power are key characteristics. RF4CE also offers star-networking capabilities (point-to-multipoint). PRO can be considered the "backbone" network layer of ZigBee, where the key characteristic is mesh-networking with the capability to cover large areas with redundant connections and therefore reliable coverage. Green Power is a feature of PRO and supports ultra-low power devices that are powered by energy harvesters or (non-replaceable) batteries. These devices are part of the network, but usually they are only included in network activity when they have to be, and otherwise they are completely shut down.

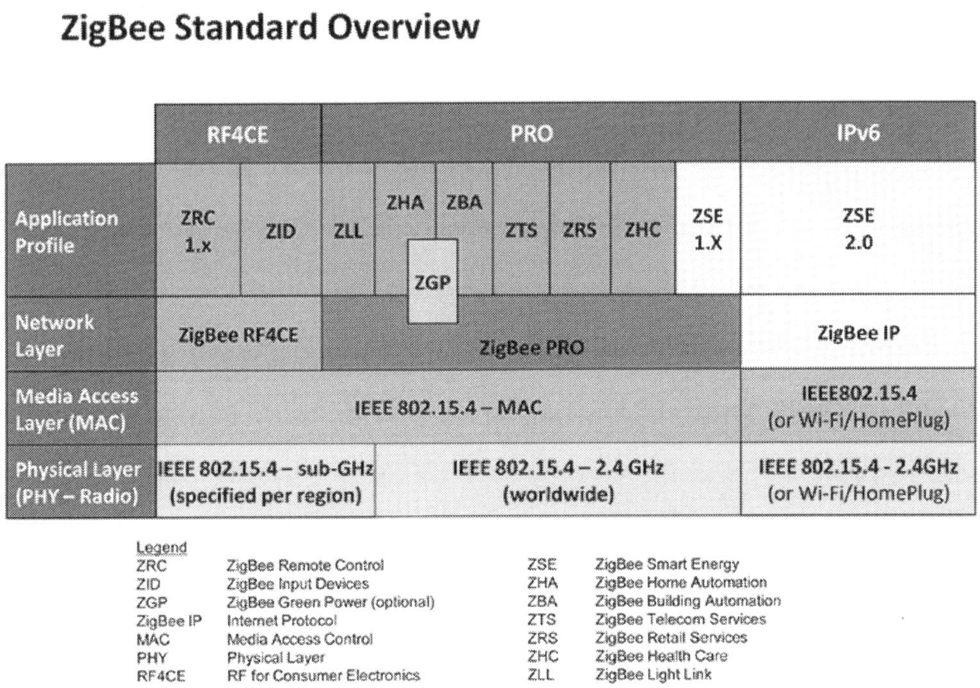

ZigBee Standard Overview

	RF4CE		PRO							IPv6
Application Profile	ZRC 1.x	ZID	ZLL	ZHA / ZBA / ZGP		ZTS	ZRS	ZHC	ZSE 1.X	ZSE 2.0
Network Layer	ZigBee RF4CE		ZigBee PRO							ZigBee IP
Media Access Layer (MAC)	IEEE 802.15.4 – MAC									IEEE802.15.4 (or Wi-Fi/HomePlug)
Physical Layer (PHY – Radio)	IEEE 802.15.4 – sub-GHz (specified per region)		IEEE 802.15.4 – 2.4 GHz (worldwide)							IEEE 802.15.4 - 2.4GHz (or Wi-Fi/HomePlug)

Legend

ZRC	ZigBee Remote Control	ZSE	ZigBee Smart Energy
ZID	ZigBee Input Devices	ZHA	ZigBee Home Automation
ZGP	ZigBee Green Power (optional)	ZBA	ZigBee Building Automation
ZigBee IP	Internet Protocol	ZTS	ZigBee Telecom Services
MAC	Media Access Control	ZRS	ZigBee Retail Services
PHY	Physical Layer	ZHC	ZigBee Health Care
RF4CE	RF for Consumer Electronics	ZLL	ZigBee Light Link

Figure: ZigBee Standard Overview

As mentioned, the fourth network layer, ZigBee IP has just been released. Based on its success in the data world IPv6, has been considered for ZigBee as well, but as the focus for IPv6 is on massive data volumes (and high data rates). This technology is currently focusing on Smart Energy/Smart Grid applications.

16.2.1.4 IEEE 802.11 AH

IEEE 802.11ah is a light (low-energy) version of the original IEEE 802.11 wireless medium access standard. It has been designed with less overhead to meet IoT requirements. IEEE 802.11 standards (also known as Wi-Fi) are the most commonly used wireless standards. They have been widely used and adopted for all digital devices including laptops, mobiles, tablets, and digital TVs. However, the original WiFi standards are not suitable for IoT applications due to their frame overhead and power consumption. Hence, IEEE 802.11 working group initiated 802.11ah task group to develop a standard that supports low overhead, power friendly communication suitable for sensors and motes.

802.11ah operating in the 900MHz range, with data rates from 150 Kbit/s with a 1 MHz band to as much as 40 Mbit/s over an 8 MHz band, lower power consumption, and a least double of the range of a typical 802.11n device, capable of covering an area of about 1 km2. The target applications are sensors networks, backhaul networks for sensor and meter, and extended range Wi-Fi, as the standard allows long range and more clients at low bitrates.

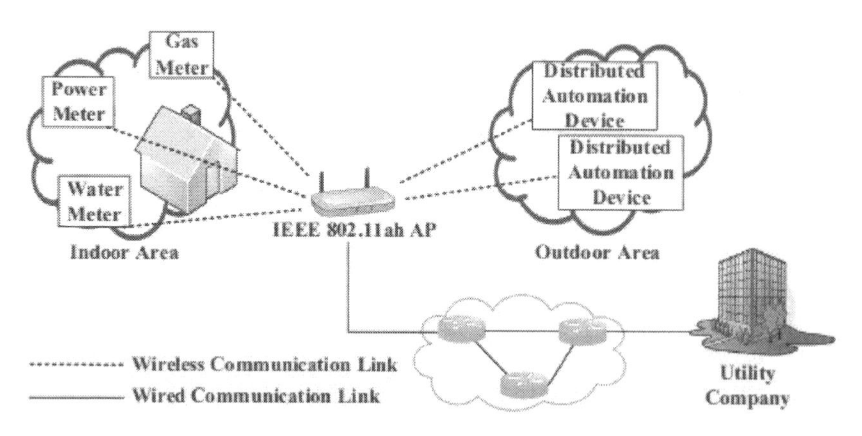

Figure: Smart Grid with 802.11ah – Source: Seoul National University

16.2.1.5 WirelessHART

WirelessHART is a datalink protocol that operates on the top of IEEE 802.15.4 PHY and adopts Time Division Multiple Access (TDMA) in its MAC. It is a secure and reliable MAC protocol that uses advanced encryption to encrypt the messages and calculate the integrity in order to offer reliability. The architecture, as shown in below Figure consists of a network manager, a security manager, a gateway to connect the wireless network to the wired networks, wireless devices as field devices, access points, routers and adapters. The standard offers end-to-end, per-hop or peer-to- peer security mechanisms. End to end security mechanisms enforce security from sources to destinations while per-hop mechanisms secure it to next hop only.

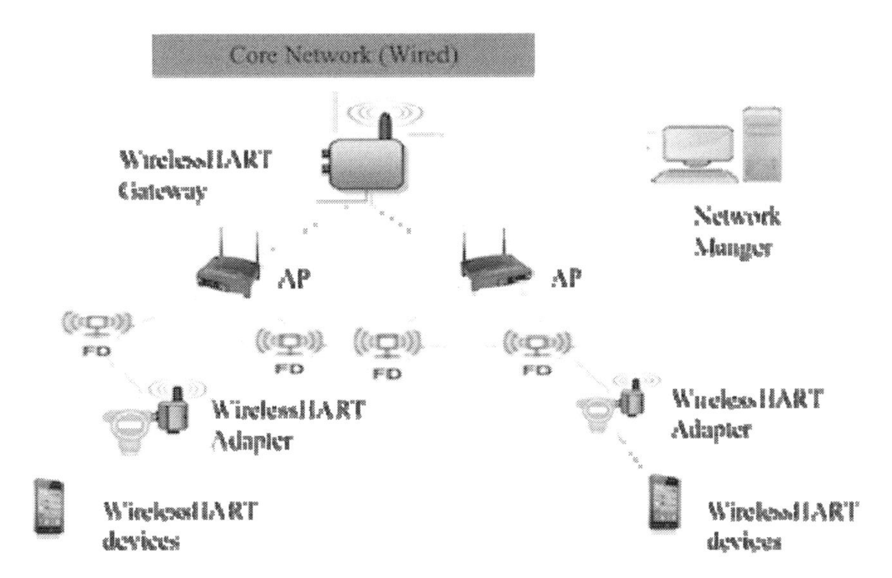

Figure : WirelessHART Architecture

16.2.1.6 Z-Wave

Z-Wave is a low-power MAC protocol designed for home automation and has been used for IoT communication, especially for smart home and small commercial domains. It covers about 30-meter point-to-point communication and is suitable for small messages in IoT applications, like light control, energy control, wearable healthcare control and others.

Figure: Z-Wave

The Z-wave network consists of controllers (one primary controller and more than one secondary controllers) and slaves. Controller devices are the nodes in a z-wave network which initiates control commands. It also sends out the commands to other nodes. The slave devices are the nodes which replies based on command received and execute the commands. Slave nodes also forward the commands to other nodes in the network.

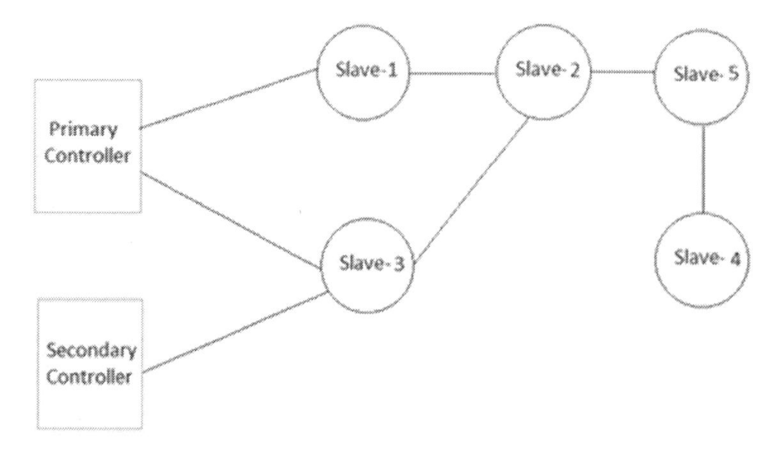

Figure: Z-Wave Network

16.2.1.7 DASH7

DASH7 is a wireless communication protocol for active RFID that operates in globally available Industrial Scientific Medical (ISM) band and is suitable for IoT requirements. It is mainly designed for scalable, long range outdoor coverage with higher data rate compared to traditional ZigBee. It is a low-cost solution that supports encryption and IPv6 addressing. It supports a master/slave architecture and is designed for burst, lightweight, asynchronous and transitive traffic.

16.2.1.8 HomePlug

HomePlug is the family name for various power line communications specifications under the HomePlug designation, with each offering unique performance capabilities and coexistence or compatibility with other HomePlug specifications.

HomePlug GreenPHY (HomePlugGP) is another MAC protocol developed by HomePlug Powerline Alliance that is used in home automation applications. HomePlug suite covers both PHY and MAC layers and has three versions: HomePlug-AV, HomePlug-AV2, and HomePlugGP. HomePlug-AV is the basic power line communication protocol which uses TDMA and CSMA/CA as MAC layer protocol, supports adaptive bit loading which allows it to change its rate depending on the noise level and uses Orthogonal Frequency Division Multiplexing (OFDM) and four modulation techniques.

HomePlugGP is designed for IoT generally and specifically for home automation and smart grid applications. It is basically designed to reduce the cost and power consumption of HomePlug-AV while keeping its interoperability, reliability and coverage. Hence, it uses OFDM, as in HomePlug, but with one modulation only. In addition, HomePlugGP uses Robust OFDM coding to support low rate and high reliability transmission. HomePlug-AV uses only CSMA as a MAC layer technique while HomePlugGP uses both CSMA and TDMA. Moreover, HomePlugGP has a power-save mode that allows nodes to sleep much more than Home Plug by synchronizing their sleep time and waking up only when necessary.

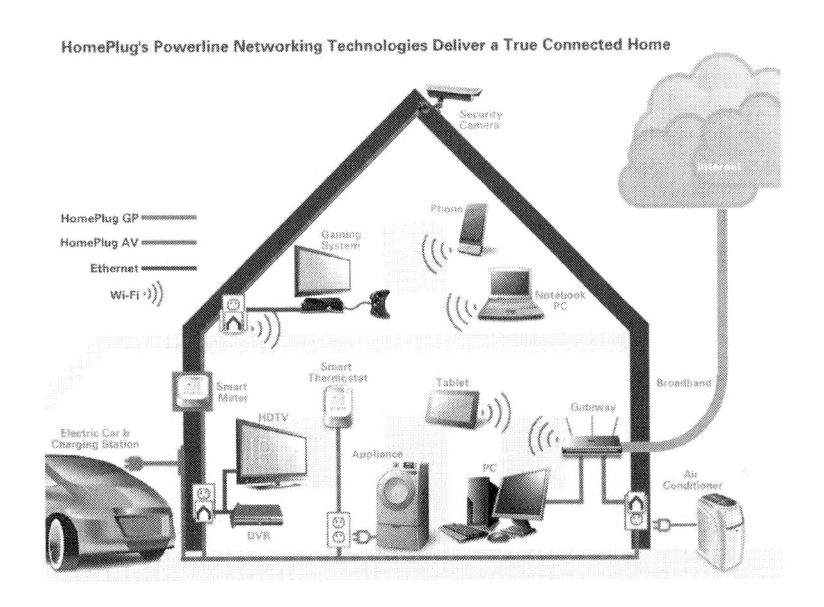

Figure: As interoperable technologies, HomePlug AV and Green PHY bring digital entertainment and smart energy/grid applications under one interoperable umbrella in a true connected home.

The first HomePlug specification, HomePlug 1.0, was released in June 2001. The HomePlug AV (for audio-video) specification, released in 2005, increased physical layer (PHY) peak data rates from approximately 13.0 Mbit/s[2] to 200 Mbit/s. The HomePlug Green PHY specification was released in June 2010 and targets Smart Energy and Smart Grid applications as an interoperable "sibling" to HomePlug AV with lower cost, lower power consumption and decreased throughput.

In 2010, the IEEE 1901 was approved and HomePlug AV, as baseline technology for the FFT-OFDM PHY within the standard, was now an international standard. The HomePlug Powerline Alliance is a certifying body for IEEE 1901 products. The three major specifications published by HomePlug (HomePlug AV, HomePlug Green PHY and HomePlug AV2) are interoperable and compliant.

16.2.1.9 G.9959

G.9959 is a MAC layer protocol from ITU, designed for low bandwidth and cost, half-duplex reliable wireless communication. It is designed for real-time applications where time is really critical, reliability is important, and low power consumption is required. The MAC layer characteristics include: unique network identifiers that allow 232 nodes to join one network, collision avoidance mechanisms, backoff time in case of collision, automatic retransmission to guarantee reliability, dedicated wakeup pattern that allows nodes to sleep when they are out of communication and hence saves their power. G9959 MAC layer features include unique channel access, frame validation, acknowledgments, and retransmission.

16.2.1.10 Other Wireless Connectivity alternatives

16.2.1.10.1 Unlicensed LPWA

New proprietary radio technologies, provided by, for example, LoRa, Weightloss ,Sigfox etc.. have been developed and designed solely for machine-type communication (MTC) applications addressing the ultra-low-end sensor segment.

16.2.1.10.1.1 LoRaWAN

LoRaWAN is a newly arising wireless technology designed for low-power WAN networks with low cost, mobility, security, and bi- directional communication for IoT applications. It is a low-power consumption optimized protocol designed for scalable wireless networks with millions of devices. It supports redundant operation, location free, low cost, low power and energy harvesting technologies to support the future needs of IoT while enabling mobility and ease of use features.

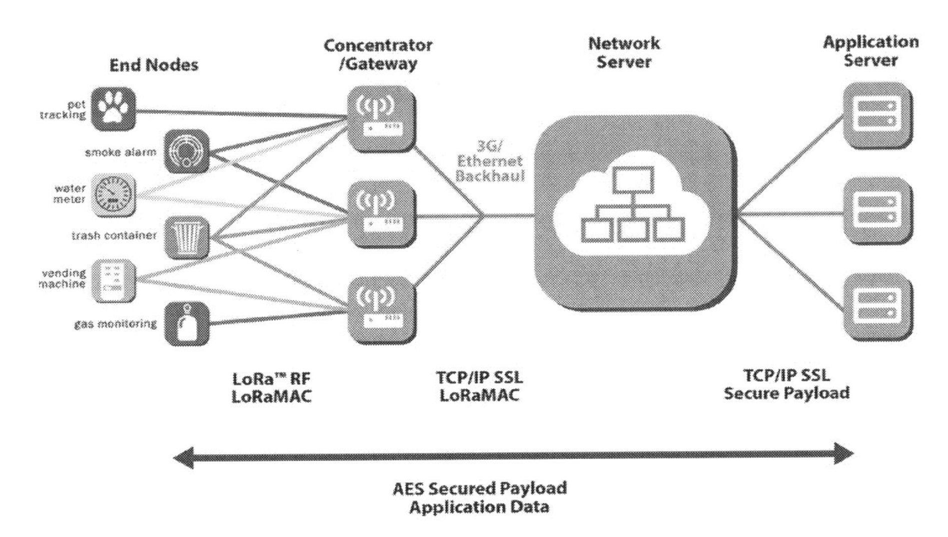

Figure: LoRaWAN Network Architecture

LoRaWAN uses star topology as it increases battery lifetime when long-range connectivity is used.

LoRa network consists of several elements.

LoRa Nodes / End Points: LoRa end points are the sensors or application where sensing and control takes place. These nodes are often placed remotely. Examples, sensors, tracking devices, etc.

LoRa Gateways: Unlike cellular communication where mobile devices are associated with the serving base stations, in LoRaWAN nodes are associated with a specific gateway. Instead, any data transmitted by the node is sent to all gateways and each gateway which receives a signal transmits it to a cloud based network server.

Typically the gateways and network servers are connected via some backhaul (cellular, Wi-Fi, ethernet or satellite).

Network Servers: The networks server has all the intelligence. It filters the duplicate packets from different gateways, does security check, send ACKs to the gateways. In the end if a packet is intended for an application server, the network server sends the packet to the specific application server.

Using this type of network where all gateways can send the same packet to the network server, the need of hand-off or handover is removed. This is useful for asset-tracking application where assets move from one location to another.

16.2.1.10.2 *Weightless*

Weightless is another wireless WAN technology for IoT applications designed by the Weightless Special Interest Group (SIG) - a non-profit global organization. It has two sets of standards: Weightless-N and Weightless-W. Weightless-N was first developed to support low cost, low power M2M communication using time division multiple access with frequency hopping to minimize the interference. It uses ultra- narrow bands in the sub-1GHz ISM frequency band. On the other hand, Weightless-W provides the same features but uses television band frequencies.

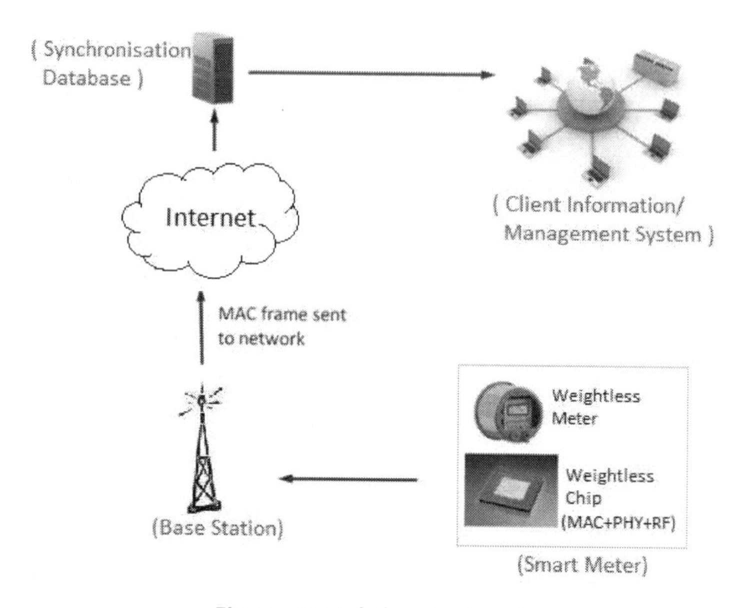

Figure: Weightless Network

As shown in the figure above, terminal (e.g. smart meter) will pass the reading information to the weightless PHY/radio module which encodes and transmits the information over weightless air interface. There are various modules in the weightless terminal transmitter. Typically, it provides modulation, spreading and forward error correction functionalities to the data. Base station receiver receives the downlink frame and does the decoding and passes the frame to the core internet backbone network as per appropriate format. The information transmitted by the terminals are routed via synchronization database to the client information/management system.

16.2.1.10.3 *Sigfox*

SIGFOX provides a cellular style network operator that provides a tailor-made solution for low-throughput Internet of Things and M2M applications.

For a host of applications from smart meters to control nodes that need connectivity over long ranges the only option until recently has been to use a cellular connection.

This option has several disadvantages because cellular phone systems are focussed on voice and high data rates. They are not suited to low data rate connections as the radio interface is complex and this adds cost and power consumption - too much for most M2M / IoT applications.

In view of the low data rates used for IoT connections, the SIGFOX network employs Ultra-Narrow Band, UNB technology. This enables very low transmitter power levels to be used while still being able to maintain a robust data connection.

The SIGFOX radio link uses unlicensed ISM radio bands. The exact frequencies can vary according to national regulations, but in Europe the 868MHz band is widely used and in the US it is 915MHz.

The density of the cells in the SIGFOX network is based on an average range of about 30-50km in rural areas and in urban areas where there are usually more obstructions and noise is greater the range may be reduced to between 3 and 10km. Distances can be much higher for outdoor nodes where SIGFOX states line of sight messages could travel over 1000km.

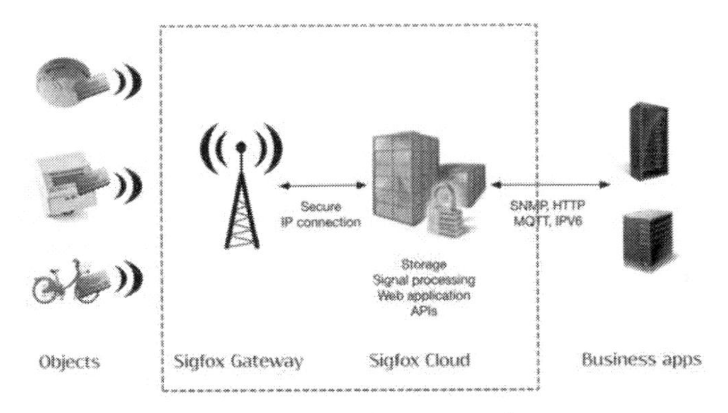

Figure: Sigfox Network

The overall SIGFOX network topology has been designed to provide a scalable, high-capacity network, with very low energy consumption, while maintaining a simple and easy to rollout star-based cell infrastructure.

16.2.1.11 Cellular Technologies

With each new generation, cellular systems (3GPP technologies like GSM, WCDMA, LTE) have evolved to provide support for new services and applications. Currently the fifth generation (5G) radio access network is under development and an important pillar in this project is support for the Internet of Things (IoT). These WANs operate on licensed spectrum and historically have primarily targeted high-quality mobile voice and data services. Now, however, they are being rapidly evolved with new functionality and the new radio access technology narrowband IoT (NB-IoT) specifically tailored to form an attractive solution for emerging low power wide area (LPWA) applications.

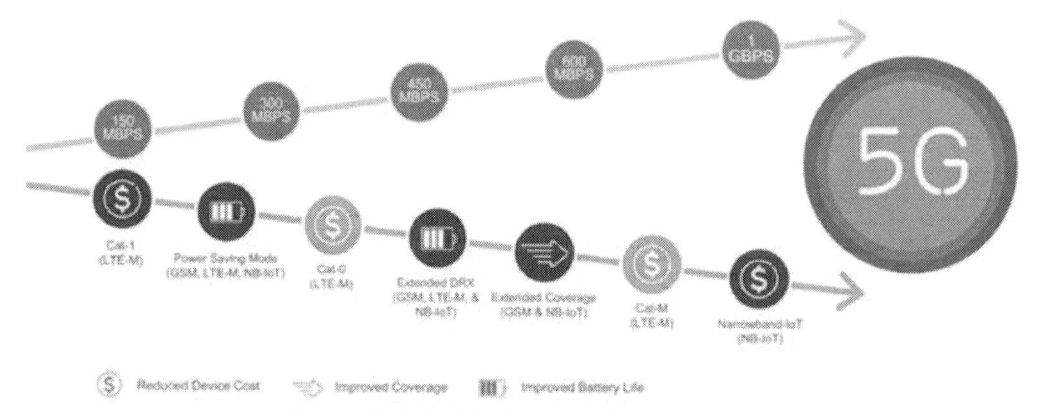

Figure: Cellular Technology Development

16.2.1.11.1 Extended Coverage GSM Internet of Things (EC-GSM-IoT)

EC-GSM-IoT stands for Extended Coverage GSM IoT. This is a low-power wide area (LPWA) cellular technology based on eGPRS and designed as a high capacity, long range, low energy cellular system for IoT applications.

GSM is still the dominant mobile technology in many markets, and the vast majority of cellular M2M applications today use GPRS/EDGE for connectivity. GSM is likely to continue playing a key role in the IoT well into the future, due to its global coverage footprint, time to market and cost advantages.

Recognizing this – and identifying the requirements for IoT– an initiative was undertaken in 3GPP Release 13 to further improve GSM.

EC-GSM-IoT can be deployed on existing GSM networks with a simple software update. This will allow for EC-GSM-IoT networks to co-exist with 2G, 3G, and 4G mobile networks. It will also benefit from all the security and privacy mobile network features, such as support for user identity confidentiality, entity authentication, confidentiality, data integrity, and mobile equipment identification.

The resulting EC-GSM functionality enables coverage improvements of up to 20dB with respect to GPRS on the 900MHz band. This coverage extension is achieved for both the data and control planes by utilizing the concept of repetitions and signal combining techniques. It is handled in a dynamic manner with multiple coverage classes to ensure optimal balance between coverage and performance.

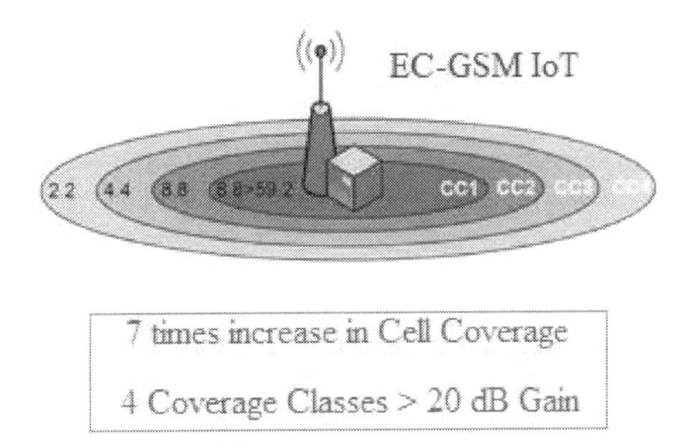

Figure: There are four coverage classes supported by EC-GSM

EC-GSM is achieved by defining new control and data channels mapped over legacy GSM. It allows multiplexing of new EC-GSM devices and traffic with legacy EDGE and GPRS. No new network carriers are required: new software on existing GSM networks is sufficient and provides combined capacity of up to 50,000 devices per cell on a single transceiver.

16.2.1.11.2 LTE-M

LTE-M is the simplified industry term for the LTE-MTC(Machine Type Communication) low power wide area (LPWA) technology standard published by 3GPP in the Release 13 specification. It specifically refers to LTE CatM1, suitable for the IoT. LTE-M is a low power wide area technology which supports IoT through lower device complexity and provides extended coverage, while allowing the reuse of the LTE installed base. This allows battery lifetime as long as 10 years or more for a wide range of use cases, with the modem costs reduced to 20-25% of the current EGPRS modems.

Supported by all major mobile equipment, chipset and module manufacturers, LTE-M networks will co-exist with 2G, 3G, and 4G mobile networks and benefit from all the security and privacy features of mobile networks, such as support for user identity confidentiality, entity authentication, confidentiality, data integrity, and mobile equipment identification.

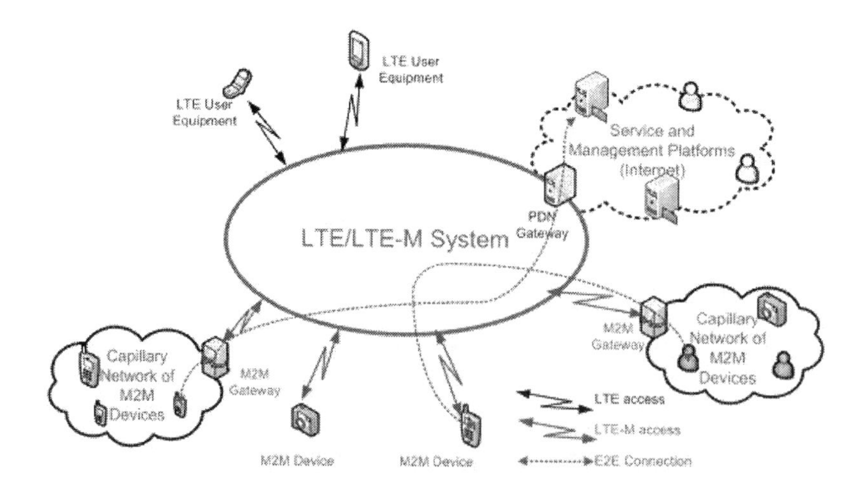

Figure: Typical LTE/LTE-M M2M/IoT Network Architecture

16.2.1.11.3 NB-IoT

Narrowband Internet of Things (NB-IoT) is to a large extent a new radio access technology reusing components from LTE. It can operate in a system bandwidth as narrow as 200 kHz and supports deployment both in spectrum originally intended for GSM or LTE. It also supports a minimum device bandwidth of only 3.75 kHz. This design gives NB-IoT a high deployment flexibility and high system capacity. Just as EC-GSM-IoT and LTE-M the technology also supports energy efficient operation, ultra-low device complexity and ubiquities coverage.

NB-IoT provides lean setup procedures, and a capacity evaluation indicates that each 200kHz NB-IoT carrier can support more than 200,000 subscribers. The solution can easily be scaled up by adding multiple NB-IoT carriers when needed. NB-IoT also comes with an extended coverage of up to 20dB, and battery saving features, Power Saving Mode and eDRX for more than 10 years of battery life.

NB-IoT can be deployed in 3 different modes; namely In-band, Guard-band and Stand-Alone.

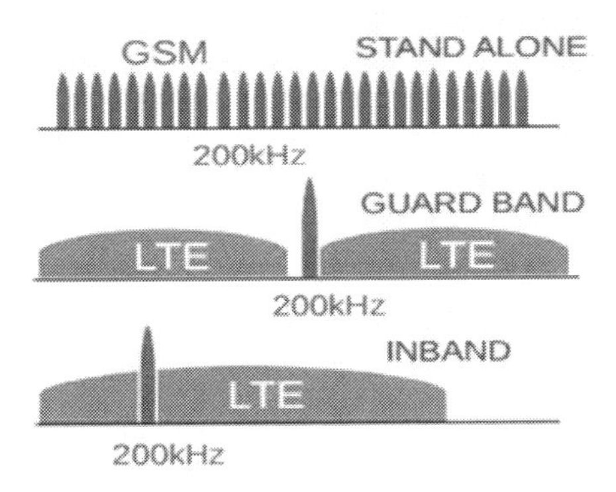

Figure: NB-IoT Deployment

NB-IoT is designed to be tightly integrated and interwork with LTE, which provides great deployment flexibility. The NB-IoT carrier can be deployed in the LTE guard band, embedded within a normal LTE carrier, or as a standalone carrier in, for example, GSM bands.

16.2.2 Network Layer Routing Protocol

In this section, we discuss some standard and non-standard protocols that are used for routing in IoT applications. It should be noted that we have partitioned the network layer in two sublayers: routing layer which handles the transfer the packets from source to destination, and an encapsulation layer that forms the packets. Encapsulation mechanisms will be discussed in the next section.

16.2.2.1 RPL

RPL is the IPv6 Routing Protocol for LLN (Low-Power and Lossy Networks) as defined in RFC6550. It builds a Destination Oriented Directed Acyclic Graph (DODAG) that has only one route from each leaf node to the root in which all the traffic from the node will be routed to. At first, each node sends a DODAG Information Object (DIO) advertising itself as the root. This message is propagated in the network and the whole DODAG is gradually built. When communicating, the node sends a Destination Advertisement Object (DAO) to its parents, the DAO is propagated to the root and the root decides where to send it depending on the destination. When a new node wants to join the network, it sends a DODAG Information Solicitation (DIS) request to join the network and the root will reply back with a DAO Acknowledgment (DAO-ACK) confirming the join. RPL nodes can be stateless, which is most common, or stateful. A stateless node keeps tracks of its parents only. Only root has the complete knowledge of the entire DODAG. Hence, all communications go through the root in every case. A stateful node keeps track of its children and parents and hence when communicating inside a sub-tree of the DODAG, it does not have to go through the root.

16.2.2.2 CORPL

An extension of RPL is CORPL, or cognitive RPL, which is designed for cognitive networks and uses DODAG topology generation but with two new modifications to RPL. CORPL utilizes opportunistic forwarding to forward the packet by choosing multiple forwarders (forwarder set) and coordinates between the nodes to choose the best next hop to forward the packet to. DODAG is built in the same way as RPL. Each node maintains a forwarding set instead of its parent only and updates its neighbor with its changes using DIO messages. Based on the updated information, each node dynamically updates its neighbor priorities in order to construct the forwarder set.

16.2.2.3 CARP

Channel-Aware Routing Protocol (CARP) is a distributed routing protocol designed for underwater communication. It can be used for IoT due to its lightweight packets. It considers link quality, which is computed based on historical successful data transmission gathered from neighboring sensors, to select the forwarding nodes. There are two scenarios: network initialization and data forwarding. In network initialization, a HELLO packet is broadcasted from the sink to all other nodes in the networks. In data forwarding, the packet is routed from sensor to sink in a hop- by-hop fashion. Each next hop is determined independently. The main problem with CARP is that it does not support reusability of previously collected data. In other words, if the application requires sensor data only when it changes significantly, then CARP data forwarding is not beneficial to that specific application. An enhancement of CARP was done in E-CARP by allowing the sink node to save previously received sensory data. When new data is needed, E-CARP sends a Ping packet which is replied with the data from the sen sors nodes. Thus, E-CARP reduces the communication overhead drastically.

16.2.2.4 Comparison of RPL, CORPL & CARP

Following table mentions comparison between RPL, CORPL and CARP protocols.

Features	RPL	CORPL	CARP
Full Form	Routing Protocol for Low-Power and Lossy Networks	Cognitive RPL	Channel-Aware Routing Protocol
Server technologies	Supported	Supported	Not supported
Security	Not supported	Not supported	Not supported
Storage management	Supported	Not supported	Supported
Data management	Supported	Supported	Supported

Table : Comparison of RPL, CORPL & CARP

Three routing protocols in IoT were discussed in above section. RPL is the most commonly used one. It is a distance vector protocol designed by IETF in 2012. CORPL is a non-standard extension of RPL that is designed for cognitive networks and utilizes the opportunistic forwarding to forward packets at each hop. On the other hand, CARP is the only distributed hop based routing protocol that is designed for IoT sensor network applications. CARP is used for underwater communication mostly. Since it is not standardized and just proposed in literature, it is not yet used in other IoT applications.

16.2.3 Network Layer Encapsulation Protocols

16.2.3.1 IPv6 in IoT

The creation of IPv6, and its slow replacement of IPv4, has been a huge and critical innovation for the future of internet communications. The primary function of IPv6 is to allow for more unique TCP/IP address identifiers to be created, now that we've run out of the 4.3 billion created with IPv4. This is one of the main reasons why IPv6 is such an important innovation for the Internet of Things (IoT). Internet-connected products are becoming increasingly popular, and while IPv4 addresses couldn't meet the demand for IoT products, IPv6 gives IoT products a platform to operate on for a very long time.

While IPv6 is an excellent and necessary upgrade from IPv4, it's certainly not an be all, end all solution for the IoT.

One problem in IoT applications is that IPv6 addresses are too long and cannot fit in most IoT datalink frames which are relatively much smaller. Some need very slim packet headers to maximize payload allocations in very small datagrams. For example, if you are limited to sending only 20 numbers at a time, and 18 are needed just for addressing, you don't have much room left for useful information. This means that long-range messages need as little of the message to be used for addressing as possible. When this is the case, IPv6 isn't a good solution because it has too much overhead. Hence, IETF is developing a set of standards to encapsulate IPv6 datagrams in different datalink layer frames for use in IoT applications. In coming sections, we review these mechanisms briefly.

16.2.3.2 6LoWPAN

IPv6 over Low power Wireless Personal Area Network (6LoWPAN) is the first and most commonly used standard in this category. It efficiently encapsulates IPv6 long headers in IEEE802.15.4 small packets, which cannot exceed 128 bytes. The specification supports different length addresses, low bandwidth, different topologies including star or mesh, power consumption, low cost, scalable networks, mobility, unreliability and long sleep time. The standard provides header compression to reduce transmission overhead, fragmentation to meet the 128-byte maximum frame length in IEEE802.15.4, and support of multi-hop delivery. Frames in 6LoWPAN use four types of headers: No 6loWPAN header (00), Dispatch header (01), Mesh header (10) and Fragmentation header (11). In No 6loWPAN header case, any frame that does not follow 6loWPAN specifications is discarded. Dispatch header is used for multicasting and IPv6 header compressions. Mesh headers are used for broadcasting; while Fragmentation headers are used to break long IPv6 header to fit into fragments of maximum 128-byte length.

16.2.3.3 6TiSCH

6TiSCH working group in IETF is developing standards to allow IPv6 to pass through Time-Slotted Channel Hopping (TSCH) mode of IEEE 802.15.4e datalinks. It defines a Channel Distribution usage matrix consisting of available frequencies in columns and time-slots available for network scheduling operations in rows. This matrix is portioned into chucks where each chunk contains time and frequencies and is globally known to all nodes in the network. The nodes within the same interference domain negotiate their scheduling so that each node gets to transmit in a chunk within its interference domain. Scheduling becomes an optimization problem where time slots are assigned to a group of neighboring nodes sharing the same application. The standard does not specify how the scheduling can be done and leaves that to be an application specific problem in order to allow for maximum flexibility for different IoT applications. The scheduling can be centralized or distributed depending on application or the topology used in the MAC layer.

16.2.3.4 6Lo

IPv6 over Networks of Resource-constrained Nodes (6Lo) working group in IETF is developing a set of standards on transmission of IPv6 frames on various datalinks. Although, 6LowPAN and 6TiSCH, which cover IEEE 802.15.4 and IEEE 802.15.4e, were developed by different working groups, it became clear that there are many more datalinks to be covered and so 6Lo working group was formed. At the time of this writing most of the 6Lo specifications have not been finalized and are in various stages of drafts. For example, IPV6 over Bluetooth Low Energy Mesh Networks, IPv6 over IEEE 485 Master-Slave/Token Passing (MS/TP) networks, IPV6 over DECT/ULE, IPV6 over NFC, IPv6 over IEEE 802.11ah, and IPv6 over Wireless Networks for Industrial Automation Process Automation (WIA-PA) drafts are being developed to specify how to transmit IPv6 datagrams over their respective datalinks. Two of these 6Lo specifications IPv6 over G.9959 and IPv6 over Bluetooth Low Energyâ€ have been approved as RFC and are described in next sections.

16.2.3.4.1 IPv6 over G.9959

RFC 7428 defines the frame format for transmitting IPv6 packet on ITU-T G.9959 networks. G.9959 defines a unique 32-bit home network identifier that is assigned by the controller and 8-bit host identifier that is allocated for each node. An IPv6 link local address must be constructed by the link layer derived 8-bit host identifier so that it can be compressed in G.9959 frame. Furthermore, the same header compression as in 6lowPAN is used here to fit an IPv6 packet into G.9959 frames. RFC 7428 also provides a level of security by a shared network key that is used for encryption. However, applications with a higher level of security requirements need to handle their end-to-end encryption and authentication using their own higher layer security mechanisms.

16.2.3.4.2 IPv6 over Bluetooth Low Energy

Bluetooth Low Energy is also known as Bluetooth Smart and was introduced in Bluetooth V4.0 and enhanced in V4.1. RFC 7668 [RFC7668], which specifies IPv6 over Bluetooth LE, reuses most of the 6LowPAN compression techniques. However, since the Logical Link Control and Adaptation Protocol (L2CAP) sublayer in Bluetooth already provides segmentation and reassembly of larger payloads in to 27 byte L2CAP packets, fragmentation features from 6LowPAN standards are not used. Another significant difference is that Bluetooth Low Energy does not currently support formation of multi-hop networks at the link layer. Instead, a central node acts as a router between lower-powered peripheral nodes.

To Summaries this section , First, two standards for IPv6 over 802.15.4 and 802.15.4e were discussed. Such protocols are important as 802.15.4e is the most widely use encapsulation framework designed for IoT. Following that, 6Lo specifications are briefly and broadly discussed just to present their existence in IETF standards. These drafts handle passing IPv6 over different channel access mechanism using 6LoWPAN standards. Then, two of 6Lo Specifications which became IETF RFCs are discussed in more details. The importance of presenting these standards is to highlight the challenge of interoperability between different MAC standards which is still challenging due to the diversity of protocols.

16.2.4 Application Layer Protocols

This section reviews standards and protocols for message passing in IoT session layer proposed by different standardization organizations. Most of the IP applications, including IoT applications use TCP or UDP for transport. However, there are several message distribution functions that are common among many IoT applications; it is desirable that these functions be implemented in an interoperable standard ways by different applications.

16.2.4.1 MQTT

Message Queue Telemetry Transport (MQTT) was introduced by IBM in 1999 and standardized by OASIS in 2013. It is designed to provide embedded connectivity between applications and middleware™ on one side and networks and communications on the other side. It is used in IoT wireless technologies such as zigbee, LoRaWAN etc.

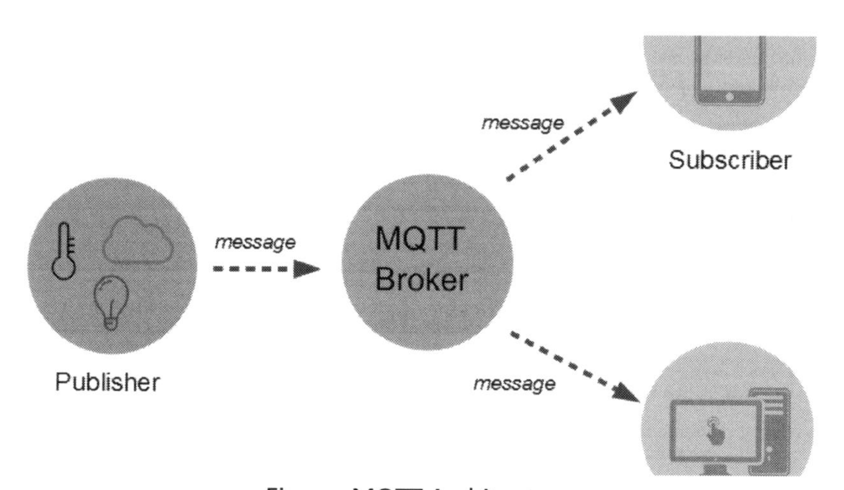

Figure: MQTT Architecture

It follows a publish/subscribe architecture, as shown in Figure, where the system consists of three main components: publishers, subscribers, and a broker. From IoT point of view, publishers are basically the lightweight sensors that connect to the broker to send their data and go back to sleep whenever possible. Subscribers are applications that are interested in a certain topic, or sensory data, so they connect to brokers to be informed whenever new data are received. The brokers classify sensory data in topics and send them to subscribers interested in the topics.

16.2.4.2 SMQTT

An extension of MQTT is Secure MQTT (SMQTT) which uses encryption based on lightweight attribute-based encryption. The main advantage of using such encryption is the broadcast encryption feature, in which one message is encrypted and delivered to multiple other nodes, which is quite common in IoT applications.

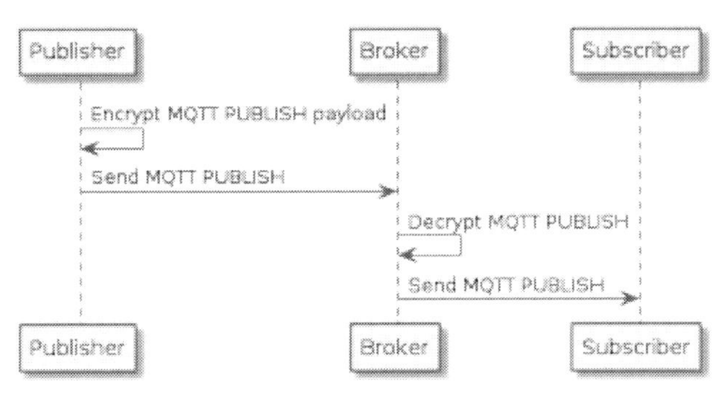

Figure : SMQTT End-to-End (E2E) Encryption

In general, the algorithm consists of four main stages: setup, encryption, publish and decryption. In the setup phase, the subscribers and publishers register themselves to the broker and get a master secret key according to their developer™ choice of key generation algorithm. Then, when the data is published, it is encrypted, published by the broker which sends it to the subscribers and finally decrypted at the subscribers which have the same master secret key. The key generation and encryption algorithms are not standardized. SMQTT is proposed only to enhance MQTT security feature.

16.2.4.3 AMQP

The Advanced Message Queuing Protocol (AMQP) is another Application/session layer protocol that was designed for financial industry. It runs over TCP and provides a publish/ subscribe architecture which is like that of MQTT. The difference is that the broker is divided into two main components: exchange and queues, as shown in Figure.

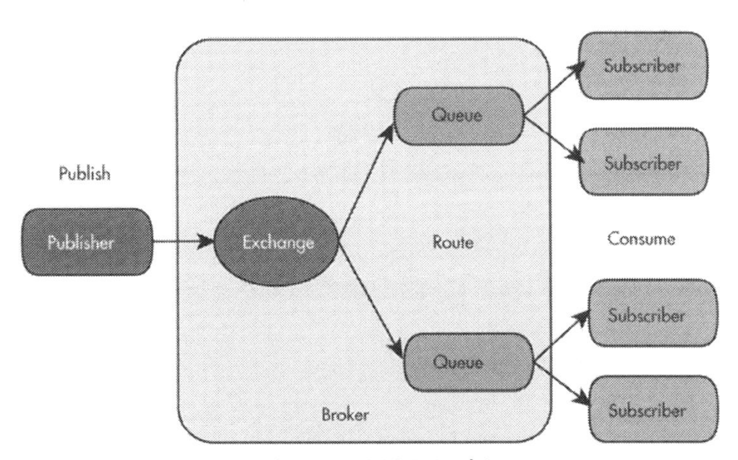

Figure: AMQP Architecture

The exchange is responsible for receiving publisher messages and distributing them to queues based on pre-defined roles and conditions. Queues basically represent the topics and subscribed by subscribers which will get the sensory data whenever they are available in the queue.

16.2.4.4 CoAP

The Constrained Application Protocol (CoAP) is another session layer protocol designed by IETF Constrained RESTful Environment (Core) working group to provide lightweight RESTful (HTTP) interface. Representational State Transfer (REST) is the standard interface between HTTP client and servers. However, for lightweight applications such as IoT, REST could result in significant overhead and power consumption. CoAP is designed to enable low-power sensors to use RESTful services while meeting their power constrains. It is built over UDP, instead of TCP commonly used in HTTP and has a light mechanism to provide reliability.

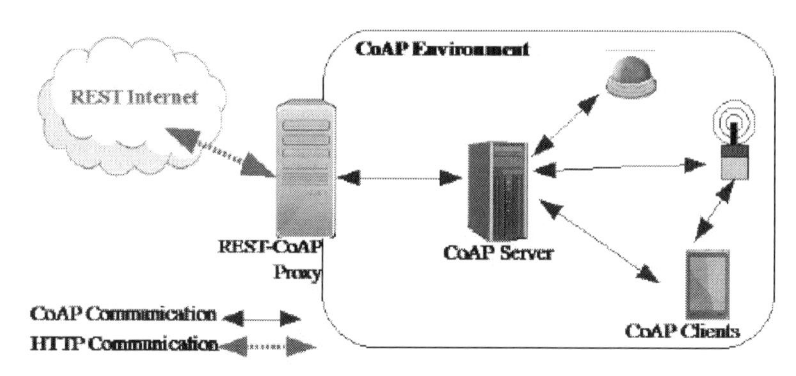

Figure: CoAP Architecture

CoAP architecture is divided into two main sublayers: messaging and request/response. The messaging sublayer is responsible for reliability and duplication of messages while the request/response sublayer is responsible for communication.

As shown in below Figure, CoAP has four messaging modes: confirmable, non-confirmable, piggyback and separate. Confirmable and non-confirmable modes represent the reliable and unreliable transmissions, respectively while the other modes are used for request/response. Piggyback is used for client/server direct communication where the server sends its response directly after receiving the message, i.e., within the acknowledgment message. On the other hand, the separate mode is used when the server response comes in a message separate from the acknowledgment, and may take some time to be sent by the server.

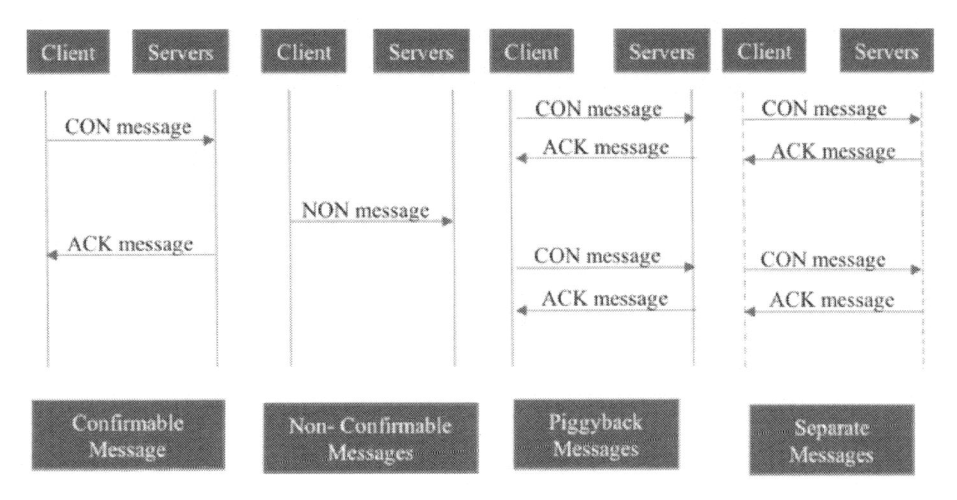

Figure CoAP messages

As in HTTP, CoAP utilizes GET, PUT, PUSH, DELETE messages requests to retrieve, create, update, and delete, respectively.

16.2.4.5 XMPP

Extensible Messaging and Presence Protocol (XMPP) is a messaging protocol that was designed originally for chatting and message exchange applications. It was standardized by IETF more than a decade ago. Hence, it is well known and has proven to be highly efficient over the internet. Recently, it has been reused for IoT applications as well as a protocol for SDN. This reusing of the same standard is due to its use of XML which makes it easily extensible. XMPP supports both publish/ subscribe and request/ response architecture and it is up to the application developer to choose which architecture to use. It is designed for near real-time applications and, thus, efficiently supports low-latency small messages. It does not provide any quality of service guarantees and, hence, is not practical for M2M communications. Moreover, XML messages create additional overhead due to lots of headers and tag formats which increase the power consumption that is critical for IoT application. Hence, XMPP is rarely used in IoT but has gained some interest for enhancing its architecture in order to support IoT applications.

16.2.4.6 DDS

Data Distribution Service (DDS) is another publish/subscribe protocol that is designed by the Object Management Group (OMG) for M2M communications [DDS]. The basic benefit of this protocol is the excellent quality of service levels and reliability guarantees as it relies on a broker-less architecture, which suits IoT and M2M communication. It offers 23 quality-of-service levels which allow it to offer a variety of quality criteria including: security, urgency, priority, durability, reliability, etc. It defines two sublayers: data-centric publish- subscribe and data-local reconstruction sublayers. The first takes the responsibility of message delivery to the subscribers while the second is optional and allows a simple integration of DDS in the application layer. Publisher layer is responsible for sensory data distribution. Data writer interacts with the publishers to agree about the data and changes to be sent to the subscribers. Subscribers are the receivers of sensory data to be delivered to the IoT application. Data readers basically read the published data and deliver it to the subscribers and the topics are basically the data that are being published. In others words, data writers and data reader take the responsibilities of the broker in the broker-based architectures.

16.2.4.7 IoT Application/Session Layer Standards Comparison

IoT has many standardized session layer protocols which were briefly highlighted in this section. These session layer protocols are application dependent and the choice between them are very application specific. It should be noted that MQTT is the most widely used in IoT due to its low overhead and power consumption. It's an organizational and applications specific to choose between these standards. For example, if an application has already been built with XML and can, therefore, accept a bit of overhead in its headers, XMPP might be the best option to choose among session layer protocols. On the other hand, if the application is really overhead and power sensitive, then choosing MQTT would be the best option, however, that comes with the additional broker implementation. If the application requires REST functionality as it will be HTTP based, then CoAP would be the best option if not the only one. Table below summarizes comparison points between these different application/session layer protocols.

Protocols	UDP/TCP	Architecture	Security and QoS	Header Size (bytes)	Max Length(bytes)
MQTT	TCP	Pub/Sub	Both	2	5
AMQP	TCP	Pub/Sub	Both	8	-
CoAP	UDP	Req/Res	Both	4	20 (typical)
XMPP	TCP	Both	Security	-	-
DDS	TCP/UDP	Pub/Sub	QoS	-	-

Table: IoT Application/Session Layer Standards Comparison

16.2.5 IoT Management Protocols

This section discusses two main management standards for IoT that provide heterogeneous communication i.e. communication between different datalinks. Management protocols play an important role in IoT due to the diversity of protocols and standards at different layers of networking. The need for heterogeneous and easy communication between different protocols at the same or different layers is critical for IoT applications. Existing standards mainly facilitate communication between protocols at the same layer; however, it is still a challenge to facilitate communication at different layers in IoT.

16.2.5.1 Interconnection of Heterogeneous Datalinks

As IoT environments rely on many different MAC protocols, interoperability among all these technologies in a challenge that needs to be handled. IEEE 1905.1 standards offer such interoperability by providing an abstraction layer that is built in top on all these heterogeneous MAC protocols [1905]. This abstraction hides the diversity of the different protocols without requiring any change to the design of each MAC, as illustrated earlier Section of Data Link Layer. The basic idea behind this protocol is the abstraction layer which is used to exchange messages, called Control Message Data Units (CMDUs) among all standards compatible devices. As shown in below Figure, All IEEE 1905.1 compliant devices understand a common Abstraction Layer Management Entity (ALME) protocol which offers different services including: neighbor discovery, topology exchange, topology change notification, measured traffic statistics exchange, flow forwarding rules, and security associations.

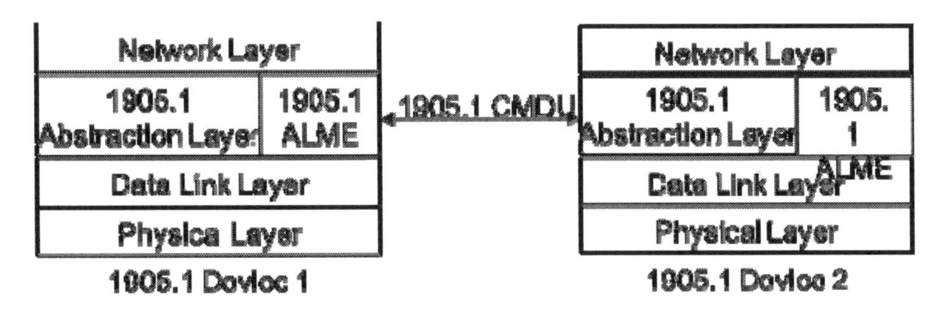

Figure: IEEE 1905.1 Protocol Structure

16.2.5.2 Smart Transducer Interface

IEEE 1451 is a set of standards developed to allow management of different analog transducers and sensors. The basic idea of this standard is the use of plug and play identification using standardized Transducer electronic data sheets (TEDSs). Each transducer contains a TEDS which includes all the information needed by the measurement system including device ID, characteristics and interface beside the data coming from the sensors. Data sheets are stored embedded memory within the transducer or the sensor and have a defined encoding mechanism to understand a broad number of sensor types and applications. The memory usage is minimized by utilizing the small XML based messages which are understood by different manufactures and different applications.

16.2.6 Security in IoT Protocols

Security is another aspect of IoT applications which is critical and can be found in all almost all layers of the IoT protocols. Threats exist at all layers including datalink, network, session, and application layers. In this section, we briefly discuss the security mechanisms built in the IoT protocols.

16.2.6.1 MAC 802.15.4

MAC 802.15.4 offers different security modes by utilizing the Security Enabled Bit in the Frame Control field in the header. Security requirements include confidentiality, authentication, integrity, access control mechanisms and secured Time-Synchronized Communications.

16.2.6.2 6LoWPAN

6LoWPAN by itself does not offer any mechanisms for security. However, relevant documents include discussion of security threats, requirement and approach to consider in IoT network layer. For example, RFC 4944 discusses the possibility of duplicate EUI-64 interface addresses which are supposed to be unique . RFC 6282 discusses the security issues that are raised due to the problems introduced in RFC 4944. RFC 6568 addresses possible mechanisms to adopt security within constrained wireless sensor devices. In addition, a few recent drafts in discuss mechanisms to achieve security in 6loWPAN.

16.2.6.3 RPL

RPL offers different level of security by utilizing a Security field after the 4-byte ICMPv6 message header. Information in this field indicates the level of security and the cryptography algorithm used to encrypt the message. RPL offers support for data authenticity, semantic security, protection against replay attacks, confidentiality and key management. Levels of security in RPL include Unsecured, Preinstalled, and Authenticated. RPL attacks include Selective Forwarding, Sinkhole, Sybil, Hello Flooding, Wormhole, Black hole and Denial of Service attacks.

16.2.6.4 Application Layer

Applications can provide additional level of security using TLS or SSL as a transport layer protocol. In addition, end to end authentication and encryption algorithms can be used to handle different levels of security as required. For further discussion on security.

It should be noted that a number of new security approaches are also being developed that are suitable for resource constrained IoT devices. Some of these protocols are listed in Figure IoT Protocols.

16.2.6.5 IoT Security Protocols

16.2.6.5.1 OAuth 2.0 (Open Authorization)

OAuth 2.0 is an authorization framework, that allows a third party to access a resource owned by a resource owner without giving unencrypted credentials to the third party. For example a mobile app wants to access your Facebook profile to post status updates. You don't need to give your Facebook password to the app, instead you log into Facebook and as a result the app is authorized to use Facebook on your behalf. You can also revoke this authorization any time by deleting the privilege in the Facebook settings.

OAuth 2.0 has been standardized by the Internet Engineering Taskforce (IETF) and is published as RFC 6749. The IETF is now looking at specifying the use of OAuth 2.0 specifically for IoT applications, in its Authentication and Authorization in Constrained Environments working group.

This work is detailed in *RFC 7744 Use Cases for Authentication and Authorization in Constrained Environments (https://www.rfc-editor.org/rfc/rfc7744.txt),* published in January 2016. It details numerous use cases: container monitoring, home automation, personal health monitoring, building automation, smart metering and industrial control systems. In short, API security for IoT is a work in progress with promising prospects.

16.2.6.5.2 SMACK (Short message authentication check)

As we know ,IoT devices are supposed to be directly connected to the Internet, and many of them are likely to be battery powered. Hence, they are particularly vulnerable to Denial of Service (DoS) attacks specifically aimed at quickly draining battery and severely reducing device lifetime. SMACK, a security service which efficiently identifies invalid messages early after their reception, by checking a short and lightweight Message Authentication Code (MAC). So doing, further useless processing on invalid messages can be avoided, thus reducing the impact of DoS attacks and preserving battery life.

16.2.6.5.3 SASL (Simple Authentication and Security Layer)

As far as security, MQTT only supports the Simple Authentication and Security Layer (SASL) and Transport Layer Security (TLS) mechanisms respectively for the authentication and for the encryption of the communication channel.

The Simple Authentication and Security Layer (SASL) described in IEEE RFC 4422, is a framework for providing authentication and data security services in connection-oriented protocols via replaceable mechanisms. It provides a structured interface between protocols and mechanisms. The resulting framework allows new protocols to reuse existing mechanisms and allows old protocols to make use of new mechanisms. The framework also provides a protocol for securing subsequent protocol exchanges within a data security layer.

16.2.6.5.4 DTLS (Datagram Transport Layer Security)

DTLS allows datagram-based applications to communicate in a way that is designed to prevent eavesdropping, tampering, or message forgery. The DTLS protocol is based on the stream-oriented Transport Layer Security (TLS) protocol and is intended to provide similar security guarantees. The DTLS protocol datagram preserves the semantics of the underlying transport — the application does not suffer from the delays associated with stream protocols, but has to deal with packet reordering, loss of datagram and data larger than the size of a datagram network packet.

16.3 IoT Alliances and Organizations

16.3.1 Organizations:

- **ETSI** (European Telecommunications Standards Institute)
  (http://www.etsi.org/) - Connecting Things Cluster
  (http://www.etsi.org/technologies-clusters/clusters/connecting-things)
- **IETF** (Internet Engineering Task Force) (https://www.ietf.org/)
  - CoRE working group (Constrained RESTful Environments)
  (https://datatracker.ietf.org/wg/core/charter/)
  - 6lowpan working group (IPv6 over Low power WPAN)
  (https://datatracker.ietf.org/wg/6lowpan/charter/)
  - ROLL working group (Routing Over Low power and Lossy networks)
  (https://datatracker.ietf.org/wg/roll/charter/)
- **IEEE** (Institute of Electrical and Electronics Engineers)
  - IoT "Innovation Space" (http://standards.ieee.org/innovate/iot/)
- **OMG** (Object Management Group) (http://omg.org/)
  - Data Distribution Service Portal (http://portals.omg.org/dds/)

- **OASIS** (Organization for the Advancement of Structured Information Standards) (https://www.oasis-open.org/)
  - MQTT Technical Committee (https://www.oasis-open.org/committees/tc_home.php?wg_abbrev=mqtt)
- **OGC** (Open Geospatial Consortium) (http://www.opengeospatial.org/ogc)
  - Sensor Web for IoT Standards Working Group (http://www.ogcnetwork.net/IoT)

- **IoT-A** (http://www.iot-a.eu/public) :"The European Lighthouse Integrated Project addressing the Internet-of-Things Architecture, proposes the creation of an architectural reference model together with the definition of an initial set of key building blocks."
- **OneM2M** (http://www.onem2m.org/) :"The purpose and goal of oneM2M is to develop technical specifications which address the need for a common M2M Service Layer that can be readily embedded within various hardware and software, and relied upon to connect the myriad of devices in the field with M2M application servers worldwide."
- **OSIOT**(http://osiot.org/) :"An organization with the single focus to develop and promote royalty-free, open source standards for the emerging Internet of Things."
- **IoT-GSI** (Global Standards Initiative on Internet of Things) (http://www.itu.int/en/ITU-T/gsi/iot/Pages/default.aspx)
- **ISA** (International Society of Automation) (http://www.isa.org/)
- **W3C** (https://www.w3.org/)
  - Semantic Sensor Net Ontology (C:\Madhu\Pers\Goals\My books\0.2Student Guide Instrumentation\W3C (https:\www.w3.org\)
  - Web of Things Community Group (https://www.w3.org/community/wot/)
- EPC Global (http://www.gs1.org/epcglobal)
- **The IEC** (International Electrotechnical Commission), and ISO (International Organization for Standardization), through the JTC (Joint Technical Committee). Committee Page : (http://www.iso.org/iso/home/standards_development/list_of_iso_technical_committees.htm)
- **RRG** (Routing research group) (https://trac.tools.ietf.org/group/irtf/trac/wiki/RoutingResearchGroup)
- **HIPRG** (Host identity protocol research group) (https://www.ietf.org/proceedings/60/264.htm)

Eclipse Paho Project (https://www.eclipse.org/proposals/technology.paho/) : "The scope of the Paho project is to provide open source implementations of open and standard messaging protocols that support current and emerging requirements of M2M integration with Web and Enterprise middleware and applications. It will include client implementations for use on embedded platforms along with corresponding server support as determined by the community."

OpenWSN
(https://openwsn.atlassian.net/wiki/pages/viewpage.action?pageId=688187) :"Serves as a repository for open-source implementations of protocol stacks based on Internet of Things standards, using a variety of hardware and software platforms."

CASAGRAS (http://www.iot-casagras.org/page/standards-and-regulations) :"We are a key group of international partners representing Europe, the USA, China, Japan and Korea who has joined a strategic EU funded 7th Framework initiative that will look at global standards, regulatory and other issues concerning RFID and its role in realising an "Internet of Things."

16.3.2 Alliances

AllSeen Alliance (https://allseenalliance.org/) : "The AllSeen Alliance is a nonprofit consortium dedicated to enabling and driving the widespread adoption of products, systems and services that support the Internet of Everything with an open, universal development framework supported by a vibrant ecosystem and thriving technical community'

IPSO (http://www.ipso-alliance.org/) : "The Alliance is a global non-profit organization serving the various communities seeking to establish the Internet Protocol as the network for the connection of Smart Objects by providing coordinated marketing efforts available to the general public."

Wi-SUN Alliance (http://www.wi-sun.org/) : The Wi-SUN Alliance seeks to "advance seamless connectivity by promoting IEEE 802.15.4g standard based interoperability for global regional markets."

OMA (Open Mobile Alliance) (http://openmobilealliance.org/about-oma/work-program/m2m-enablers/) :"OMA is the Leading Industry Forum for Developing Market Driven, Interoperable Mobile Service Enablers"

Industrial Internet Consortium (http://www.iiconsortium.org/) :"Founded in 2014 to further development, adoption and wide-spread use of interconnected machines, intelligent analytics and people at work'

Z-Wave Alliance (https://z-wavealliance.org/): Established in 2005, the Z-Wave Alliance is comprised of industry leaders throughout the globe that are dedicated to the development and extension of Z-Wave as the key enabling technology for 'smart' home and business applications.

17 Chapter: Top Technology Trends in Future

17.1 Top Technologies in Sensors and Instrumentation

17.1.1 Biosensors

A biosensor comprises a transducer that measures the electrical signals generated through interaction of an analyte and a biological element.

A biosensor integrated with a signal processor measures the electrical signals and converts them into digital form, which can then be read on a local or remote computer.

17.1.2 LiDAR Sensor

LiDAR (light detection and ranging) is a remote, optical sensing technology, which uses laser pulses to illuminate the target object and photo detector to measure the return time of the laser pulse back to the source.

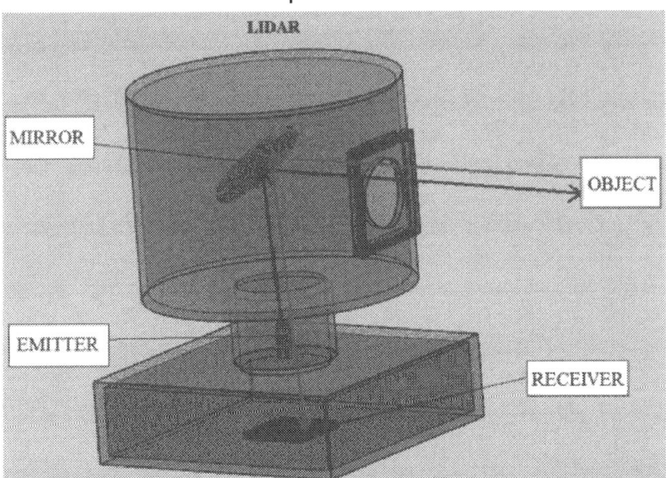

Figure: Lidar Sensor

The data received from the laser pulse can provide a high-resolution detailed picture of the object"s surroundings.

17.1.3 Sensor Fusion

Sensor fusion refers to a combination of sensor data from different and disparate sources, with the aim of providing a more meaningful, comprehensive, and relevant set of information.

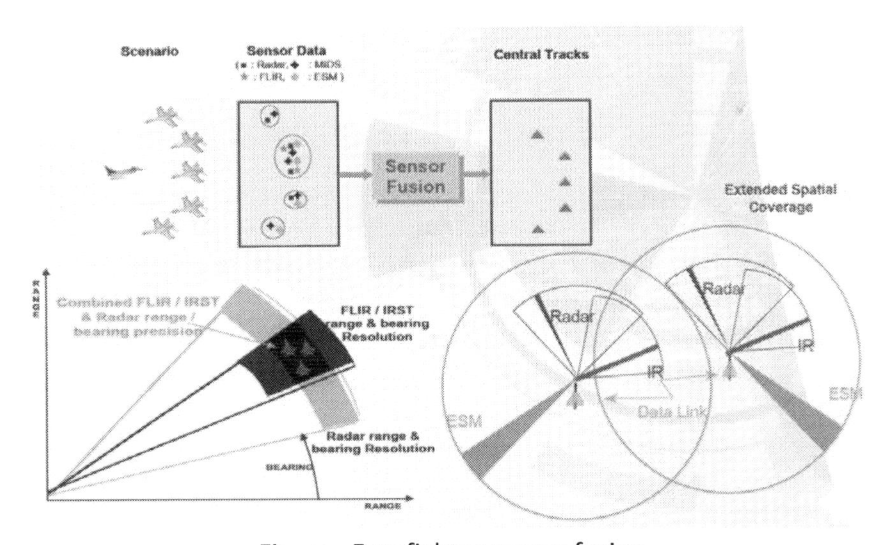

Figure : Eurofighter sensor fusion

Fusion of data from multiple sensors corrects the limitations of individual sensors. Developments in sensor fusion involve both hardware and software technologies and platforms.

17.1.4 Advanced Driver Assistance System (ADAS)

Advanced Driver Assistance System (ADAS) is a sophisticated device that assists vehicle drivers with enhanced and automated driving experience, while ensuring high safety and convenience.

Sensor technologies play a vital role in ensuring safety and also help to avoid collisions and fatal accidents, and entering into hazardous areas.

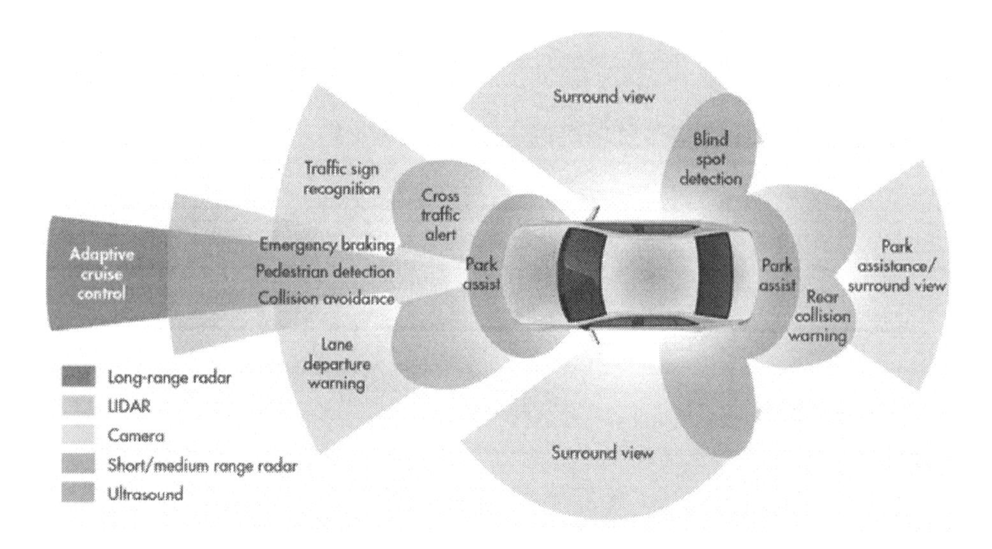

Figure : Advanced Driver Assistance System (ADAS)

17.1.5 Smart Sensors

The definition for Smart Sensors has changed over the years; the benchmark to be smart is ever increasing. A Smart Sensor needs to be able to capture information, process the information, and transmit the processed information through a designated location. With technology advancements, sensors have become interoperable with various systems and can tell their own health, control their operating conditions, and operate alongside sophisticated analytics systems.

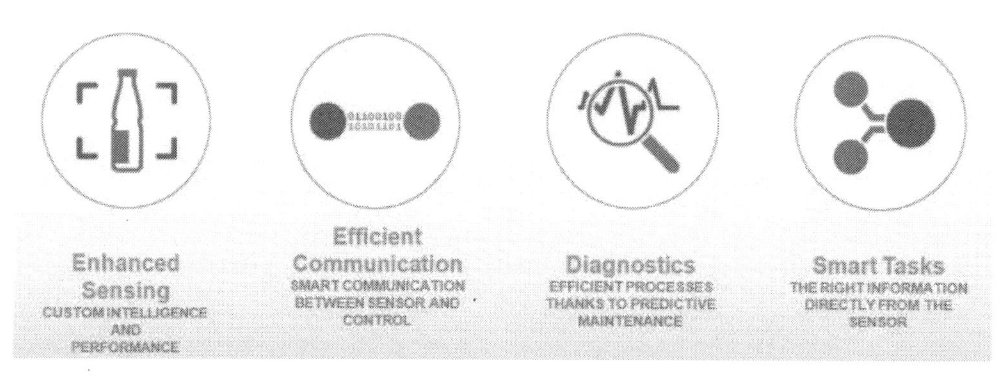

Figure: The Four Pillars of Smart Sensors

17.1.6 Photonic Sensors

A photonic sensor is a sensory device that is used to detect light and convert it into electricity.

Figure: Illustration of Photonic Sensor

The photonic sensor includes light transmission, light emission, light refraction, light amplification, and light detection capabilities through electro-optical instrumentation, optical components, and sophisticated nanophotonic systems, thereby causing flow of current.

17.1.7 Gesture Recognition

Gesture recognition (also includes touchless sensing) is an advanced form of human-machine interaction, where users can control devices in a touchless manner.

Figure: Gesture Recognition

The technology is gaining increased adoption in the consumer electronics and automotive industries for control of smart devices. R&D on the integration of gesture recognition technology in new products is expected to bolster technology development in a wide array of applications.

17.1.8 Large Area Sensors

Large area sensor comprises of an array of sensors, which are fabricated over a stretchable or a flexible substrate, unlike traditional sensors, which are usually fabricated on a single chip. This makes large area sensors a major innovation in the sensors segment owing to the flexibility they provide.

Large area sensors can be used for various applications in major industrial segments such as healthcare, robotics, retail, packaging, and defense owing to their thin, stretchable, and flexible form factors while exhibiting multiparameter sensing with high efficiency and accuracy. This makes them a unique sensor technology for the future.

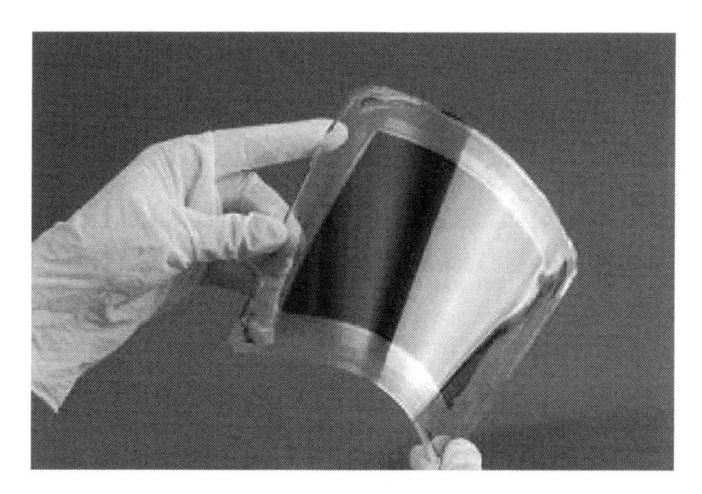

Figure : Large Area Sensors ,*The prime application for thin-film-transistors are backplanes for active-matrix displays, including in particular flexible displays. They are well-suited for integration with temperature or chemical sensors and more.*

Large area sensors will play a major role in offering flexible sensors for emerging applications such as wearables. These sensors will also give a new perspective to devices resulting in innovations in this domain.

17.1.9 Energy Harvesting

Energy harvesting (EH) involves harvesting energy from ambient sources and converting it into useful energy.

Advancements in materials are driving innovations in the EH area. Converting kinetic energy into electrical energy for wearable devices is emerging as a key trend for consumer electronics appliances.

The Internet-of-Things (IoT) demands endless machine-to-machine connectivity. Energy harvesting technologies facilitate self-powered wireless sensor networks, enabling IoT applications, such as building automation and industrial process control.

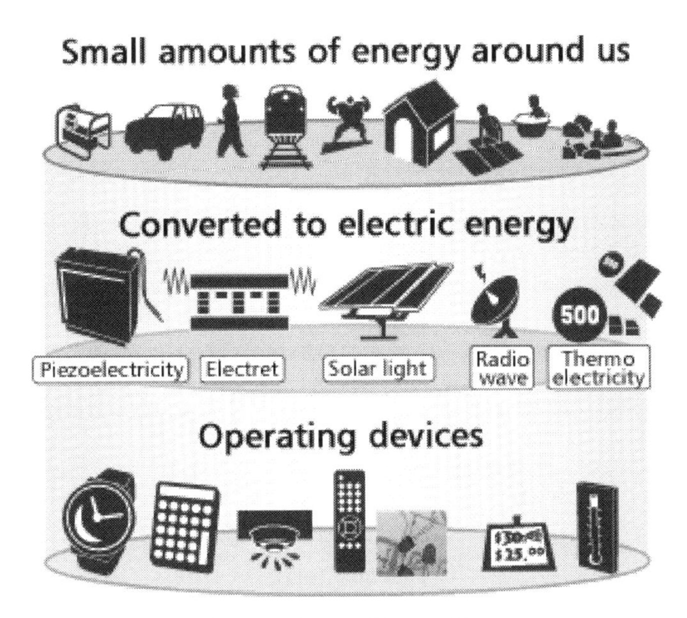

Figure: Energy Harvesting

Harvesting renewable energy sources, such as solar for self-powered sensor networks is a key driving force for notable advancements in the energy harvesting sector.

17.1.10 Electronic Skin/Skinput

With skinput technology, the electronic skin (E-skin) is utilized as an input medium and it forms an integral part of measuring human body electrical activities. E-skin is an ultrathin, flexible electronics technology which can be worn on human/animal skin. It can be used to measure and monitor the body's internal vital and physiological signs. Advancements in E-skin support technological developments in the field of flexible and stretchable displays which can be imprinted on the skin.

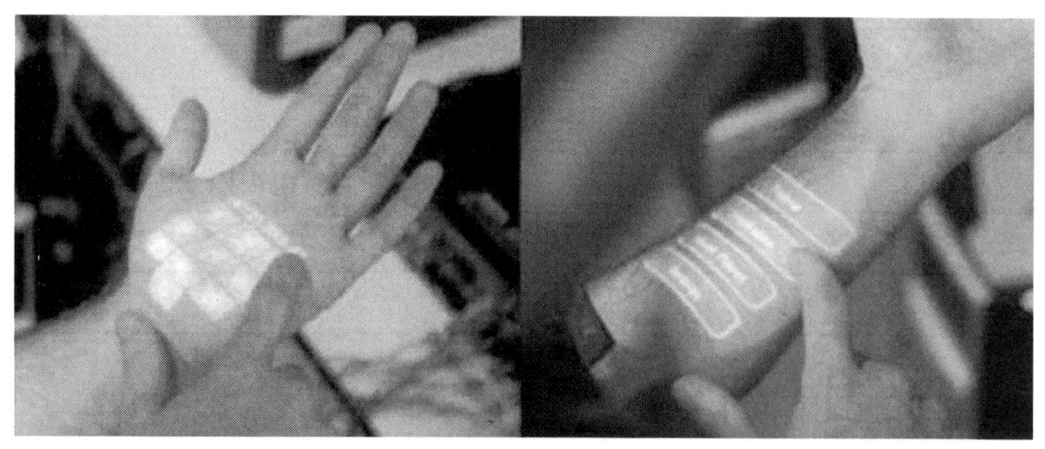

Figure: Skinput

Currently, the E-skin market is niche and technology developments are at a relatively nascent stage. The technology eliminates the need for the use of bulky devices, (especially for health monitoring applications) such as vital signs monitoring, drug delivery systems, and external controller devices including smart televisions, smartphones, and other smart computing devices.

17.2 Top Technologies in Manufacturing and Automation

Industries today are in the midst of the Fourth Industrial Revolution (4IR) driven by the use of smart connected devices. IoT solutions are moving towards integrating machines, services, and products leveraging big data analytics to obtain useful data through intelligent networks. Key benefits include: customization, improved productivity and efficiency, and reduced cost.

17.2.1 Collaborative Robots

Collaborative robots or cobots are robots that can work alongside humans. Their ability to ensure worker safety and integrate with the existing environment is enabling them to gain steady traction in recent years. Their interaction with humans is more natural. They share the same workspace and can be easily reprogrammed.

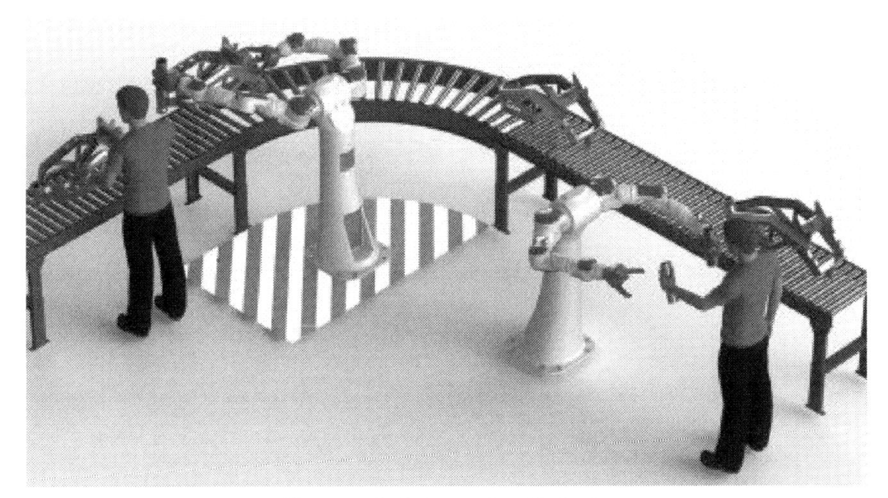

Figure: Collaborative Robots

Enabling technologies, such as artificial intelligence (AI), sensor technologies, facial recognition, and other technologies supporting Internet of Things have also contributed to advancements in collaborative robots.

17.2.2 Cognitive Manufacturing

Cognitive manufacturing enables interaction between humans and machines in an industrial environment. It provides machines with cognitive capabilities and enables them to perform tasks that are normally possible only by humans. This includes cognitive technologies such as speech recognition, decision making, and translation between languages.

Enabling technologies, such as deep learning, machine learning, predictive analytics, Natural Language Processing (NLP), pattern recognition, robotics supporting Industrial Internet of Things have also contributed to advancements in cognitive manufacturing.

17.2.3 Metal 3D Printing

Metal 3D printing (Metal 3DP) builds a three-dimensional metal object by adding layer-upon-layer of material. This novel technology is currently being used for rapid prototyping and manufacturing low-volume metal parts. Aluminum, titanium, stainless steel, cobalt chrome, copper, bronze, Inconel, and iron are some of the materials that can be printed using metal 3DP technologies.

3D printing of metal components will have mechanical properties equivalent to parts produced using conventional processes such as casting.

Selective Laser Melting (SLM) and Direct Metal Laser Sintering (DMLS) are the two commonly used processes that use granular materials for printing 3D metal parts.

Electron beam melting and ultrasonic additive manufacturing are other 3D printing techniques that can be used for printing dense metal parts.

17.2.4 Nanofabrication

Nanofabrication involves building or assembling of nanoscale (1 to 100 nanometer) electronic devices, structures, and producing nanoscale materials.

Materials, devices, and systems are built at the nanoscale level using top-down or bottom up approach and manipulating matter at scale less than 100 nm. Nanofabrication comprises techniques and tools for fabrication of one-, two-, or three-dimensional nanostructures.

Applications range from aerospace (for an airplane wing) to transportation (for laying miles of concrete on highways).

17.2.5 Hybrid Manufacturing

Hybrid manufacturing (HM) involves integration of additive manufacturing (AM) technology and conventional subtractive technology, which enables each process to work on the same machine. It can also involve integrating Laser beam machining and CNC or milling and 3D laser cladding in a single machine.

Combining high productivity machining (subtractive machining) and material deposition (additive manufacturing) is beginning to gain traction in the repair of blades, hardfacing, and other opportunities, where it offers reduced costs on repairs.

17.2.6 Big Area Additive Manufacturing

Big Area Additive Manufacturing, (or BAAM) is a type of 3D printing method aimed at production manufacturing applications for developing parts with dimensions in yards (and not inches). BAAM was developed by Cincinnati Inc., Oak Ridge National Laboratory, and Lockheed Martin, and was introduced in 2014.

Figure: Cincinnati BAAM Printer (Installed at Cincinnati)

BAAM initially was to print carbon-filled ABS and carbon fiber composites. The technology also can print thermoplastics, such as ABS, PLA, polycarbonate, polyphenyl sulfide, polyetherimide. With BAAM, cars, a house, wind turbine blade molds, were printed. The technique is used in rapid prototyping or tooling applications.

Other materials such as metal and concrete are also compatible with BAAM, but adoption appears to be limited.

R&D is geared towards development of large scale metal parts and concrete mold which has impact on aerospace and the building and construction industries.

17.2.7 Automated Optical Metrology

Optical metrology, which involves the science of measurement using light, is one method that is gaining major attraction due to its speed of inspection in comparison to other methods.

Another advantage is that it is a non-contact method of measuring without marking or deforming surfaces, such as sheet metal, composite parts, or surgical implants.

Automated optical metrology, a type of dimensional metrology, combines inspection and metrology for detecting a defective point in a part using 3D laser scanners or white light scanners.

17.2.8 Self-piercing Rivet

The self-piercing rivet (SPR) is a mechanical joining process which is used for joining of two or more sheets of material. SPR uses a rivet to pierce through the top sheet into the bottom or middle sheets to form a mechanical joint. SPR is a single step technique that uses tubular rivets to clinch sheets.

SPR is a widely used mechanical process for joining of materials, such as steel and aluminum. SPR process is illustrated in the figure below.

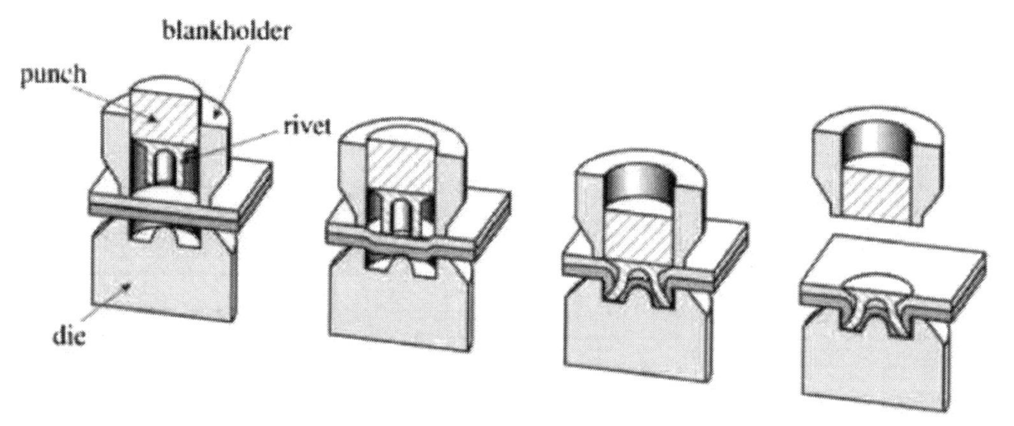

Figure: SPR Process

The technology has the advantage of using no heat for joining. Limited fumes, low noise, and low energy consumption are the major benefits, making this joining process an economical and viable solution for major sheet metal joining firms.

17.2.9 Advanced Lithography

Lithography is one of the key manufacturing technologies for realizing miniaturized feature sizes in component level, especially in electronics wherein the size of the chips and ICs (integrated circuits) has seen drastic reduction in size in recent years.

It has been established as the major fabrication technique in the semiconductor market. Prominent advanced lithography techniques include:

- Extreme ultraviolet lithography (EUV-lithography)
- X-ray lithography
- Electron beam lithography
- Focused ion beam lithography
- Nanoimprint lithography

17.2.10 Agile Robots

Agile robots are smart robots with artificial intelligence capabilities. They can work in tandem with humans without any interference.

Their agility, flexibility, and maneuverability are encouraging adoption in applications where human intervention is risky and unsafe (for example space exploration, surveillance, and military).

The increased agility and movement, enhanced efficiency and reduced downtime offered by agile robots are expected to encourage adoption in various applications.

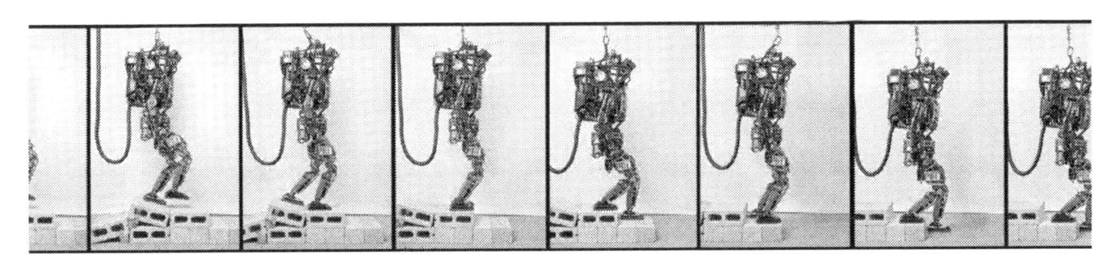

Image: Courtesy MIT *(Researchers at MIT, led by Seth Teller and Russ Tedrake, replaced the dynamic balance software that comes with Atlas with their own version. This lets the robot walk relatively quickly over uneven and unfamiliar ground)*

18 Chapter: General Interview Questions

18.1 Automation

1. What is Automation?

Automation is delegation of human control functions to technical equipment for increasing productivity, better quality, reduce cost & increased in safety working conditions.

2. What are the different components used in automation?

The components of automation system include

- Sensors for sensing the input parameters
- Transmitters for transmitting the raw signal in electrical form
- Control system which includes PLC, DCS & PID controllers
- Output devices/ actuators like drives, control valves.

3. What are the different control systems used in Automation?

- PID Controller based control system
- PLC based control system
- DCS based Control system
- PC Based automation system

4. Explain PID based control system.

PID (Proportional Integral Derivative) is the algorithm widely used in closed loop control. The PID controller takes care of closed loop control in plant. A number of PID controller with single or multiple loop can be taken on network. PID Controllers are widely for independent loops. Although some logic can be implemented but not much of sequential logic can be implemented in PIDs.

5. Difference between PLC & Relay?

- PLC can be programmed whereas a relay cannot.
- PLC works for analog I/Os such a PID loops etc. whereas a relay cannot.

- PLC is much more advanced as compared to relay.

6. Difference between PLC & DCS?

DCS: The system uses multiple processors, has a central database and the functionality is distributed. That is the controller sub system performs the control functions, the history node connects the data, the IMS node gives reports, the operator station gives a good HML, the engineering station allows engineering changes to be made.

PLC: The system has processor & I/O's and some functional units like basic modules and so on. Uses a SCADA for visualization. Generally, the SCADA does not use a central database.

7. What is Encoder?

A feedback device which converts mechanical motion into electronic signals. Usually an encoder is a rotary device that outputs digital pulses which correspond to incremental angular motion. The encoder consists of a glass or metal wheel with alternating clear and opaque stripes that are detected by optical sensors to produce the digital outputs.

18.2 Instrumentation System

8. What types of sensors are used for measuring different parameters?

- Temperature sensors – RTD, Thermocopule, Thermister
- Pressure Sensor – Borden Tube, Bellows, Strain gauge
- Flow – sensor – Pitot tube
- Level, Conductivity, Density, Ph

9. What is transmitter?

A transmitter is an electronic device that is generally mounted in the field in close proximity to a sensor. The sensor (also known as a transducer) measures a physical variable such as temperature or pressure and outputs a very low level electronic signal. The basic function of the transmitter is to provide the correct electrical power to turn on (or excite) the sensor then to read the low-level sensor signal, amplify it to a higher level electrical signal and send that signal a long distance to a control or read-out device.

Since low-level electrical signals do not transmit long distances with great accuracy, installing a transmitter generally gives a tremendous improvement in the accuracy of the information delivered to a larger control system. Typically, the output form the transmitter is 4-20 mA or 0-10 V

10. Why 4-20 mA preferred over 0-10 V signal?

The 0-10 V signal has tendency to drop because of line resistance. If the distance between sensor and input card is more the signal will not properly represent the field value. The 4-20 mA will travel a long distance without dropping signal value.

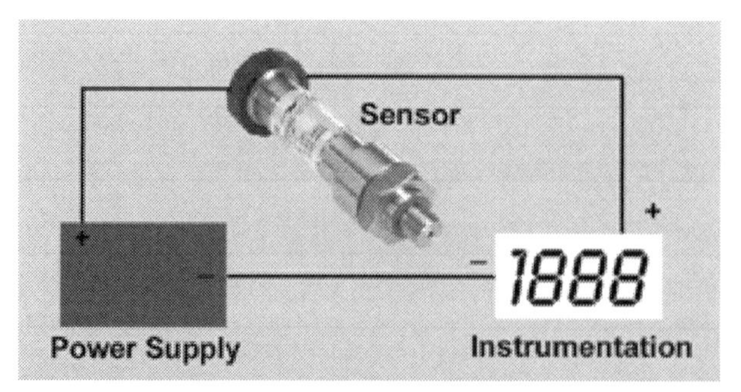

11. Why 4-20 mA preferred over 0-20 mA signal?

With 0-20 mA you can not distinguish between minimum field value and connection break. With 2-20 mA, internal circuit can distinguish between connection break of minimum value. Normally when the value is minimum the transmitter will give you 4 mA while in case of connection breakage it will give 0 mA.

12. Deference between 2 wire, 3 wire and 4 wire transmitter.

In 2 wire transmitter the power and signal are transmitted through same cable.

In 3 wire transmitter the data signal and power are with respect to common ground.

In 4 wire transmitter two wires for power supply and two for signals.

Only current transmitters can be used as 2 wire transmitters.

13. What is a "Smart" Transmitter?

A "Smart" transmitter is a transmitter that uses a microprocessor as the heart of the electronics. in addition, a "Smart" transmitter will output some type of remote digital communications allowing you to read and set-up the device from a remote position.

14. What is Field bus?

Field bus is a general term for a digital only, high speed communications protocol. The Key attribute to Field bus communications is higher speed communications with the possibility of addressing multiple transmitters all on the same field wiring. The foundation Field bus is a specific digital protocol that is often shortened to just be called field bus. Other digital only communications such as Prefabs are also Field bus protocols.

15. What is Actuator?

In a closed- loop control system, the part of the final control element that translates the control signal into action by the control device.

16. Explain Working of RTDs

Resistance Temperature Device works on the principles that the resistance of the material changes as its temperature changes Temperature is determined by measuring resistance and then using the RTD Resistance vs Temp characteristic to detect temperature. Typical elements used for RTD are Nickel, copper and Platinum, Platinum is widely used in RTDs because of accuracy. PT 100 means at 0 deg temp 100 ohms resistance, A typical RTD consists of a fine platinum wire wrapped around a mandrel and covered with a protective coating (glass or ceramic).

17. Temperature measurement range supported by RTDs?

The RTD work on temperature range between-250 to 850 deg C.ddfkm

18. Explain Working of Thermocouple

Thermocouple consists of two strips or wires made up of different metals and joined at one end. The temperature at that juncture induces and electromotive force (emf) between the other ends. As the temperature goes up the emf also increases. Through standard charts and tables the corresponding temperature can be found out.

The relationship between the thermocouple output and the temperature is quite non linear. Different metallurgies produce different outputs. The different metallurgies and different linearities result in different thermocouple designations as "J" "K" "N" "L", etc.

19. What is Cold junction compensation?

The industry accepted standard for the temperature at open end is 0 deg C. Therefore, tables and chart make the assumption that the temp open end is 0 deg C. In industry the open ends are always at actual room temperature and not0 deg C. The emf adjustment because of difference between the temp and 0 deg C is referred as Cold Correction (CJ Correction)

20. Temperature measurement range supported by thermocouple?

The thermocouple work on board temperature range ie- 270 to 2300.

21. Can split my one T/C signal to two separate instruments?

No. The T/C signal is a very low- level millivolt signal and should only be connected to the device. Splitting to two devices may result in bad readings or loss of signal. The is to use a " dual" T/C probe, or convert one T/C output to a 4-20 mA signal by using a transmitter or signal conditioner, then the new signal can be sent can be sent more than one instrument.

22. What are the flow measuring instruments used in flow measurement?

- Different pressure meters
- Positive displacement
- Velocity meters

23. Explain working of differential pressure measurement?

Suitable restriction placed in flow stream causes a different pressure across it. As flow depends upon different pressure (Head) & area so any of them or both can be varied for varying flow.

24. What are the components of different flow sensor?

For creating different pressure: Orifice plate, Venturi Tube, flow Nozzle, pitot tube For measuring pressure: U- Tube Manometers, Ring- Balance Manometer, P. Cell

25. What type of pressure sensors used in pressure measurement?

- Manometers
- Bourdon tubes
- Bellow elements
- Diaphragm elements

- DP transmitters

26. Explain working of different pressure transmitters.

Process pressure is transmitted through isolating diaphragms and oil fill fluid to a sensing diaphragm. The sensing diaphragm is a stretched spring element that deflects in response to differential pressure across it. The displacement of the sensing diaphragm, a maximum deflection f 0.004 inch (0.10mm), is proportional to the applied pressure, Capacitor plates on both sides of the sensing diaphragm detect the position of the diaphragm. The transmitter electronics convert the different capacitance between the sensing diaphragm and the capacitor plates into a two- wire mA signal and a digital output signal.

27. **What is Control Valves?**

The control valve, commonly named the final control element of control contains a pneumatic device that converts the control signal from the controller in action, regulation the flow.

28. What type of control valves used in the industry?

- ON – OFF SERVICES:- Gate, Ball, Diaphragm, Plug, Butterfly valves.
- THROTTILING SERVICES:- Globe, Butterfly, Diaphragm, Pinch valves.
- NON – REVERSE FLOW:- Check valves.

29. What are specifications of the control valve?

Following specifications are used for control valve

- Flow medium and operating temperature
- Flow rate kg/hr or Nm3/hr Max/Min/Normal
- Inlet and Outlet pressure : kg/cm2 Max/Min/Normal
- Max. allowable diff. Pressure : kg/cm2
- Density of medium : kg/m3
- Viscosity
- Cv: Valve Flow Coefficient

30. What are the components of control valve?

- Actuator, Body, Trim, Diaphragm, Diaphragm plate, Actuator stem
- Actuator spring, Seat, Travel Indicator, Valve stem, Gaskets, Yoke, Hand wheel

31. What is flow coefficient?

It is the flow of water (G=1, T=6 to 34 deg. C) through the valve at full lift in U.S. gallon per minute with a pressure drop across the valve of 1 psi.

18.3 SCADA Software : Wonderware In Touch

32. What is SCADA ? Role of MMI/HMI/SCADA in Industrial Automation

SCADA : Supervisory control and data acquisition

MMI : Man Machine Interface

HMI : Human machine Interface

This acts as an operator station. The operator can monitor as well as control the process parameters from this stations. Apart from online process data the operator will have access to historical and real-time trends, alarms and reports. The operator can give commands to control hardware for opening the valve, change the set point, start the pump etc.

33. What are Features of SCADA software?

The common features of SCADA include Dynamic process mimic, Tends, alarm, Connectivity with hardware, Recipe management etc.

34. Applications of SCADA.

SCADA systems has many applications right industrial automation, power distribution to water management.

35. Some of the leading SCADA companies

- Invensys Wonder ware In Touch
- Siemens WinCC (Earlier COROS)
- Allen Bradley RS View (Earlier Control View)
- Intellution ifix (Earlier Fix DMACS)
- GE Fanuc Simplicity

36. Types of Wonder ware SCADA packages

- No. of I/Os- Wondeware In Touch comes is 64, 128,256,1000, and 64,000 tags package.
- Development+ Runtime + Network (DRN)/ Runtime + Network (R+N) and View Node

D+R+N: With this packers development and editing of the application is possible, Runtime monitoring and control of the plant is possible and Networking is possible.

R+N: With this package development and editing of the application is NOT possible, Runtime monitoring and control of the plant is possible and Networking is possible.

Factory Focus: With this package developments and editing of the application is N OT possible, Runtime monitoring is possible but control of the plant is NOT possible and Networking is possible. This package is used a view node

37. What type of licensing patterns used in the SCADA software

Typically two types of licenses are used in the SCADA software

- Dongle Key : It is a hardware lock which can be put on the communication port of the PC.
- Software Lock: Hare the software code is the license. Typically can put the code while installation or transfer the code from Floppy to hard- disk.

38. Various EXE files used in Touch Software an there role

InTouch : It is an application manager. Using this you can create new application. Move between various application.

View: Window viewer. This will start Runtime application. From this you can monitor and control the plant.

WM: WM.XE is Window maker. This will start the development package in InTouch. Using you can you can develop the application.

39. Types of Window?

Replace: Automatically closes any window (s) it intersect when it appears on the screen including popup other replace type windows.

Overlay: Appears on top of currently displayed window (s) and can be larger than the window (s) it is overlaying. When an overlay window is closed, any window (s) that ware hidden behind it will reappear. Clicking on any on any visible portion of a window behind an overlay window will bring that window to the foreground as the active window.

Popup: Similar to an overlay window except, it always stays on top of all other open windows (even if another window is clicked) Popup window usually require a response from the user in order to be removed.

40. What is Symbol Factory?

Symbol Factory contains symbols which cab be readily used in the application. The symbol is contains include various Tanks, Reactor, Pipes Icons, and Flags.

41. What type of user input in InTouch?

User inputs include data entry Discrete, Analog, String/ Message, Sliders and Pushbuttons.

42. What type of animation can be given in InTouch?

Colour fill, % Fill Blinding, Size Control, Location, Orientation, Visibility, Action, Hide Show window

43. What are trends?

Trends means graphical representation of data.

"Real- time and " Historical. You can configure both trend objects to display graphical representations of multiple tag names over time. Real- time trends allow you allow you to chart up to four pens (data values). While Historical trends allow to chart up to pens.

Real-time trends are dynamic. They are updated continuously during runtime with whatever time span given in configuration. You can not scroll the real- time trends to see previous data.

Historical trends provide you with a " snapshot" of data from a time and data in the past. They are not dynamic Unlike real- time trends historical trends are only updated when they instructed to do so either through the execution of a Quick- Script or an action by the operator, for example, clicking a button, You can zoom in/ zoom out the trends. You can also access the pervious data.

44. What are Alarm and Events in InTouch?

Alarms and Events are the notifications used to inform operator of process activity. Alarms represent warnings of process conditions that could cause problems, and

require an operator response. A typical Alarms is triggered when a process value exceeds a user defined limit. InTouch uses for types of alarm LOLO LO HI and HIHI.
Events represent normal system status message, and do not require an operator response. A typical event is triggered when a certain system condition takes place, such as an operator logging into InTouch.

45. Security management

Security provides the ability to control whether or not specific operators are allowed to perform specific functions within an application. Security is based on the concept of the operator "logging on" to the application and entering a "User Name" and "Password." (The application developer sets up each operator with a "User Name," a pre-assigned "Password" and an "Access Level" via the Special/Security/Configure Users command wither in WindowMaker or Windows Viewer).
When a new application is created, the default "User Name" is "Administrator" with an access level of 1999 (which allows access to all security commands).

46. What is the use of Scripts in InTouch?

Is a way of writing logic in InTouch. InTouch has its own instructions and way of writing program.

Application : Linked to the entire application.
Windows : Linked to a specific window.
Key : Linked to a specific key or key combination on the keyboard.
Condition : Linked to a discrete tagname or expression.
Data Change : Linked to a tagname and/or tagname field only.

47. What is driver?

A software which allows a computer to access the external devices using com ports or communication cards.

48. What is DDE?

Dynamic Data Exchange is the facility developed by Microsoft for exchanging the data between various programs.
DDE has three important settings
Application / Server name, Topic Name and Item Name

49. Communication with software (Excel)

DDE settings for Excel

- App. Name – Excel,
- Topic Name – [book.xls]sheet1
- Item Name – Cell Address ie (R1C1)

DDE settings for InTouch

- App. Name – View
- Topic Name – TAGNAME
- Item Name – Name of tag

50. How to monitor the data from InTouch to Excel? ie One way communication from InTouch to Excel.

In Excel worksheet write a formula ie = View [TAGNAME A1

A1 is name of tag in InTouch

51) How to have two way communication between Excel and InTouch?

18.4 PLC (Programmable Logic Controller)

51. What is PLC?

PLC means Programmable Logic Controller. It is a class of industrially hardened devices that provides hardware interface for input sensors and output control element. The field I/p include element like limit switches, sensors, push button and the final control elements like actuator, solenoid/control valves, drives, hooters etc.

PLC Senses the input through I/p modules, Processes the logic through CPU and memory and gives output through output module.

52. Applications of PLC

PLC can be used in almost all industrial application solutions right from small machine to large manufacturing plants. Even it caters applications of redundant systems at critical process plants.

53. Role of PLC in Automation?

PLC plays most important role in automation. All the monitoring as well as the control actions are taken by PLCs. PLC Senses the input through I/P modules, Processes the logic through CPU and memory and gives output through output module.

54. Role of CPU ?

This component act as a brain of the system.

CPU consist of Arithmetic Logic Unit, Program memory, Process image memory, Internal timers and counters, flags.

It receives information from I/P device, makes decisions depending upon the information and logic written and sends information through the O/P devices.

The CPU's are distinguished with following features

Memory capacity, Instruction set supported, communication option, time required to execute the control program.

55. Role of Power supply in PLC system?

Power supply provides system power requirement to processor, I/O and communication modules. Typically the power supply has input voltage 120 V – 230 V AC or 24 V DC and back plane output current 2 A to 5 A at 5 V DC

56. Role of Rack or Chassis in PLC system?

A hardware assembly, which houses the processor, communication and I/O modules. It does following functions.

- Power distribution

- Containment of I/O modules
- Communication path between I/O module and CPU

The chassis are available in different slots in various PLC systems. Additional chassis can be connected using chassis interconnecting cable.

57. What is role of I/O modules?

Electronic plug in units used for interfacing the i/p and o/p device in the machine or process to be controlled.

I/P module receives data from i/p devices (Pushbutton, Switches, Transmitters) and send it to processor. The O/P module receives data from processor and send it to output device (Relay, Valves).

Digital/Discrete:- Sends and Receives On/Off signal

Analog:- Sends and receives variable input or output signals

58. Role of EEPROM memory module?

This module is inserted into processor system for maintaining a copy of project (PLC program). This is helpful in case of memory corruption or Extended power loss.

59. Communication module

communication modules are used either for communication between external hardware or software. The hardware can be PLCs (same or other make), Controller, I/O module, smart transmitters. The software can be SCADA software, MIS system or programming software.

60. Difference Between Fixed and Modular PLCs?

In non modular PLCs the processor will have inbuilt power supply and I/Os in one unit.

The modular PLC, will have separate slots for components like Power supply, I/O modules. You can select the I/Os or power supply as per the need.

61. What are the Types of I/Os?

- Local – These are the I/Os placed in the PLC main rack containing CPU. These I/Os are connected to CPU through jackplane.

Distributed – These are the I/O placed at remote location from the main rack containing the CPU. These I/O's are to be connected on communication bus like control net, device net or FIP I/O.

62. What is meaning of resolution in I/O cards in PLCs?

It is the minimum change in i/p parameter which can sensed by the i/p card. As far as Digital I/O is concerned it takes only one bit for operation. In case of analog input the resolution determines how much bits are used for input or output. For example a 12 bit resolution card means the input will come as 0 to 4095 count (2^12). For 16 bit data the counts will be from 0 – 65536(2^16). More the resolution the data will be more accurate.

63. What is an Analog Input Module?

An I/O module that contains circuits that convert analog input signals to digital values that can be manipulated by the processor. The signals for pressure, flow, level, temperature transmitters are connected to this module. Typically the input signal in 4-20 mA, 0-10 V

64. What is Analog Output Module?

An I/O module that contains circuits that output an analog dc signal proportional to a digital value transferred to the module from the processor. By implication, these analog outputs are usually direct (i.e., a data table value directly controls the analog signal value).

65. What is meaning of universal analog input card?

Normally there are different cards for different signals. But in universal input card the same channels can be configured for RTD, Thermocouple, Current or voltage input.

66. Give examples of I/P and Output connected to PLCs

- Digital I/P (Pushbutton, Switches)
- Analog I/P (Temp, Pressure, Flow, Level)
- Digital O/P (Solenoids valves, Contactors)
- Analog O/P (Control Valves, Speed)

67. Explain Source and Sink Concept?

Sinking Source type modules gives out the current to the field digital devices while sins type modules draw current when the device is in high sate.

Sinking – When active the output allows the current to flow to a common ground.

Sourcing – When active, current flows from a supply, through the output device and to ground.

68. What is forcing of I/O?

Forcing the I/Os means making the desired status of I/O in PLCs irrespective of its status coming from the field.

In certain cases when there is problem in receiving field input/ output, we can force the i/Os so that the logic takes desired state.

69. Scan cycle of PLC

PLC's can cycle follows following path Scan cycle of PLC

- Input Image Updation
- Process Logic Execution

Output Updation

70. What is meaning of scan time in PLC?

Scan time is the Time required to read the I/P, Process the logic and update the output in one cycle.

71. What is typical scan time in PLCs? What effects scan time?

Typically it is less than 10 ms. It depends on the complexity of logic, PID algorithm etc.

72. How to program PLCs?

Every PLC manufacturer have their own software for programming the PLC. For example Siemens uses Simantic S7 Manager, Allen Bradley uses RS Logix and Modicon uses PLC pro programming software. The programming language used is Ladder Logic (LD), Statement List (STL), Functional Block Diagram (FBD), Sequential Foundation Chart (SFC), Instruction List (IL) etc.

73. What is ladder diagram?

This is a programming language, which expresses a program as a series of "coils" and "contacts", simulating the operation of electromechanical relays. The resultant program is the equivalent of an equation, which is executed continuously in a combinatorial manner. The advantage of this language is the familiarity many electricians have with the simple operation of relays. Disadvantages include the complexity of large, cross-connected programs, and the difficulty of expressing such non-binary functions as motion control and analog I/P.

74. What is redundancy?

The capacity to switch from primary equipment to standby equipment automatically without affecting the process under control. Redundancy means provision for standby module. In case of failure of one module is running process, the standby module takes over. Hot redundancy means the changeover of control from active processor to standby processor in less than 1 scan time.

75. Need of redundancy

in critical processes, it is important to run the plant without failure. In such case it is important to have redundancy so that even in one system fails the redundant system can take care without affecting plant.

76. Types of redundancy

CPU redundancy : In case of CPU failure the standby CPU takes care of the plant.

Power Supply redundancy : In case the power supply fails the standby power supply takes control of the situation.

Communication : Multiple communication channels are provided to take care of communication failure.

I/O Redundancy : Multiple I/O channels are provided to take care of input or output failure.

77. What are components of redundant PLC system?

Typical component on Schneider Redundant PLCs

The backplane used is either 4 slot o66 slot with

- Power Supply
- Controller with built-in Modbus Plus and Modbus ports
- Optional dual cable Modbus Plus
- Optional fiber optic Modubs Plus
- CHS Hot Standby module
- Dual cable Remote I/O Head

The master and Standby configuration must be identical

78. Commonly used Instructions in PLCs

Examine if Closed (XIC):- | | -	Examines if the bit is in ON condition. If the bit is ON the instruction is tue.
Examine if Open (XIO):- |/| -	Examines if the bit is in OFF condition. If the bit is OFF the instruction is true.
One short rising (OSR):- [OSR]-	When the conditions preceding the instruction is true, makes the rung run for one program scan.

Note retentive Output instruction

Output Entergies (OTE):-- ()--	If the rung is true, it turns on the bit. If the rung goes or a power cycle occurs the bit turns off.

Retentive Output instruction

Output Latch (OTL):-- (L)---	If the rung is true, turns ON a bit. The bit stays ON until the rung containing an OUT with the same address goes true.
Output Latch (OUT):-- (U)---	If the rung is true, turns OFF a bit. The bit stays OFF until the rung containing an OTL with the same address goes true.

79. Timers

Timer and Counter are used to control operation based on time or number of events
Types of timers

TON – (Timer ON delay)	An output instruction that can be used to turn an output ON or OFF after the timer has been timing for a preset time interval.
TOF – (Timer OFF delay)	An output instruction used to turn an output ON or OFF after its rung has been off for a preset time interval.
RTO – (Retentive Timer)	An output instruction that can be used to turn an output ON or OFF after the timer has been timing for a preset time interval. Once it has begin timing, it holds the count of time even when the rung continuity is lost.

Timer Status Bit
DeviceNet: A low-cost communication link that connects industrial devices to a network It is based on broadcast – oriented communication protocol- the Controller Area Network (CAN).
Ethernet: The standard for local communications network developed jointly by Digital Equipment Corp, Xerox, and Intel Ethernet base band coaxial cable transmits data at speed up to 10 megabits per second. Ethernet is used as the underlying transport vehicle by several upper-level protocols, including TCP/ IP.
Programmable Logic Controller – Allen Bradley

80. What are the PLC ranges available in Rockwell?

- Pico : Non modular small PLCs
- Micrologix 1000, 1200 and 1500 Series
- SLC : SLC 5/01, 5/02, 5/03 ....
- Control Logix Flex Logic and Soft PLC

Diff b/w Micro Logic and SLC
Micrologix

1. Has limited 1/0 Large capacity of 1/0
2. Use DFI only Use PID , DH+

81. What are the software Used with AB?

For Pico soft for Pico PLC programming
RS Logix 500 for Micrologix and SLC PLCs programming
RS Logix 5000 for Control Logix PLCs programming
SCADA – RS View earlier Control View

82. What is use of RS Linx software?

RS Linx software is used to perform following tasks.

- Configure communication drivers
- View configured drivers and active nodes
- Enable communication tasks such as uploading, downloading, going online, updating firmware and sending messages.

83. What is use of RS Logix Software?

RS Logix is a PLC programming software. It contains all the instructions needed for PLC programming. We can develop the program, down load/upload the program, work on line/ off line and force the I/Os using the software.
RS Logix 500 is used for Micrologix and SLCs
RS Logix 5000 is used for Control Logix PLCs

84. What file gets created in PC for RS Logix PLC program?

The extension of the file will be.rss. So if you create an application with "Reliance" name the file created will be "Reliance.rss". By default it will be stored in "c:\Program Files\Rockwell Software/RS Logix 500 Eng location.

85. How to creating linkage between PLCs and PLC programming software?

The RS Linx software is used for linking the PLC and software
Either you can manually configuration the communication settings or By using Auto Configure facility the software will detect the communication settings automatically.

86. Hardware configuration (PLC and I/Os)

- Manual configuration
- Auto detect

87. What is the meaning of Upload and Download?

Upload means transferring the program data from PLC to PC
Download means transferring the program data from PC to PLC

88. What are the various communication interfaces supported by AB?

The commonly used communication protocol in AB includes DH+, DH485, ETH, Modbus, Device Net and Control Net.

89. Give information about DH, Control Net, DeviceNet and Ethernet protocol.

Data Highway : The proprietary data network used by Allen Bradley PLCs to communicate information to and from other PLCs on the network or to and from host computers attached to the network.

ControlNet : A real-time, control-layer network providing high-speed transport of both time-critical IO data and messaging data, including upload/ download of programming and configuration data and peer-to-peer messaging, on a single physical media link.

FC Functions : It is a logic block without memory. An FC is always executed by calling in another block. FC is used either for returning a function value to a calling function or executing a technological function. Temporary variable belonging to FC are saved in local stack and this data is lost when the FC has been executed.

Function Blocks (FBs) : A function block Is block with a memory. A FB contains a program that is always executed when a different logic block calls the FB. FB make it much easier to program frequently occurring complex functions.

90. What is latency in communication?

The delay time between the end of one communication and the start of another. During this time, the processes associated with the communication are hung up and cannot continue. The latency to be minimum.

91. How the communication protocols are distinguished?

The protocols are distinguished with following specifications

- No. of nodes supported, total network length, speed of communication.

92. Comparison between various Protocols used with AB

	DH+	DH485	Device Net	Control Net
Baud rate max	230.4 kbits/s	19.2 kbit/s	500 kbit/s	5 Mbit/s
No. of max. nodes	64	32	64	99
Network Length	3.048 Km	1.2 km	0.487 Km	30 km

Programmable Logic Controller- Siemens

93. What are the various PLC system in SIMATIC range?

Siemens has broadly 3 PLC ranges ie Siemens S7 200, 300 and 400

94. What are the software used with Siemens?

For S7 200 PLC programming Micro win
For S7 300 and 400 system: Simantic S7 manager
The SCADA software used by Siemens is Win CC. Earlier Siemans use to supply COROS LS /B

95. Components of Siemens S7 300 Series PLC system?

CPUs (312 IFM, 313,314,IFM,314,315,2DP,316-2DP,318
Single Modules (SM), Digital I/O (SM321/322/323), Analog I/O (SM 331/332/334)
Function modules (FM) ex Positioning Modules, Closed Loop
Communication Processor ex CP 342-5 DP for Profibus
Interface module- For interconnecting individual racks (IM 360/361, IM 365 S/R)

96. What are the Communication Protocol used in Siemens?

Multi- Point Interface (MPI):
Data Transfer-187,5 kbits to 15 Mbit/ s,
Distance-50 m without RS 485 repeater /10 Km with repeater
Number of nodes- up to 32
Profibus
Data Transfer-12 Mbit/s,
Distance-23 Km with fibre optic cable
Number of nodes – up to 125

97. What are the blocks used in Siemens?

Simantic S7 manger uses DB, OB, FC, PB, and FB
Obs: Determine the structure of the user program

Data Black: These are the blocks used by logic blocks in CPU program for storing the data. DB's does not contain any instructions and it take up space in the user memory. The user program and access a data block with bit, byte, word or double word operations. Global data block: These contain information that can be accessed by all the logic block in the user program.
Instance data block: These DBs are always assigned to a particular FB.
UN : (enable) Specifies whether or not the timer instruction is enabled
DN : (Done) Specifies whether or not the accumulated value of the timer equals to the preset value of the timer.
Programming instruction (Advance)
MCR, Compute, PID, STI, Sequencer, Register, RTC

98. What are SFCs and SFBs?

SFBs and SFCs are integrated in the S7 CPU and allow you access to some important system functions.

99. What is Statement List?

Statement List (STL) is a textual programming language that can be used to create the code section of logic blocks. Its syntax for statements is similar to assembler language and consists of instructions followed by addresses on which the instructions act.

100. What happen to PLC when it goes to Fatal Error?

The PLC performs the following tasks when a Fatal Error is detected.

- Changes to STOP mode
- Turns on both the System Fault LED and the STOP LED

The PLC remains in this condition until the fatal error is corrected.

18.5 IOT

101. What is IoT?

Answer: IoT stands for Internet of Things. It is basically a network using which things can communicate with each other using internet as means of communication between them. All the things should be IP protocol enabled to have this concept possible. Not one but multiple technologies are involved to make IoT an enormous success.

102. There may be some questions on Linux OS, as it is most popular in IoT domain.

Answer: One can refer the same on net on very basics such as what are the qualities of linux OS? What are the features of linux OS over other Operating Systems etc.

This set of IoT(Internet Of Things) interview questions and answers are useful for freshers and experienced level of job positions.

103. What impacts will the Internet of Things (IoT) have on Infrastructure and Smart Cities Sector?

Answer: The capabilities of the smart grid, smart buildings, and ITS combined with IoT components in other public utilities, such as roadways, sewage and water transport and treatment, public transportation, and waste removal, can contribute to more integrated and functional infrastructure, especially in cities.

For example, traffic authorities can use cameras and embedded sensors to manage traffic flow and help reduce congestion. IoT components embedded in street lights or other infrastructure elements can provide functions such as advanced lighting control, environmental monitoring, and even assistance for drivers in finding parking spaces. Smart garbage cans can signal waste removal teams when they are full, streamlining the routes that garbage trucks take.

This integration of infrastructure and service components is increasingly referred to as smart cities, or other terms such as connected, digital, or intelligent cities or communities. A number of cities in the United States and elsewhere have developed smart-city initiatives.

4.What role does the network play in the Internet of Everything?

Answer: The network plays a critical role in the Internet of Everything ? it must provide an intelligent, manageable, secure infrastructure that can scale to support billions of context-aware devices.

5.How Wireless Communications might affect the Development and Implementation of the Internet of Things (IoT)?

Answer: Many observers believe that issues relating to access to the electromagnetic spectrum will need to be resolved to ensure the functionality and interoperability of IoT devices. Access to spectrum, both licensed and unlicensed, is essential for devices and objects to communicate wirelessly. IoT devices are being developed and deployed for new purposes and industries, and some argue that the current framework for spectrum allocation may not serve these new industries well.

6.How does the Internet of Everything relate to the Internet of Things?

Answer: The "Internet of Everything" builds on the foundation of the "Internet of Things" by adding network intelligence that allows convergence, orchestration and visibility across previously disparate systems.

7.How is Industrial Internet of Things (IIoT) different from the Internet of Things (IoT)?

Answer: There are two perspectives on how the Industrial IoT differs from the IoT.

The first perspective is that there are two distinctly separate areas of interest. The Industrial IoT connects critical machines and sensors in high-stakes industries such as aerospace and defense, healthcare and energy. These are systems in which failure often results in life-threatening or other emergency situations. On the other hand, IoT systems tend to be consumer-level devices such as wearable fitness tools, smart home thermometers and automatic pet feeders. They are important and convenient, but breakdowns do not immediately create emergency situations.

The second perspective sees the Industrial IoT as the infrastructure that must be built before IoT applications can be developed. In other words, the IoT, to some extent, depends on the Industrial IoT.

For example, many networked home appliances can be classified as IoT gadgets, such as a refrigerator that can monitor the expiration dates of the milk and eggs it contains, and remotely-programmable home security systems. On the Industrial Internet side, utilities are enabling better load balancing by taking power management decisions down to the neighborhood level. What if they could go all the way down to individual appliances? Suppose users could selectively block power to their devices during high-demand scenarios? Your DVR might power down if it wasn't recording your favorite show, but your refrigerator would continue to work, resulting in less food spoilage. You could set your washer and dryer to be non-functional, and make an exception with a quick call from your smartphone. Rolling blackouts could be a thing of the past.

Innovators are only beginning to imagine the possibilities that may be achieved by taking advantage of devices and systems that can communicate and act in real time, based on information they exchange amongst themselves. As the Industrial IoT becomes better defined and developed, more impactful IoT applications can and will be created.

Internet of Things:

Everyday consumer-level devices connected to one another and made smarter and slightly self-aware.

Examples: consumer cell phone, smart thermostat

Industrial Internet of Things:

Equipment and systems in industries and businesses where failures can be disastrous.
Examples: individual health monitors and alert systems in hospitals

How will Internet of Things (IoT) impact sustainability of Environment or Business?

Answer: Internet of Things (IoT) can significantly reduce carbon emissions by making business and industry more efficient. "By managing street lights more efficiently you can save approximately 40% of energy used to make them run," Will Franks says.

Bill Ruh, vice-president of GE Software, agrees. "We have created 40 applications," says Ruh. "One of these, PowerUp, uses sensors to collect weather and performance data from wind turbines to enable operators to generate up to 5% more electricity without physically changing it, which generates 20% more profit for our customers."

9. What is the difference between the Internet of Things (IoT) and the Sensor Business?

Sensors can be used in lots of different ways, many of which don't need to be internet connected.

IoT also includes the control side, not just the sensing side.

10.What impacts will the Internet of Things (IoT) have on Economic Growth?

Answer: Several economic analyses have predicted that the IoT will contribute significantly to economic growth over the next decade, but the predictions vary substantially in magnitude. The current global IoT market has been valued at about $2 trillion, with estimates of its predicted value over the next five to ten years varying from $4 trillion to $11 trillion. Such variability demonstrates the difficulty of making economic forecasts in the face of various uncertainties, including a lack of consensus among researchers about exactly what the IoT is and how it will develop.

11. Why Internet of Things(IoT) will be successful in the coming years?

As the telecommunication sector is becoming more extensive and efficient, broadband internet is widely available. With technological advancement it is now much cheaper to produce necessary sensors with built-in Wi-Fi capabilities making connecting devices less costly.

Most important, the smart phone usage has surpassed all the predicted limits and telecommunication sector is already working on its toes to keep their customers satisfied by improving their infrastructure. As IoT devices need no separate communication than the existing one building IoT tech is very cheap and highly achievable.

12.How Cybersecurity might affect the Development and Implementation of the Internet of Things (IoT), especially in USA?

Answer: The security of devices and the data they acquire, process, and transmit is often cited as a top concern in cyberspace. Cyberattacks can result in theft of data and sometimes even physical destruction. Some sources estimate losses from cyberattacks in general to be very large, in the hundreds of billions or even trillions of dollars. As the number of connected objects in the IoT grows, so will the potential risk of successful intrusions and increases in costs from those incidents.

Cybersecurity involves protecting information systems, their components and contents, and the networks that connect them from intrusions or attacks involving theft, disruption, damage, or other unauthorized or wrongful actions. IoT objects are potentially vulnerable targets for hackers. Economic and other factors may reduce the degree to which such objects are designed with adequate cybersecurity capabilities built in. IoT devices are small, are often built to be disposable, and may have limited capacity for software updates to address vulnerabilities that come to light after deployment.

The interconnectivity of IoT devices may also provide entry points through which hackers can access other parts of a network. For example, a hacker might gain access first to a building thermostat, and subsequently to security cameras or computers connected to the same network, permitting access to and exfiltration or modification of surveillance footage or other information. Control of a set of smart objects could permit hackers to use their computing power in malicious networks called botnets to perform various kinds of cyberattacks.

Access could also be used for destruction, such as by modifying the operation of industrial control systems, as with the Stuxnet malware that caused centrifuges to self-destruct at Iranian nuclear plants. Among other things, Stuxnet showed that smart objects can be hacked even if they are not connected to the Internet. The growth of smart weapons and other connected objects within DOD has led to growing concerns about their vulnerabilities to cyberattack and increasing attempts to prevent and mitigate such attacks, including improved design of IoT objects. Cybersecurity for the IoT may be complicated by factors such as the complexity of networks and the need to automate many functions that can affect security, such as authentication. Consequently, new approaches to security may be needed for the IoT.

IoT cybersecurity will also likely vary among economic sectors and subsectors, given their different characteristics and requirements. Each sector will have a role in developing cybersecurity best practices, unique to its needs. The federal government has a role in securing federal information systems, as well as assisting with security of nonfederal systems, especially critical infrastructure. Cybersecurity legislation considered in the 114th Congress, while not focusing specifically on the IoT, would address several issues that are potentially relevant to IoT applications, such as information sharing and notification of data breaches.

13.What impacts will the Internet of Things (IoT) have on Health Care Sector?

Answer: The IoT has many applications in the health care field, in both health monitoring and treatment, including telemedicine and telehealth. Applications may involve the use of medical technology and the Internet to provide long-distance health care and education. Medical devices, which can be wearable or nonwearable, or even implantable, injectable, or ingestible, can permit remote tracking of a patient's vital signs, chronic conditions, or other indicators of health and wellness.36 Wireless medical devices may be used not only in hospital settings but also in remote monitoring and care, freeing patients from sustained or recurring hospital visits. Some experts have stated that advances in healthcare IoT applications will be important for providing affordable, quality care to the aging U.S. population.

14.What are the main Social and Cultural Impacts of Internet Of Things (IoT)?

Answer: The IoT may create webs of connections that will fundamentally transform the way people and things interact with each other. The emerging cyberspace platform created by the IoT and SMAC has been described as potentially making cities like "computers" in open air, where citizens engage with the city "in a real-time and ongoing loop of information."

Some observers have proposed that the growth of IoT will result in a hyperconnected world in which the seamless integration of objects and people will cause the Internet to disappear as a separate phenomenon. In such a world, cyberspace and human space would seem to effectively merge into a single environment, with unpredictable but potentially substantial societal and cultural impacts.

15.Will IoT actually work over the internet or will it have its own dedicated wide area network?

Answer: Interoperability between various wireless and networking standards is still an issue and something that forums and standards bodies are trying to address. According to Franks, businesses have to collaborate on standards to create strong ecosystems for a range of industries, otherwise the industry will remain fragmented.

"The IoT is a whole myriad of different ways of connecting things ? it could be fixed, Wi-Fi NFC, cellular, ultra-narrow band or even Zigbee. You have to mix and match what is best for each task," he says. "Interoperability is essential, for economies of scale."

16.What will happen in terms of jobs losses and skills as IoT makes devices and robots more intelligent?

Answer: A Digital Skills Select Committee report to the House of Lords in February estimated that 35% of UK jobs would be lost to automation in the next 20 years. It echoes the sort of thinking that Erik Brynjolfsson and Andrew McAfee's The Second Machine Age: Work, Progress, and Prosperity in a Time of Brilliant Technologies predicts. Their answer is that you need to switch skills but to do this you need to switch the schools too.

Will Franks agrees. He saw when he launched his business in 2004 that a massive shortage in relevant skills can impede progress, so he was forced to look overseas. The same he says will happen with IoT unless we get schools and colleges to start gearing courses to meet the challenges of tomorrow's automated economy.

It is a huge challenge and one which is a top three priority for Chi Onwuruh MP and Labour's Digital Review. Digital inclusion, a data review and a focus on digital skills are she says essential.

Last July the Digital Skills Taskforce called for the Government to review skills development in schools and colleges. The Perkins Review in November last year also called for a review into developing engineering skills to boost the UK economy.

What is clear is that the jobs landscape will change dramatically in the next 20 years. But it will be a slow process and whether or not we are prepared to cope with it will depend as much on education policy as digital policy. The robots are definitely coming but don?t hand your notice in just yet.

17.What are the elements of the Internet of Everything?

Answer: People: People will continue to connect through devices, like smartphones, PCs and tablets, as well as through social networks, such as Facebook and LinkedIn. As the Internet of Everything emerges, the interaction of people on the Internet will evolve. For example, it may become common to wear sensors on our skin or in our clothes that collect and transmit data to healthcare providers. Some analysts even suggest that people may become individual nodes that produce a constant stream of static data.

Process: This includes evolving technology, business, organizational and other processes that will be needed in order to manage and, to a large extent, automate the explosive growth in connections and the resultant accumulation, analysis and communication of data that will be inevitable in the Internet of Everything. Processes will also play an important role in how each of these entities people, data, and things interact with each other within the Internet of Everything to deliver societal benefits and economic value.

Things: This element includes many physical items like sensors, meters, actuators, and other types of devices that can be attached to any object, that are or will be capable of connecting to the network and sharing information. These things will sense and deliver more data, respond to control inputs, and provide more information to help people and machines make decisions. Examples of things in the Internet of Everything include smart meters that communicate energy consumption, assembly line robots that automate factory floor operations, and smart transportation systems that adapt to traffic conditions.

Data: Today, devices typically gather data and stream it over the Internet to a central source, where it is analyzed and processed. Such data is expected to surpass today's largest social media data set by another order of magnitude. Much of this data has very transient value. In fact, its value vanishes almost as quickly as it is created. As a result, not all generated data can be or should be stored. As the capabilities of things connected to the Internet continue to advance, they will become more intelligent and overcome the limits of traditional batch-oriented data analysis by combining data into more useful information. Rather than just reporting raw data, connected things will soon send higher-level information and insights back to machines, computers, and people in real time for further evaluation and decision making. The intelligent network touches everything and is the only place where it's possible to build the scalable intelligence required to meet and utilize this new wave of "data in motion". This transformation made possible by the emergence of the Internet of Everything is important because it will enable faster, more intelligent decision making by both people and machines, as well as more effective control over our environment.

18.What are the important Components of Internet of Things?

Answer: Many people mistakenly think of IoT as an independent technology. Interestingly, internet of things is being enabled by the presence of other independent technologies which make fundamental components of IoT.

The fundamental components that make internet of things a reality are:-

1. Hardware-Making physical objects responsive and giving them capability to retrieve data and respond to instructions

2. Software – Enabling the data collection, storage, processing, manipulating and instructing

3. Communication Infrastructure – Most important of all is the communication infrastructure which consists of protocols and technologies which enable two physical objects to exchange data.

19.What are the main Challenges of Internet of Things (IoT)?

Answer: Like any other technology there are challenges which make the viability of IoT doubtful.

Security is one of the major concerns of experts who believe virtually endless connected devices and information sharing can severely compromise one's security and well being. Unlike other hacking episodes which compromise online data and privacy with IoT devices can open gateway for an entire network to be hacked.

One such flaw is well presented by Andy Greenberg on wired.com where he works with hackers to remotely kill his Jeep on the highway. Another very relevant example is provided by W. David Stephenson in his post Amazon Echo: is it the smart home Trojan Horse? You can estimate the amount of personal and private data the connected devices will be producing once they are on a network. The major challenge for IoT tech companies is to figure out how the communication in the internet of things realm can be made truly secure.

20.What kinds of information do Internet of Things (IoT) objects communicate?

Answer: The answer depends on the nature of the object, and it can be simple or complex. For example, a smart thermometer might have only one sensor, used to communicate ambient temperature to a remote weather-monitoring center. A wireless medical device might, in contrast, use various sensors to communicate a person's body temperature, pulse, blood pressure, and other variables to a medical service provider via a computer or mobile phone.

Smart objects can also be involved in command networks. For example, industrial control systems can adjust manufacturing processes based on input from both other IoT objects and human operators. Network connectivity can permit such operations to be performed in "real time" that is, almost instantaneously.

Smart objects can form systems that communicate information and commands among themselves, usually in concert with computers they connect to. This kind of communication enables the use of smart systems in homes, vehicles, factories, and even entire cities.

Smart systems allow for automated and remote control of many processes. A smart home can permit remote control of lighting, security, HVAC (heating, ventilating, and air conditioning), and appliances. In a smart city, an intelligent transportation system (ITS) may permit vehicles to communicate with other vehicles and roadways to determine the fastest route to a destination, avoiding traffic jams, and traffic signals can be adjusted based on congestion information received from cameras and other sensors.

Buildings might automatically adjust electric usage, based on information sent from remote thermometers and other sensors. An Industrial Internet application can permit companies to monitor production systems and adjust processes, remotely control and synchronize machinery operations, track inventory and supply chains, and perform other tasks.

IoT connections and communications can be created across a broad range of objects and networks and can transform previously independent processes into integrated systems. These integrated systems can potentially have substantial effects on homes and communities, factories and cities, and every sector of the economy, both domestically and globally.

21.Which Companies and Organizations Support the Industrial IoT?

Answer: General Electric coined the term Industrial Internet in late 2012. It is effectively synonymous with the Industrial Internet of Things, and abbreviated as Industrial IoT or IIoT.

Many other companies and organizations are realizing the potential and significance of the Industrial IoT. A recent study conducted by Appinions and published in Forbes listed RTI as the #1 most influential company for the Industrial Internet of Things. Other influencers included Google, Cisco, GE, Omron, DataLogic and Emerson Electric.

The Industrial Internet Consortium also advocates for the advancement of the Industrial IoT. It is a not-for-profit organization that manages and advances the growth of the Industrial IoT through the collaborative efforts of its member companies, industries, academic institutions and governments. Founding members include AT&T, Cisco Systems Inc., General Electric, IBM and Intel.

22.What is the Internet of Everything?

Answer: The Internet of Everything is the intelligent connection of people, process, data and things.

23.What is a "Thing" in the context of Internet of Things (IoT)?

Answer: The "Thing" commonly referred to by the concept of the Internet of Things is any item that can contain an embedded, connected computing device. A "Thing" in the IoT could be a shipping container with an RFID tag or a consumer's watch with a Wi-Fi chip that sends fitness data or short messages to a server somewhere on the Internet.

24.How the Internet of Things (IoT) makes a difference to the businesses?

Answer: Businesses focus on getting products to the marketplace faster, adapting to regulatory requirements, increasing efficiency, and most importantly, persisting to innovate. With a highly mobile workforce, evolving customer, and changing supply chain demand, the IoT can move your enterprise forward, starting today.

25.What are the major Privacy and Security Issues in case of Internet Of Things (IoT)?

Answer: Cyberattacks may also compromise privacy, resulting in access to and exfiltration of identifying or other sensitive information about an individual. For example, an intrusion into a wearable device might permit exfiltration of information about the location, activities, or even the health of the wearer.

In addition to the question of whether security measures are adequate to prevent such intrusions, privacy concerns also include questions about the ownership, processing, and use of such data. With an increasing number of IoT objects being deployed, large amounts of information about individuals and organizations may be created and stored by both private entities and governments.

With respect to government data collection, the U.S. Supreme Court has been reticent about making broad pronouncements concerning society's expectations of privacy under the Fourth Amendment of the Constitution while new technologies are in flux, as reflected in opinions over the last five years.

Congress may also update certain laws, such as the Electronic Communications Privacy Act of 1986, given the ways that privacy expectations of the public are evolving in response to IoT and other new technologies. IoT applications may also create challenges for interpretation of other laws relating to privacy, such as the Health Insurance Portability and Accountability Act and various state laws, as well as established practices such as those arising from norms such as the Fair Information Practice Principles.

26.What is Bluetooth Low Energy (BLE) Protocol for Internet of Things (IoT)?

Answer: Nokia originally introduced this protocol as Wibree in 2006. Also known as Bluetooth Smart this protocol provides the same range coverage with much reduced power consumption as the original Bluetooth. It has similar bandwidth with narrow spacing as used by ZigBee. Low power latency and lower complexity makes BLE more suitable to incorporate into low cost microcontrollers.

Low power latency and lower complexity makes BLE more suitable to incorporate into low cost microcontrollers.

As far as application is concerned BLE is in healthcare sector. As wearable health monitors are becoming prevalent the sensors of these devices can easily communicate with a smart phone or any medical instrument regularly using BLE protocol.

27.What impacts will the Internet of Things (IoT) have?

Answer: Many observers predict that the growth of the IoT will bring positive benefits through enhanced integration, efficiency, and productivity across many sectors of the U.S. and global economies.

Among those commonly mentioned are agriculture, energy, health care, manufacturing, and transportation. Significant impacts may also be felt more broadly on economic growth, infrastructure and cities, and individual consumers. However, both policy and technical challenges, including security and privacy issues, might inhibit the growth and impact of IoT innovations.

28.Why is the Internet of Everything happening now?

Answer: The explosion of new connections joining the Internet of Everything is driven by the development of IP-enabled devices, the increase in global broadband availability and the advent of IPv6.

29.What are the top 5 Machine-to-Machine (M2M) applications in the world?

Answer: 1. Asset Tracking and/or Monitoring in some form or another (Stolen Vehicles, Fleet, Construction Equipment, Wood Pellets, Tank level monitoring, etc.) seems to be the biggest. Low data requirements, high volumes of devices, etc. It isn't hot or particularly exciting, but it is changing the world in subtle ways and very quickly. New business models will spring from this.

2. Insurance Telematics is huge as if offers Insurance companies the opportunity to cut risk and drive better/more attractive pricing.

3. Utilities/Automated Meter Reading/Smart Grids – lots of regulation and investment into this at the moment. There a lot of national solutions as the requirements and business case are driven in very diverse ways.

4. Security has been an early adopter here. The requirements are quite heavy on the network as many of these applications have a fixed line legacy.

5. mHealth has been out there for a while, but hasn't really taken off. There are some exciting early adopters. Many of the established companies (largely built through acquisition) have some challenges moving quickly in this space, but when they get up to speed and the business models are established there will be a massive uptake.

6. Automotive is a big one – driven by consumers' expectation of being always connected as well as regulation.

30.What impacts will the Internet of Things (IoT) have on Transportation Sector?

Answer: Transportation systems are becoming increasingly connected. New motor vehicles are equipped with features such as global positioning systems (GPS) and in-vehicle entertainment, as well as advanced driver assistance systems (ADAS), which utilize sensors in the vehicle to assist the driver, for example with parking and emergency braking. Further connection of vehicle systems enables fully autonomous or self-driving automobiles, which are predicted to be commercialized in the next 5-20 years.

Additionally, IoT technologies can allow vehicles within and across modes including cars, buses, trains, airplanes, and unmanned aerial vehicles (drones) to "talk" to one another and to components of the IoT infrastructure, creating intelligent transportation systems (ITS). Potential benefits of ITS may include increased safety and collision avoidance, optimized traffic flows, and energy savings, among others.

31.What is the Current Federal Regulatory Role of USA Government pertinent to Internet Of Things (IoT)?

Answer: There is no single federal agency that has overall responsibility for the IoT, just as there is no one agency with overall responsibility for cyberspace. Federal agencies may find the IoT useful in helping them fulfill their missions through a variety of applications such as those discussed in this report and elsewhere. Each agency is responsible under various laws and regulations for the functioning and security of its own IoT, although some technologies, such as drones, may also fall under some aspects of the jurisdiction of other agencies.

Various agencies have regulatory, sector-specific, and other mission-related responsibilities that involve aspects of IoT. For example, entities that use wireless communications for their IoT devices will be subject to allocation rules for the portions of the electromagnetic spectrum that they use.

. The Federal Communications Commission (FCC) allocates and assigns spectrum for nonfederal entities.

. In the Department of Commerce, the National Telecommunications and Information Administration (NTIA) fulfills that function for federal entities, and the National Institute of Standards and Technology (NIST) creates standards, develops new technologies, and provides best practices for the Internet and Internet-enabled devices.

. The Federal Trade Commission (FTC) regulates and enforces consumer protection policies, including for privacy and security of consumer IoT devices.

. The Department of Homeland Security (DHS) is responsible for coordinating security for the 16 critical infrastructure sectors. Many of those sectors use industrial control systems (ICS), which are often connected to the Internet, and the DHS National Cybersecurity and Communications Integration Center (NCCIC) has an ICS Cyber Emergency Response Team (ICS-CERT) to help critical-infrastructure entities address ICS cybersecurity issues.

. The Food and Drug Administration (FDA) also has responsibilities with respect to the cybersecurity of Internet-connected medical devices.

. The Department of Justice (DOJ) addresses law-enforcement aspects of IoT, including cyberattacks, unlawful exfiltration of data from devices and/or networks, and investigation and prosecution of other computer and intellectual property crimes.

. Relevant activities at the Department of Energy (DOE) include those associated with developing high-performance and green buildings, and other energy-related programs, including those related to smart electrical grids.

. The Department of Transportation (DOT) has established an Intelligent Transportation Systems Joint Program Office (ITS JPO) to coordinate various programs and activities throughout DOT relating to the development and deployment of connected vehicles and systems, involving all modes of surface transportation. DOT mode-specific agencies also engage in ITS activities. The Federal Aviation Administration (FAA) is involved in regulation and other activities relating to unmanned aerial vehicles (UAVs) and commercial systems (UAS).

. The Department of Defense was a pioneer in the development of much of the foundational technology for the IoT. Most of its IoT deployment has related to its combat mission, both directly and for logistical and other support.

In addition to the activities described above, several agencies are engaged in research and development (R&D) related to the IoT:-

. Like NIST, the National Science Foundation (NSF) engages in cyber-physical systems research and other activities that cut across various IoT applications.

. The Networking and Information Technology Research and Development Program (NITRD), under the Office of Science and Technology Policy (OSTP) coordinates Federal agency R&D in networking and information technology. The NITRD Cyber Physical Systems Senior Steering Group "coordinates programs, budgets and policy recommendations" for IoT R&D.

Other agencies involved in such R&D include the Food and Drug Administration (FDA), the National Aeronautics and Space Administration (NASA), the National Institutes of Health (NIH), the Department of Veterans Affairs (VA), and several DOD agencies.

. The White House has also announced a smart-cities initiative focusing on the development of a research infrastructure, demonstration projects, and other R&D activities.

32.What companies are working on Internet of Things (IoT)?

Answer: At this point, the easier question might be who isn't working on an IoT product. Big names like Samsung, LG, Apple, Google, Lowes and Philips are all working on connected devices, as are many smaller companies and startups. Research group Gartner predicts that 4.9 billion connected devices will be in use this year, and the number will reach 25 billion by 2020.

33.How Does the Internet of Things (IoT) Work?

Answer: The IoT is not separate from the Internet, but rather, a potentially huge extension and expansion of it. The things that form the basis of the IoT are objects. They could be virtually anything, streetlights, thermostats, electric meters, fitness trackers, factory equipment, automobiles, unmanned aircraft systems (UASs or drones), or even cows or sheep in a field. What makes an object part of the IoT is embedded or attached computer chips or similar components that give the object both a unique identifier and Internet connectivity. Objects with such components are often called "smart" such as smart meters and smart cars.

Internet connectivity allows a smart object to communicate with computers and with other smart objects. Connections of smart objects to the Internet can be wired, such as through Ethernet cables, or wireless, such as via a Wi-Fi or cellular network.

To enable precise communications, each IoT object must be uniquely identifiable. That is accomplished through an Internet Protocol (IP) address, a number assigned to each Internet-connected device, whether a desktop computer, a mobile phone, a printer, or an IoT object. Those IP addresses ensure that the device or object sending or receiving information is correctly identified.

34.What impacts will the Internet of Things (IoT) have on Energy Sector?

Answer: Within the energy sector, the IoT may impact both production and delivery, for example through facilitating monitoring of oil wellheads and pipelines. When IoT components are embedded into parts of the electrical grid, the resulting infrastructure is commonly referred to as the "smart grid". This use of IoT enables greater control by utilities over the flow of electricity and can enhance the efficiency of grid operations. It can also expedite the integration of microgenerators into the grid.

Smart-grid technology can also provide consumers with greater knowledge and control of their energy usage through the use of smart meters in the home or office.

Connection of smart meters to a building's HVAC, lighting, and other systems can result in "smart buildings" that integrate the operation of those systems. Smart buildings use sensors and other data to automatically adjust room temperatures, lighting, and overall energy usage, resulting in greater efficiency and lower energy cost. Information from adjacent buildings may be further integrated to provide additional efficiencies in a neighborhood or larger division in a city.

35.Why is the Internet of Everything important?

Answer: The Internet of Everything brings together people, process, data and things to make networked connections more relevant and valuable than ever before – turning information into actions that create new capabilities, richer experiences and unprecedented economic opportunity for businesses, individuals and countries.

36.What impacts will the Internet of Things (IoT) have on Agriculture Sector?

Answer: The IoT can be leveraged by the agriculture industry through precision agriculture, with the goal of optimizing production and efficiency while reducing costs and environmental impacts. For farming operations, it involves analysis of detailed, often real-time data on weather, soil and air quality, water supply, pest populations, crop maturity, and other factors such as the cost and availability of equipment and labor. Field sensors test soil moisture and chemical balance, which can be coupled with location technologies to enable precise irrigation and fertilization. Drones and satellites can be used to take detailed images of fields, giving farmers information about crop yield, nutrient deficiencies, and weed locations.

For ranching and animal operations, radio frequency identification (RFID) chips and electronic identification readers (EID) help monitor animal movements, feeding patterns, and breeding capabilities, while maintaining detailed records on individual animals.

37.Can all IoT devices talk to each other? What is the Standard for Communication between these devices?

Answer: With so many companies working on different products, technologies and platforms, making all these devices communicate with each other is no small feat seamless overall compatibility likely won't happen.

Several groups are working to create an open standard that would allow interoperability among the various products. Among them are the AllSeen Alliance, whose members include Qualcomm, LG, Microsoft, Panasonic and Sony; and the Open Interconnect Consortium, which has the support of Intel, Cisco, GE, Samsung and HP.

While their end goal is the same, there are some differences to overcome. For example, the OIC says the AllSeen Alliance doesn't do enough in the areas of security and intellectual property protection. The AllSeen Alliance says that these issues have not been a problem for its more than 110 members.

It's not clear how the standards battle will play out, though many believe we'll end up with three to four different standards rather than a single winner (think iOS and Android).

In the meantime, one way consumers can get around the problem is by getting a hub that supports multiple wireless technologies, such as the one offered by SmartThings.

38.Should the consumers be concerned about security and privacy issues considering the amount of data Internet of Things (IoT) collects?

Answer: The various amounts of data collected by smart home devices, connected cars and wearables have many people worried about the potential risk of personal data getting into the wrong hands. The increased number of access points also poses a security risk.

The Federal Trade Commission has expressed concerns and has recommended that companies take several precautions in order to protect their customers. The FTC, however, doesn't have the authority to enforce regulations on IoT devices, so it is unclear how many companies will heed its advice.

For example, Apple requires that companies developing products for its HomeKit platform include end-to-end encryption and authentication and a privacy policy. The company also said it doesn't collect any customer data related to HomeKit accessories.

39.What is meant by a Smart City, in the context of Internet Of Things (IoT)?

Answer: As with IoT and other popular technology terms, there is no established consensus definition or set of criteria for characterizing what a smart city is. Specific characterizations vary widely, but in general they involve the use of IoT and related technologies to improve energy, transportation, governance, and other municipal services for specified goals such as sustainability or improved quality of life.

The related technologies include:-
. Social media (such as Facebook and Twitter),
. Mobile computing (such as smartphones and wearable devices),
. Data Analytics (big data ? the processing and use of very large data sets; and open data ? databases that are publicly accessible), and
. Cloud computing (the delivery of computing services from a remote location, analogous to the way utilities such as electricity are provided).

Together, these are sometimes called SMAC.

40.What is GainSpan's GS2000 Protocol for Internet of Things (IoT)?

Answer: GainSpan's GS2000 is one such tech which used both ZigBee and Wi-Fi. It makes optimum use of power by putting the device into energy-saving standby mode when no data transmission is taking place. Only when device is awaked or checked for connection failure the high-power consumption connection of Wi-Fi is used.

41.What risks and challenges should be considered in the Internet of Everything?

Answer: Some important considerations in the Internet of Everything include privacy, security, energy consumption and network congestion.

42.How Lack of Uniform Technical Standards might affect the Development and Implementation of the Internet of Things (IoT)?

Answer: Currently, there is no single universally recognized set of technical standards for the IoT, especially with respect to communications, or even a commonly accepted definition among the various organizations that have produced IoT standards or related documents.

Many observers agree that a common set of standards will be essential for interoperability and scalability of devices and systems. However, others have expressed pessimism that a universal standard is feasible or even desirable, given the diversity of objects that the IoT potentially encompasses. Several different sets of de facto standards have been in development, and some observers do not expect formal standards to appear before 2017. Whether conflicts between standards will affect growth of the sector as it did for some other technologies is not clear.

43.What is the difference between the Internet of Things (IoT) and Machine to Machine (M2M)?

Answer: Generally speaking, M2M could be considered a subset of IoT. M2M is like a line connecting 2 points, and IoT is like a network, a system composed of lots of M2M and triggering lots of interactions/activities.

Giving a simple definition to M2M which is transferring data from one machine to another one. It's been used everywhere in our daily life. For example, entrance security. Just like using your employee card to unlock a door. When the security detector receives the ID from the employee card and then unlock the door once the ID is approved. This is M2M.

In this case, what IoT can offer? As mentioned, IoT is a network, is a system composed of lots of M2M and algorithms. When the system knows you are the person entering the office, it can turn on the light and the air conditioner of your partition, even it can adjust the most comfortable light level and temperature that you like the most from time to time after learning your behavior for a period of time. The system can get all the data from all the sensors and machines to know, for example, who and when enters the office, how much electricity you have consumed, what the environment makes you feel most comfortable, and other applications.

IoT will make the facilities and things smarter and make people's life more convenient. Not only machine to machine, but also human to machine, machine to human, and so on.

44.How Energy Consumption might affect the Development and Implementation of the Internet of Things (IoT)?

Answer: Energy consumption can also be an issue. IoT objects need energy for sensing, processing, and communicating information. If objects isolated from the electric grid must rely on batteries, replacement can be a problem, even if energy consumption is highly efficient. That is especially the case for applications using large numbers of objects or placements that are difficult to access. Therefore, alternative approaches such as energy harvesting, whether from solar or other sources, are being developed.

45.What is Bluegiga APx4 Protocol for Internet of Things (IoT)?

Answer: BLE and Wi-Fi together can be used without interference as they are compliable to coexistence protocols. The Bluegiga APx4 is one such solution which supports both BLE and Wi-Fi and is based on 450MHz ARM9 processor.

46.Who coined the term Internet of Things (IoT) and when?

Answer: The term Internet of Things is 16 years old. But the actual idea of connected devices had been around longer, at least since the 70s. Back then, the idea was often called "embedded internet" or "pervasive computing". But the actual term "Internet of Things" was coined by Kevin Ashton in 1999 during his work at Procter&Gamble. Ashton who was working in supply chain optimization, wanted to attract senior management's attention to a new exciting technology called RFID. Because the internet was the hottest new trend in 1999 and because it somehow made sense, he called his presentation "Internet of Things".

Even though Kevin grabbed the interest of some P&G executives, the term Internet of Things did not get widespread attention for the next 10 years.

47.How High-Speed Internet might affect the Development and Implementation of the Internet of Things (IoT)?

Answer: Use and growth of the IoT can also be limited by the availability of access to high-speed Internet and advanced telecommunications services, commonly known as broadband, on which it depends. While many urban and suburban areas have access, that is not the case for many rural areas, for which private-sector providers may not find establishment of the required infrastructure profitable, and government programs may be limited.

48.How Internet Addresses (IPv6) might affect the Development and Implementation of the Internet of Things (IoT)?

Answer: A potential barrier to the development of IoT is the technical limitations of the version of the Internet Protocol (IP) that is used most widely. IP is the set of rules that computers use to send and receive information via the Internet, including the unique address that each connected device or object must have to communicate. Version 4 (IPv4) is currently in widest use. It can accommodate about four billion addresses, and it is close to saturation, with few new addresses available in many parts of the world.

Some observers predict that Internet traffic will grow faster for IoT objects than any other kind of device over the next five years, with more than 25 billion IoT objects in use by 2020,76 and perhaps 50 billion devices altogether. IPv4 appears unlikely to meet that growing demand, even with the use of workarounds such as methods for sharing IP addresses.

Version 6 (IPv6) allows for a huge increase in the number IP addresses. With IPv4, the maximum number of unique addresses, 4.2 billion, is not enough to provide even one address for each of the 7.3 billion people on Earth. IPv6, in contrast, will accommodate over 1038 addresses more than a trillion per person.

It is highly likely that to accommodate the anticipated growth in the numbers of Internet-connected objects, IPv6 will have to be implemented broadly. It has been available since 1999 but was not formally launched until 2012. In most countries, fewer than 10% of IP addresses were in IPv6 as of September 2015. Adoption is highest in some European countries and in the United States, where adoption has doubled in the past year to about 20%.

Globally, adoption has doubled annually since 2011, to about 7% of addresses in mid-2015. While growth in adoption is expected to continue, it is not yet clear whether the rate of growth will be sufficient to accommodate the expected growth in the IoT. That will depend on a number of factors, including replacement of some older systems and applications that cannot handle IPv6 addresses, resolution of security issues associated with the transition, and availability of sufficient resources for deployment.

Efforts to transition federal systems to IPv6 began more than a decade ago. According to estimates by NIST, adoption for public-facing services has been much greater within the federal government than within industry or academia. However, adoption varies substantially among agencies, and some data suggest that federal adoption plateaued in 2012. Data were not available for this report on domains that are not public-facing, and it is not clear whether adoption of IPv6 by federal agencies will affect their deployment of IoT applications.

49.What is Industrial Internet of Things (IoT)?

Answer: The Industrial Internet of Things (IIoT) is the use of Internet of Things (IoT) technologies in manufacturing.

Also known as the Industrial Internet, IIoT incorporates machine learning and big data technology, harnessing the sensor data, machine-to-machine (M2M) communication and automation technologies that have existed in industrial settings for years. The driving philosophy behind the IIoT is that smart machines are better than humans at accurately, consistently capturing and communicating data. This data can enable companies to pick up on inefficiencies and problems sooner, saving time and money and supporting business intelligence efforts. In manufacturing specifically, IIoT holds great potential for quality control, sustainable and green practices, supply chain traceability and overall supply chain efficiency.

50.How Remote Updation of Software might affect the Development and Implementation of the Internet of Things (IoT)?

Answer: Several other technical issues might impact the development and adoption of IoT. For example, if an object's software cannot be readily updated in a secure manner, that could affect both function and security. Some observers have therefore recommended that smart objects have remote updating capabilities. However, such capabilities could have undesirable effects such as increasing power requirements of IoT objects or requiring additional security features to counter the risk of exploitation by hackers of the update features.

51.What are the industrial applications for wireless sensor networks Internet of Things (IoT)?

Answer: You can easily and cheaply buy sensors that can measure a variety of variables that would be interesting in industrial applications, for example (and this is for sure not a complete list): light or sound intensity; voltage; current; pressure; temperature; rotational position; XYZ orientation; compass direction; acceleration; location; fluid flow rate and so on.

These sensors can be interrogated by microcontroller, and data stored to memory card, or communicated in real-time to other systems via Bluetooth, Zigbee, Wi-Fi , Ethernet, serial, USB, infra-red and so on.

The inexpensive nature of these microcontrollers (for example Google for ESP8266 to see a Wi-Fi-enabled microcontroller) means that you could deploy a large number of these in an industrial setting (even in hazardous environments) and gather data without a large capital investment, and without the worry of "what if it gets destroyed"?

There must be so many industrial applications of this technology that it's impossible to enumerate the possibilities. The limiting factor is really only "how can we process all of this data"?

52.Why Should we Care about the Industrial IoT?

Answer: The Industrial IoT focuses strongly on intelligent cyber-physical systems. These systems comprise machines connected to computers that interpret, analyze and make decisions almost instantly, based on sensor data from many widely distributed sources.

The Industrial IoT enables the smart system in your car that brakes automatically when it detects an obstacle in the road. It enables the patient monitoring system in hospitals to track everything from a patient's heart rate to their medication intake. It enables a mining machine or space robot to safely and efficiently operate where humans can't.

The world is building more and more intelligent machines that interact with other machines, with their environments, with data centers and with humans.

53.What is Wi-Fi Protocol for Internet of Things (IoT)?

Answer: Counted as the most mature wireless radio technology, Wi-Fi is predominant communication technology chosen for IoT applications. Already existing protocols like WPS make the integration of internet of things devices easier with the existing network. If we talk about transmission then Wi-Fi offers the best power-per-bit efficiency. However power consumption when devices are dormant is much higher with conventional Wi-Fi designs. The solution is provided by protocols like BLE and ZigBee that reduce power consumption by sensors when devices are dormant.

Most important use of Wi-Fi is in the applications where IP stack compliance is needed and there is high data transmission. For instance in applications sharing audio, video or remote device controlling.

As the prerequisites of internet of things are scaling up, companies are working on more integrated solutions. But even at present there are many solutions available for anyone who is trying to build up internet of things applications around the major three IoT components. Vendors

54.What is ZigBee Protocol for Internet of Things (IoT)?

Answer: ZigBee is a low power consuming IEEE 802.15.4(2003) standard based specification, ZigBee is a brain child of 16 automation companies. What makes it novel is the use of mesh networking which makes utilization of communication resources much more efficient. ZigBee based IoT nodes can connect to central controller making use of in-between nodes for propagating the data. It makes transmission and handling of data robust.

55.What impacts will the Internet of Things (IoT) have on Manufacturing Sector?

Answer: Integration of IoT technologies into manufacturing and supply chain logistics is predicted to have a transformative effect on the sector. The biggest impact may be realized in optimization of operations, making manufacturing processes more efficient. Efficiencies can be achieved by connecting components of factories to optimize production, but also by connecting components of inventory and shipping for supply chain optimization.

Another application is predictive maintenance, which uses sensors to monitor machinery and factory infrastructure for damage. Resulting data can enable maintenance crews to replace parts before potentially dangerous and/or costly malfunctions occur.

56.Is the Internet of Everything a Cisco or IBM architecture or trademark?

Answer: No. The Internet of Everything does not describe a specific architecture and is not solely owned by Cisco or IBM or any other company.

57.What effect will the Internet of Things (IoT) have on our daily lives?

Answer: It already is having an impact. A recent report from Gartner says there will be 4.9bn connected things in 2015, rising to 25bn by 2020. What are these things, though?

"Let us not focus on fridges," says Will Franks, who sold Ubiquisys to Cisco for ?204m in 2013. Franks, who has just helped set up the Wireless IoT Forum, lists a number of consumer touch points. "Keeping track of possessions where insurance companies could reduce premiums," he says. "Home control devices, maintenance checks for cars and white goods, healthcare and so on."

He doesn't mention robots or Facebook. Robots will be connected too in a smart home of the future, at least according to the GSMA. And Facebook? According to The

Register, it's planning to launch software development kits (SDKs) for IoT apps and devices. Heating control through systems such as Nest and Hive are just the start, it seems.

58.What is difference between Wireless Sensor Network (WSN) and Internet of Things (IoT) network (sensor)?

Answer: About WSN:

Wireless sensor network is the foundation of IoT applications.

WSN is the network of motes, formed to observe, to study or to monitor physical parameters of desired application.

For example – Motes deployed in Agriculture land, monitor Temp-Humidity or even soil moisture, who gathers data and with perfect data analysis produce results about crop yields – quality or quantity.

About IoT:

IoT is the network of physical objects controlled and monitored over internet.

Now just as WSN, in IoT application you will encounter the monitoring of physical parameters. But desired outcomes are little different.

IoT is more about M2M, it is more about bringing smartness into daily objects.

For example – Device hooked to your Thermostat monitors surrounding temperature and adjust it to most preferred setting for

59.Give few examples of the Impact of Internet of Things (IoT) on our lives?

Answer: To put things simply any object that can be connected will be connected by the IoT. This might not make sense for you on the forefront but it is of high value. With interconnected devices you can better arrange your life and be more productive, safer, smarter and informed than ever before.

For instance how easy it will be for you to start your day if your alarm clock is not only able to wake you up but also able to communicate with your brewer to inform it that you are awake at the same time notifies your geezer to start water heating. Or you wearable wrist health band keeps track of your vitals to inform you when you are most productive during the day. These are just few examples but applications of internet of things are numerous.

On large scale transportation, healthcare, defense, environment monitoring, manufacturing and every other field you can imagine of can be benefited from IoT. It is very hard to conceive the whole application domain of internet of things at the moment but you can clearly understand why it is such an interesting and hot topic at the moment.

60.How Safety issue might affect the Development and Implementation of the Internet of Things (IoT), especially in USA?

Answer: Given that smart objects can be used both to monitor conditions and to control machinery, the IoT has broad implications for safety, with respect to both improvements and risks. For example, objects embedded in pipelines can monitor both the condition of the equipment and the flow of contents. Among other benefits, that can help both to expedite shutoffs in the event of leaks and to prevent them through predictive maintenance.

Connected vehicles can help reduce vehicle collisions through crash avoidance technologies and other applications.110 Wireless medical devices can improve patient safety by permitting remote monitoring and facilitating adjustments in care.

However, given the complexities involved in some applications of IoT, malfunctions might in some instances result in catastrophic system failures, creating significant safety risks, such as flooding from dams or levees. In addition, hackers could potentially cause malfunctions of devices such as insulin pumps or automobiles, potentially creating significant safety risks.

19 Chapter: References

1. Instrumentation for engineering measurements, J.W. Dally, W.F. Riley and K.G. McConnell, Prentice Hall, 1983.
2. Process Control Instrumentation Technology, Curtis D. Johnson, Prentice Hall, 2000
3. Principles of measurements and instrumentations, A.S. Morris, Prentice Hall, 1988.
4. Industrial Instrumentation & Control,2e By S. K. Singh.
5. Industrial Control Handbook, E.A. Parr, Industrial Press, 3rd Ed., 1998.
6. Electronic instrumentations and measurements, L.D. Jones and A. Foster, Prentice Hall, 1991.
7. Industrial Instrumentation by Tony R. Kuphaldt ,Jan 2011.
8. Mechanical Measurements: Jones' Instrument Technology edited by B E Noltingk
9. Electrical Transformers and Power Equipment By Anthony J. Pansini
10. Internet of Things and Data Analytics Handbook edited by Hwaiyu Geng
11. www.InstrumentationToday.com and many other online sites.
12. http://www.fargocontrols.com/sensors.html
13. http://www.jms-se.com/thermowell.php
14. https://www.engineeringtoolbox.com/
15. http://www.fts-arg.com.ar/producto/juntas-monoliticas-pipe-pup-copiar/
16. http://www.alma-valves.ie/general-valve-products/globe-valves/globe-valves-brass-bronze-cast-iron-steel/
17. http://www.citek.co.za/buy/dwb-01-wheatstone-bridge-755814
18. https://www.electrical4u.com/instrument-transformers/
19. https://www.elprocus.com/air-circuit-breaker-acb-working-principle-application/
20. https://www.engineeringtoolbox.com/pfd-process-flow-diagram-d_465.html
21. http://plcmanual.com
22. https://advancedinternettechnologies.wordpress.com/ipv4-header/
23. Industrial Internet of Things ;A high -level architecture discussion by PCI Industrial Computer Manufacturer's Group
24. https://dzone.com/articles/an-introduction-to-restful-apis
25. https://iot.intersog.com/blog/iot-platforms-overview-arduino-raspberry-pi-intel-galileo-and-others/
26. https://www.postscapes.com/internet-of-things-protocols/

27. https://wiki.aalto.fi/download/attachments/59704179/devadiga-802-15-4-and-the-iot.pdf?version=1
28. https://blogs.vmware.com/pulseiot/2018/02/15/potential-api-iot-security/
29. http://www.mdpi.com/1424-8220/17/2/301/htm (Enabling Secure XMPP Communications in Federated IoT Clouds Through XEP 0027 and SAML/SASL SSO by Antonio Celesti , Maria Fazio and Massimo Villari)
30. https://www.iotone.com/term/datagram-transport-layer-security-dtls/t167
31. https://www.mytectra.com/interview-question/iot-interview-questions-and-answers/
32. Gartner, "Gartner's 2014 hype cycle for emerging technologies maps the journey to digital business,", August 2014 : http://www.gartner.com/newsroom/id/2819918
33. https://www.cse.wustl.edu/~jain/cse570-15/ftp/iot_prot/#chal4
34. https://iotpoint.wordpress.com/z-wave-tutorial/
35. https://www.ecnmag.com/article/2013/07/guide-zigbee-device-developers-smart-connected-home
36. http://industrial.embedded-computing.com/articles/plug-and-play-homeplug-homeplug-powerline-alliance/
37. http://www.3glteinfo.com/lora/lora-architecture/
38. http://www.rfwireless-world.com/Articles/weightless-system-overview.html
39. https://www.ericsson.com/research-blog/cellular-internet-things-technologies-standards-performance/
40. http://telecoms.com/461282/ericsson-claims-first-complete-low-power-wide-area-solution-with-att/
41. http://www.rfwireless-world.com/Terminology/RPL-vs-CORPL-vs-CARP.html
42. http://blog.mallow-tech.com/2018/03/mqtt-protocol-for-iot-a-brief-introduction/
43. 2018 Top Technologies in Sensors and Instrumentation by Frost & Sullivan
44. 2018 Top Technologies in Advanced Manufacturing and Automation by Frost & Sullivan
45. https://en.wikipedia.org/wiki/Sensor_fusion#/media/File:Eurofighter_sensor_fusion.png
46. https://roboticsandautomationnews.com/2016/09/22/advanced-driver-assistance-systems-trump-driverless-cars-by-stealth/7304/
47. https://www.tctmagazine.com/3d-printing-news/additive-engineering-solutions-big-area-additive-manufacturing/
48. https://www.twi-global.com/technical-knowledge/faqs/faq-what-is-self-piercing-riveting-and-how-does-it-work/
49. https://www.technologyreview.com/s/526536/agile-robots/

Printed in Great Britain
by Amazon